D1104764

Hydrogen Energy

Challenges and Prospects

Hydrogen Energy
Challenges and Prospects

D.A.J. Rand
CSIRO Energy Technology, Victoria, Australia

R.M. Dell
Formerly Head of Applied Electrochemistry,
Atomic Energy Research Establishment, Harwell, UK

RSC Publishing

TP
359
.H8
H868
2008

ISBN: 978-0-85404-597-6

A catalogue record for this book is available from the British Library

Published by The Royal Society of Chemistry,
Thomas Graham House, Science Park, Milton Road,
Cambridge CB4 0WF, UK

Registered Charity Number 207890

For further information see our web site at www.rsc.org

Foreword

Hydrogen has been the most prevalent element in the universe since its early beginnings thirteen billion years ago, and there is a compelling logic that makes us all believe that surely this gas can provide an almost limitless source of energy for the world.

Although abundant on Earth, not least as water in the oceans and the atmosphere, the chemical activity and physical properties of hydrogen make its isolation an energy-intensive process and its subsequent storage and transport a challenge.

In an ideal world, hydrogen would be made available through the splitting of water into its constituent elements, drawing on renewable or long-term energy sources. Following recombination with oxygen in the air to liberate this energy — typically elsewhere — water would be returned to the atmosphere as part of the natural water cycle. Such a process would be both sustainable and carbon neutral.

In reality, nearly all the hydrogen currently manufactured in the world is through the reforming of hydrocarbons, a process that has low energy-conversion efficiency and contributes to the 8.8 billion tonnes of carbon being emitted annually as a result of fuel usage. The small proportion of hydrogen produced by electrolysis uses electricity that includes the burning of fossil fuels, although renewable sources such as hydroelectric power, wind, photovoltaics and geothermal energy are being employed increasingly.

There is the prospect of wind farms, wave and tidal power facilities being linked to hydrogen generation, as well as nuclear plants, which nevertheless pose other challenges related to security and radioactive waste disposal. In one form or other and with the right leadership and entrepreneurship, these could meet, potentially, all our heating, power and transportation needs.

The world is now at a pivotal point in planning its energy provision for the future, as the effects of climate change have to be addressed and, in the much longer term, fossil fuels will become scarcer. We cannot rely on the latter solving the former! There are schools of thought that picture a Hydrogen Economy based on combustion and fuel cells supported by electrolysis and solar pyrolysis of water, while others see a domination of electricity as the principal energy vector of the future. A further component is the role of biofuels, but current rates of energy conversion to liquid fuel (less than 1% of sunlight received) pale against photovoltaic direct power generation of 20%.

This then is the context of the present book that explores the scientific, economic, fiscal, social and regulatory framework of a world economy supported by hydrogen. It is a fascinating journey that will inspire all of us who are looking for a clean, sustainable future.

Richard Pike
Chief Executive, Royal Society of Chemistry

Preface

The early history of the proposal to harness hydrogen as a clean and abundant form of energy has been well documented.[1] In brief, on 27 November 1820, the Revd. William Cecil presented a lecture to the Cambridge Philosophical Society in which he advanced the idea of replacing steam engines by a novel form of engine in which hydrogen was combusted in a cylinder. The resultant change in pressure could be converted into mechanical energy to perform work. He suggested that this would overcome certain disadvantages of the steam engine, notably the long start-up time from cold. There is no record that Cecil ever built such an engine. Later in the 19th century, the novelist Jules Verne wrote *The Mysterious Island* (1874): a story in which he proposed replacing coal as a fuel by hydrogen derived, via electrolysis, from inexhaustible supplies of water. In this, Verne showed considerable prescience, but failed to explain that a source of primary energy was required to generate the electricity. A hundred years after Cecil's seminal lecture, J.B.S. Haldane made the following statement in a paper read to the 'Heretics' Society of Cambridge University on 4 February 1923:[2]

> '*Personally, I think that four hundred years hence the power question in England may be solved somewhat as follows. The country will be covered with rows of metallic windmills working electric motors which in their turn supply current at a very high voltage to great electric mains. At suitable distances, there will be great power stations where during windy weather the surplus power will be used for the electrolytic decomposition of water into oxygen and hydrogen. These gases will be liquefied and stored in vast vacuum jacketed reservoirs, probably sunk in the ground.*'

Many scholars consider Haldane's prophesy to be the first allusion to a 'Hydrogen Economy' – a term that, as will be seen below, was not coined until some fifty years later.

Hydrogen does not occur freely in nature; it is predominantly found in combination with oxygen as water and with carbon as fossil fuels. Energy has to be expended to extract hydrogen from these sources. It is therefore not a new form of primary energy, but merely an energy vector (or carrier) that is environmentally clean when combusted and therefore attractive, although not especially convenient to handle and use.

Little happened to advance the deployment of hydrogen energy until the 1930s and 1940s when Rudolf Erren, a German engineer, modified internal combustion engines to run on fuel–hydrogen mixtures or even on pure hydrogen. Commercial enthusiasm for hydrogen engines faded, however, after the *Hindenburg* airship disaster in 1937. It should be noted that later research has shown that the *Hindenburg* fire was more to do with the flammability of the fabric chosen for the airship's envelope than with the fact that it was filled with hydrogen.

Renewed interest in hydrogen energy was shown in the 1950s following the demonstration of the first practical fuel cell by the British scientist Francis T. Bacon, who, like Cecil and Haldane, also worked in Cambridge. Fuel cells of the Bacon type found an important application in space missions, particularly in manned flights of extended duration. Recently, the perfection of fuel cell technology for both the stationary generation of electricity and for powering road transportation has been seen as an important step towards energy security. Accordingly, much research and development of fuel cells is in progress.

Given the relatively low cost of fossil fuels, scant attention was paid to the use of hydrogen for terrestrial applications until 1973, when the shock of the first 'oil crisis' reverberated around the world. It was then widely understood that reserves of fossil fuels, especially petroleum, were finite and that, sooner or later, demand would exceed supply. Fossil fuels are laid down over geological time and, once used, cannot be replaced. Moreover, there are inequalities in the geographic distribution of these natural resources that make some countries wealthy and others relatively poor; unabated consumption will intensify political, economic and social tension. Thus, it has been deemed necessary to turn increasingly to 'renewable' sources of energy such as solar, wind, wave and tidal energy. These are mostly periodic or irregular in output and require some type of storage as a smoothing device, together with a vector to convey the primary energy from where it is produced to where it is needed. Many of these renewable forms of energy serve to generate electricity directly. While electricity is a highly versatile carrier of energy, it is not easy to store and is of limited value as a fuel for road and air transportation, although it is employed widely for rail traction.

The dawning realization of the future importance of hydrogen as an energy vector and new fuel led, in 1974, to the formation of the International Association for Hydrogen Energy (IAHE) and to its biennial World Hydrogen Energy Conference (WHEC) that is still taking place today. The latter is held in cities around the world. In 1975, the *International Journal of Hydrogen Energy* was launched and its publication continues. Also, in the same year, one of the present authors was a contributor to a paper entitled 'Hydrogen the Ultimate Fuel', in which the attributes of hydrogen as an energy vector and fuel were examined.[3] To quote from that review:

> *'In the long term, when world supplies of natural gas and petroleum are largely exhausted and society is increasingly dependent upon nuclear power or new sources of energy, hydrogen will appear attractive as an energy vector and energy storage medium. Key questions are how and over what*

timescale should the transition be made from the present situation, where we have neither the technical ability nor the financial incentive to introduce hydrogen as a fuel, to the distant future where it may be a vital component of the world energy scene.

To answer these questions it will be necessary to estimate in some detail the magnitude of the problems involved in implementing the Hydrogen Economy, in technical, industrial and financial terms, and also the lead times for society to make a smooth transition from one energy basis to another. It is then necessary to compare these findings with some estimate of when hydrogen will be required in bulk quantities to replace fossil fuels. Even without this analysis it is evident that the lead times will be measured in decades and the capital requirements will be enormous, so that planning on a national or international scale will be vital.'

The 1975 paper went on to estimate the number of 1000 MW_e nuclear power plants, for example, that would be necessary to produce (by means of electrolysis at 60% efficiency) the hydrogen needed to fuel various aspects of the UK industrial and transportation markets as they then existed. To substitute hydrogen for all the fossil fuel consumed in the UK in 1974 – a hypothetical concept – would have required 377 such power stations. This gives some idea of the magnitude of the problems that would be encountered in implementing a full Hydrogen Economy based on nuclear electricity. The challenge today for a hydrogen future centred on renewables would be even greater, especially given the growth in overall energy consumption since 1974.

In the intervening years, there has been comparatively little progress in facing up to the task of establishing a Hydrogen Economy, again thanks largely to an abundance of cheap fossil fuels. The situation is now changing with world production of petroleum expected to peak in the near future and that of natural gas within a few decades – just as the global demand for energy is growing rapidly.[4] Coupled with this concern over finite resources, there are two other pressing threats to society: (i) climate change, which is attributed largely to carbon dioxide formed by the burning of fossil fuels, and (ii) worries over the security of energy supplies in a world where the major reserves are in the hands of comparatively few countries.

There is much debate among geologists and petroleum engineers over exactly how much oil and gas remain to be exploited and how many decades these reserves will last as world consumption rises. The precise reserves are not too relevant in the greater scheme of things; even children alive today face an uncertain energy future unless totally new technology is developed. The growth in world population and the expectation of a rising standard of living in developing countries make a greater use of fossil fuels inevitable. The World Bank has stated (2002) that:[5]

'Reliable energy is a key component of economic and social development. . . lack of energy is among the key forces slowing down poverty reduction and growth of the rural sector.'

Increasing the availability, affordability and use of energy is vital if poor nations are to achieve their potential. This is a short-term objective for the next decade or two.

Looking further ahead, the world has an even more challenging goal that was set in 1987 by the World Commission on Environment and Development (the Brundtland Commission) when it defined 'sustainable development' as a process that: 'meets the needs of the present without compromising the ability of future generations to meet their own needs'.[6] In energy and environmental terms, this reduces to 'devising a set of energy technologies that meets human-ity's needs on an indefinite basis without producing irreversible environmental effects'. It is now generally accepted that mean global temperatures are rising steadily and are resulting in a change of climate. This, in turn, is thought to account for the increased incidence of violent storms and flooding, while in other parts of the world more droughts and crop failures are occurring.

Fortunately, there are unlimited supplies of renewable energy potentially available in many different forms. The challenge is therefore to find practical ways to exploit them as efficiently and cheaply as possible on a world-wide scale. Hydrogen is seen as one such option; it presents a means to convey renewable energy from the site of generation to the end-user, it serves as an energy store and it also offers the prospect of a new, emission-free fuel for transportation applications. This is a vision that is extremely attractive to environmentalists, politicians and informed members of the public. Never-theless, any decision to adopt hydrogen as an energy vector has to be backed up by hard numerical analysis of the associated energetics and by rigorous evaluation of the capital and operating costs.

Hydrogen is a clean form of energy that can be produced from water using many different primary sources of energy, *e.g.*, various fossil fuels, assorted renewables and nuclear power. This flexibility is attractive politically since it reduces the chances of a hydrogen cartel being established. During the early stages of transition to a Hydrogen Economy, it is likely that the hydrogen will derive mostly from fossil fuels while awaiting the establishment of further renewable energy systems.

The title of this book – *Hydrogen Energy: Challenges and Prospects* – has been chosen carefully. Essentially, the work presents an account of the state-of-the-art of hydrogen energy and the challenges to be met if it is to play an important role in the world's energy scene, together with an assessment of the prospects for success in such a venture. Just as coal and the steam engine gave rise to the Industrial Revolution and petroleum and the internal combustion engine brought about the transportation revolution of the 20th century, enthusiasts for a Hydrogen Economy see this as the start of a major technological and social revolution that will determine our way of life in the 22nd century. Inevitably, much will change in society during the course of the present century and forecasting events so far ahead can only be an exercise in crystal-ball gazing, albeit one based on the science and techno-logy in place today. Nevertheless, we have endeavoured to peer into that crystal ball, with due regard for physical principles, numerical values and economics.

There are particular difficulties in writing a book that addresses the future. For example, it is not easy to evaluate the prospects for new forms of energy without sound information on the attendant costs, which are hard to determine in advance of large-scale prototype demonstrations. Costs tend to vary from place to place and certainly from time to time, as technology progresses and the benefits of replication are felt. Also, when looking decades ahead, one does not know what advances will be made in the utilization of competing fuels, or the environmental regulations that will exist. For these reasons, we have confined ourselves mostly to an evaluation of the technical aspects and, where appropriate, have made just a few generalizations on costs.

A further problem is that the overall subject of hydrogen energy is multifaceted, with many component considerations, all of which are highly interactive. These include the primary energy source (or the fuel) employed, the means of conversion to hydrogen, the scale of operations, the mode of distributing and storing the hydrogen and the intended end-use. With so many contributing factors, it is difficult to subdivide the book into discrete, self-contained chapters. We have been obliged, therefore, to cross-refer the reader between chapters.

Finally, the world literature on energy production and consumption is plagued by a proliferation of measurement units. Variously, data are presented in terms of the International System of Units (SI, *e.g.*, metres, pascals, joules), traditional industry-based units (*e.g.*, barrels of oil, kilowatt hours of electricity, million tonnes of oil equivalent) and, especially in the USA, Imperial units (*e.g.*, miles, British thermal units of heat, quads of energy, cubic feet of natural gas, bars of pressure). For the expression of time, however, units of days and years are generally more appropriate than the SI unit (seconds) in this field. In order to assist readers in translating units into those with which they are familiar, a set of conversion factors has been included.

Acknowledgements

The authors are grateful to holders of copyright who have kindly consented to the use of their illustrations. Should any omissions have inadvertently occurred, sincere apologies are offered.

The authors are indebted to Dr Andrew Dicks (University of Queensland), Dr Greg Duffy (CSIRO Energy Technology) and Professor Emeritus Sten-Eric Lindquist (Uppsala University), who have generously provided technical and scientific information. The dedication and skills of Ms Trina Dearricott (CSIRO Minerals) in assisting with the production of the illustrations are also much appreciated.

Finally, the authors wish to acknowledge with gratitude the patience of their wives, Gwen Rand and Sylvia Dell, who provided encouragement and were deprived of much companionship during the preparation of this book.

References

1. P. Hoffmann, *Tomorrow's Energy: Hydrogen, Fuel Cells and the Prospects for a Cleaner Planet*, MIT Press, Cambridge, MA, 2001.
2. J.B.S. Haldane, *Daedalus; or, Science and the Future*, E. P. Dutton, New York, 1924.
3. R.M. Dell and N.J. Bridger, *Appl. Energy*, 1975, **1**, 279–292.
4. R.M. Dell and D.A.J. Rand, *Clean Energy*, Royal Society of Chemistry, Cambridge, 2004.
5. D. Lallement, *Roundtable on Energy for Sustainable Development: Partnerships for Action*. Seminar: *Energy for Sustainable Development*, United Nations Development Programme – European Commission, Brussels, 25–26 April 2002.
6. G. Bruntland (Ed.), *Our Common Future*, World Commission on Environment and Development, Oxford University Press, Oxford, 1987.

D.A.J. Rand
R.M. Dell

Biographical Notes

The authors are both senior research chemists who have spent their entire professional careers working in the energy field.

Ronald Dell Ph.D. D.Sc. CChem. FRSC graduated from the University of Bristol. He lived for several years in the USA where he worked as a research chemist in the petroleum industry. Upon returning to Britain, he joined the Atomic Energy Research Establishment at Harwell. During a tenure of 35 years, Ron investigated the fundamental chemistry of materials used in nuclear power and managed projects in the field of applied electrochemistry, especially electrochemical power sources. Since retiring in the mid-1990s, he has interested himself in the developing world energy scene, particularly with regard to environmental factors, and in technical writing.

David Rand Ph.D. Sc.D. FTSE was educated at the University of Cambridge. Shortly after graduating, he emigrated to Australia and has spent his research career working at the government's CSIRO laboratories in Melbourne. During the 1970s, he conducted investigations on noble-metal electrocatalysts for fuel cell systems. In the late 1970s, David also set up, and subsequently managed, the CSIRO battery research group. Since its inception in 1987, he has served as the Chief Battery Technical Officer of the World Solar Challenge, the trans-Australia event for cars powered exclusively by solar photovoltaic electricity. David has been the Asia-Pacific Editor of the *Journal of Power Sources* since 1983 and was elected a Fellow of the Australian Academy of Technological Sciences and Engineering in 1998. His present work is directed towards the use of hydrogen as a global energy vector.

Both authors have published their research findings extensively in the scientific literature and have also co-authored three previous books:
Batteries for Electric Vehicles, (Research Studies Press, 1998), ISBN 0-86380-205-2.
Understanding Batteries, (RSC Paperbacks, 2001), ISBN 0-85404-605-4.
Clean Energy, (RSC Clean Technology Monographs, 2004), ISBN 0-85404-546-5.

Contents

Abbreviations, Symbols and Units Used in Text

Abbreviations

a.c.	alternating current
AFC	alkaline fuel cell
ATP	adenosine 5′-triphosphate
BEV	battery electric vehicle
CB	conduction band
CBM	coal-bed methane
CCGT	combined-cycle gas turbine
CCS	carbon capture and storage
CFC	chlorofluorocarbon
CHP	combined heat and power
CNG	compressed natural gas
CSIRO	Commonwealth Scientific and Industrial Research Organisation (Australia)
CUTE	Clean Urban Transport for Europe
DAFC	direct alcohol fuel cell
DBFC	direct borohydride fuel cell
d.c.	direct current
DCFC	direct carbon fuel cell
DMFC	direct methanol fuel cell
DOE	Department of Energy (USA)
DSSC	dye-sensitized solar cell
e	(subscript) electrical
ECBM	enhanced coal-bed methane
ECTOS	Ecological City Transport System
EOR	enhanced oil recovery
EU	European Union
EV	electric vehicle

| FC | fuel cell |
| FCV | fuel cell vehicle |

GDP	gross domestic product
GHG	greenhouse gas
GTL	gas-to-liquids
GWP	global warming potential

HCFC	hydrochlorofluorocarbon
HEV	hybrid electric vehicle
HFC	hydrofluorocarbon
HHV	higher heating value
HTGCR	high-temperature gas-cooled reactor

IAHE	International Association for Hydrogen Energy
ICE	internal combustion engine
ICEV	internal-combustion-engined vehicle
IEA	International Energy Agency
IGCC	integrated gasification combined-cycle
IPCC	Inter-governmental Panel on Climate Change
IPHE	International Partnership for the Hydrogen Economy
IT	information technology

LH_2	liquid hydrogen
LHV	lower heating value
LNG	liquefied natural gas
LPG	liquid petroleum gas

M_{ox}	redox mediator in oxidized state
M_{red}	redox mediator in reduced state
MCFC	molten carbonate fuel cell
MEA	membrane–electrode assembly
MEMS	microelectromechanical systems
MIT	Massachusetts Institute of Technology
MOC	Meridional overturning cycle
MOF	metal–organic framework
MPV	multi-purpose vehicle

NADP	nicotinamide adenine dinucleotide phosphate
NASA	National Aeronautics and Space Administration (USA)
NGL	natural gas liquids
NO_x	nitrogen oxides
NTP	normal temperature (273.15 K) and pressure (101.325 kPa)

| OECD | Organisation for Economic Co-operation and Development |
| OPV | organic (polymer-based) photovoltaic |

PAFC	phosphoric acid fuel cell
PEC	photo-electrochemical
PEM	proton-exchange membrane
PEMFC	proton-exchange membrane fuel cell
PF	pulverized fuel
PFC	perfluorocarbons
PSA	pressure swing adsorption
PTFE	polytetrafluoroethylene
PV	photovoltaic
RAPS	remote-area power supply
SHE	standard hydrogen electrode
SI	Système International
SO_x	sulfur oxides
SOFC	solid oxide fuel cell
SPE	solid polymer electrolyte
SPEFC	solid polymer electrolyte fuel cell (same as PEMFC)
STEP	Sustainable Transport Energy Project
STP	standard temperature (298.15 K) and pressure (101.325 kPa)
SUV	sports utility vehicle
Syngas	synthesis gas
th	(subscript) thermal
UAV	unmanned aerial vehicle
USCAR	United States Council for Automotive Research
UTC	United Technologies Corporation
VB	valence band
WGS	water-gas shift reaction
YSZ	yttria-stabilized zirconia

Symbols and Units

Sub-units

d	deci	10^{-1}
c	centi	10^{-2}
m	milli	10^{-3}
μ	micro	10^{-6}
n	nano	10^{-9}

Multiple units

k	kilo	10^{3}
M	mega	10^{6}
G	giga	10^{9}
T	tera	10^{12}
P	peta	10^{15}

atm	atmosphere (= 101.325 kPa)
A	ampere
Ah	ampere-hour
b	barrel of oil
bar	unit of pressure (= 100 kPa)
bhp	brake horsepower (= 745.7 W)
c	speed of light in vacuum (= 2.998×10^8 m s^{-1})
C	coulomb (= 1A s)
C	carbon
cal	calorie (= 4.184 J)
cm	centimetre
°C	degree Celsius
e^-	electron
η	electrode overpotential (V)
η_+	overpotential at a positive electrode (V)
η_-	overpotential at a negative electrode (V)
eV	electron volt (= 1.602×10^{-19} J)
E	electrode potential (V)
E°	standard electrode potential (V)
E	energy (expressed in eV)
$E_{CB,b}$	energy at bottom of conduction band (eV)
E_F	Fermi level in a semiconductor (eV)
E_g	band gap energy (eV)
$E_{VB,t}$	energy at top of valence band (V)
ft	foot (linear measurement = 305 mm)
F	Faraday constant (= 96 458 C mol^{-1})
g	gram
G	Gibbs free energy (J mol^{-1})
ΔG	change in Gibbs free energy (J mol^{-1})
ΔG°	standard change in free energy (J mol^{-1})
h	Planck's constant (6.626×10^{-34} J s)
$h\nu$	energy of a photon
h	hour
h$^+$	electron hole in valence band
H	enthalpy (J mol^{-1})
ΔH	change in enthalpy (J mol^{-1})
ΔH_f°	standard heat (enthalpy) of formation (J mol^{-1})
in	inch (linear measurement = 2.54 cm)
I	current

IR'_e	resistive losses in electrolyte (V)
IR'_t	total resistive losses in electrodes (V)
J	joule ($= 1\,W\,s$)
K	kelvin (a measure of absolute temperature)
λ	wavelength of electromagnetic radiation
L	litre
m	metre
mol	mole, *i.e.*, mass of 6.022×10^{23} elementary units (atoms, molecules, *etc.*) of a substance
Mtoe	million tonnes of oil equivalent
ν	frequency of electromagnetic radiation
n	number of units (electrons, atoms, molecules) involved in a chemical or electrochemical reaction
N	newton (unit of force $= 1\,kg\,m\,s^{-2}$)
$N\text{-}m^3$	normal cubic metre of gas (*i.e.*, that measured at NTP)
Ω	ohm
ppmv	parts per million by volume
psi	pounds per square inch (1 psi $\approx 6.895\,kPa$)
P	pressure
Pa	pascal ($1\,Pa = 1\,N\,m^{-2} = 9.869\times10^{-6}\,atm$)
R'	resistance (Ω)
R	gas constant ($= 8.1345\,J\,K^{-1}\,mol^{-1}$)
s	second
S	entropy ($J\,K^{-1}\,mol^{-1}$)
ΔS	change in entropy ($J\,K^{-1}\,mol^{-1}$)
t	tonne
T	temperature
T_c	critical temperature
vol.%	volume percent
V°	reversible cell voltage (V) under standard conditions of temperature (298.15 K) and pressure (101.325 kPa); also known as the standard cell voltage
V_p	practical cell voltage (V)

V_r	reversible cell voltage (V)
V	volt
W	watt
W_e	watt, electrical power
W_p	peak watt (for solar cells)
W_{th}	watt, thermal power
Wh	watt-hour
wt.%	weight percent
x	variable in stoichiometry

Glossary of Terms

Adiabatic process
A process that takes place without heat entering or leaving the system.

Alternating current
Electric current that flows for an interval of time (half-period) in one direction and then flows for the same time in the opposite direction. The normal waveform is sinusoidal.

Anion
Ion in an electrolyte that carries a negative charge and that migrates towards the anode under the influence of a potential gradient. See **Anode, Ion**.

Anode
An electrode at which an oxidation process, *i.e.*, loss of electrons, is occurring. During electrolysis, the anode is the positive electrode. In a fuel cell, the anode is the negative electrode where hydrogen is consumed.

Aquifer
Underground water-bearing, porous rock strata that yield economic supplies of water, sometimes heated, to wells or springs.

Band gap energy
The energy gap, generally measured in electron volts (eV), between the top of the valence band and the bottom of the conduction band in a crystalline solid. See **Conduction band, Energy band, Valence band**.

Barrel
A measure of crude oil (petroleum), approximately 159 litres.

Battery electric vehicle (BEV)
A vehicle driven by an electric motor that is powered by rechargeable batteries.

Battery pack
A number of batteries connected together to provide the required power and energy for a given application.

Bio-energy
Energy derived from combustible waste materials or crops.

Biofuel
A gaseous, liquid or solid fuel that is derived from a biological source.

Biogas
A mixture of methane and carbon dioxide that results from the anaerobic decomposition of waste matter.

Biomass
The term used to describe all biologically produced matter at the end of its life. This includes both waste matter and crops that are specially grown as a source of energy.

Bipolar plates
The components of a fuel cell that act as the current collector for the positive electrode in one cell and for the negative electrode in the adjacent cell and also serve to join the cells electrically. The cells in a stack are series-connected and so allow the voltage to be built up.

Carbon footprint
A measure of the impact of human activities on the environment in terms of the amount of greenhouse gases produced; expressed as tonnes of carbon dioxide or carbon emitted, usually on a yearly basis.

Carnot cycle
The most efficient cycle of operation for a reversible heat engine. It consists of four operations, as in the four-stroke internal combustion engine, namely: isothermal expansion, adiabatic expansion, isothermal compression and adiabatic compression to the initial state.

Catalyst
A substance that increases the rate of a chemical reaction, but that is not itself permanently changed.

Cathode
An electrode at which a reduction process, *i.e.*, gain of electrons, is occurring. During electrolysis the cathode is the negative electrode. In a fuel cell, the cathode is the positive electrode where oxygen is consumed.

Cation
Ion in an electrolyte that carries a positive charge and that migrates towards the cathode under the influence of a potential gradient. See **Cathode**, **Ion**.

Chemical reactor
Engineering equipment in which a chemical reaction takes place.

Clathrates
Crystalline compounds formed between certain gases (*e.g.*, carbon dioxide, hydrogen sulfide, methane) and water at low temperatures and high pressures.

Climate change
A change of climate, attributed directly or indirectly to human activity, that alters the composition of the global atmosphere and is in addition to natural climate variability observed over comparable time periods. See **Global warming**.

Coal gas
A fuel gas, which is usually rich in hydrogen, produced when coal is heated in the absence of air (so-called 'destructive distillation' or 'pyrolysis'). It is a by-product in the preparation of coke and coal tar. Coal gas was a major source of energy in the late 19th and early 20th centuries and was also known as 'town gas'. The use of this gas declined with the increasing availability of natural gas. See **Pyrolysis**.

Cogeneration
See **Combined heat and power (CHP) system**.

Combined-cycle gas turbine (CCGT)
The technology employed in a natural gas-fired power station. The gas is first burnt in a gas turbine and the waste heat contained in the exhaust gases is then recovered and used to raise steam to drive a steam turbine.

Combined heat and power (CHP) system
An installation where there is simultaneous generation of usable heat and power (usually electricity) in a single process. The term is synonymous with cogeneration.

Conduction band
Partially filled or empty energy levels in a crystalline solid where electrons are free to move and thus allow the solid to conduct an electrical current.

Critical point (of a gas/liquid)
The temperature and pressure at and above which the gaseous and liquid states of a substance are indistinguishable.

Current density
In an electrochemical cell, the current flowing per unit electrode area.

Direct current
Electric current that flows in one direction only, although it may have appreciable pulsations in its magnitude.

Drive-train
The elements of the propulsion system that deliver mechanical energy from the power source to drive the wheels of a given vehicle.

Electric vehicle (EV)
A vehicle propelled by an electric motor.

Electrochemical capacitor
A capacitor that stores charge in the form of ions (rather than electrons), adsorbed on materials of high surface area. The ions undergo redox reactions

during charge and discharge. Also known as electrochemical double-layer capacitors, supercapacitors and ultracapacitors. See **Redox reaction**.

Electrode
An electronic conductor that acts as a source or sink of electrons that are involved in electrochemical reactions.

Electrode potential
The voltage developed by a single electrode, either positive or negative. The algebraic difference in voltage of any pair of electrodes of opposite polarity equals the cell voltage.

Electrolysis
The production of a chemical reaction by passing a direct electric current through an electrolyte.

Electrolyte
A chemical that conducts electricity by means of positive or negative ions. Electrolytes are solids, molten ionic compounds or solutions containing ions, *i.e.*, solutions of ionic salts or of compounds that ionize in solution.

Electrolytic cell
An electrochemical cell, which consists of a positive and a negative electrode and an electrolyte, through which an externally generated electric current is passed in order to produce an electrochemical reaction. See **Positive electrode**, **Negative electrode**, **Electrolyte**.

Electrolyzer
An electrochemical plant designed to effect the process of electrolysis.

Energy efficiency
The ratio of the energy output from a device to the energy input.

Endothermic reaction
A chemical reaction in which heat is absorbed. See **Enthalpy**.

Energy
The ability to do work or produce heat (measured in joules).

Energy band
The range of energies that electrons can have in a solid. In a single atom, electrons exist in discrete energy levels. In a crystal, where large numbers of atoms are held closely together in a lattice, electrons are influenced by a number of adjacent nuclei and the sharply defined energy levels of the atoms become bands of allowed energy that are separated by bands of forbidden values. In a metal, there is a continuous energy band.

Energy crops
Trees, grasses and other vegetation grown specially for use as a fuel or for extracting plant oils or alcohols that may be used as fuels in internal combustion engines.

Energy density
Stored energy per unit volume, usually expressed in MJ m^{-3} or kWh m^{-3}.

Enthalpy
A thermodynamic quantity (H) equal to the total energy of a system when it is at constant pressure. The gain or loss of energy by a system when it reacts at constant pressure is expressed by the change in enthalpy, symbolized by ΔH. When all the energy change appears as heat (Q), the change in enthalpy is equal to the heat of reaction at constant pressure, *i.e.*, $\Delta H = Q$. The values of ΔH and Q are negative for exothermic reactions (heat evolved from system) and positive for endothermic reactions (heat absorbed by system).

Entropy
A thermodynamic quantity representing the amount of energy in a system that is no longer available to do useful work. When a closed system undergoes a reversible change, the entropy change (ΔS) equals the energy lost from or transferred to, the system by heat (Q) divided by the absolute temperature (T) at which this occurs, *i.e.*, $\Delta S = Q/T$. At constant pressure, the amount of heat (Q) is equal to the change in enthalpy (ΔH). For a more detailed explanation, see **Box 4.2, Chapter 4**.

Equilibrium potential
See **Reversible potential**.

Equilibrium voltage
See **Reversible voltage**.

Exothermic reaction
A chemical reaction in which heat is evolved. See **Enthalpy**.

Fermi level
The energy level in a semiconductor where there is a 0.5 probability of finding an electron.

Fossil fuels
Carbonaceous deposits (solid, liquid or gaseous) that derive from the decay of vegetable matter over geological time spans.

Free energy of formation
See **Gibbs free energy** and **Box 4.2, Chapter 4**.

Fuel cell
An electrochemical device for generating low-voltage direct current electricity from a fuel (often hydrogen) and air or oxygen.

Galvanic cell
An electrochemical cell in which chemical energy is converted into electrical energy on demand; more commonly known as a 'battery'. See **Fuel cell**.

Gasification
A special type of pyrolysis where thermal decomposition takes place in the presence of a small amount of air or oxygen. See **Coal gas, Pyrolysis**.

Gettering
The removal of residual gas from a moderate vacuum by absorption into, or reaction with, a reactive species such as a metal.

Gibbs free energy
The energy liberated or absorbed in a reversible process at constant pressure and constant temperature. The change in free energy, ΔG, in a chemical reaction is given by $\Delta G = \Delta H - T\Delta S$, where ΔH is the change in enthalpy, ΔS is the change in entropy and T is the temperature. This is known as the Gibbs equation. See **Enthalpy**, **Entropy**.

Global warming
The observed and projected increases in the average temperature of the Earth's atmosphere and oceans.

Green electricity
Electricity generated from renewable energy sources, which include carbon-neutral biomass. (Note: nuclear energy, which does not liberate carbon dioxide, is sometimes counted as a form of green electricity.)

Greenhouse gases
Those gaseous constituents of the Earth's atmosphere, both natural and anthropogenic, that absorb and re-emit infrared radiation.

Higher heating value of a fuel (HHV)
The maximum heat of combustion (MJ kg^{-1}) of a fuel, based on complete combustion to carbon dioxide and water at 25 °C.

Hybrid electric vehicle (HEV)
A vehicle that has two power sources, often a conventional engine together with a battery or fuel cell that provides electric propulsion through an electric motor.

Hydrothermal reservoir
An aquifer containing water or brine at more than 100 °C and therefore pressurized.

Integrated gasification combined cycle (IGCC)
A technology employed in some coal-fired power stations. Instead of feeding pulverized coal directly to the boilers to raise steam, it is first converted to a gaseous mixture of carbon monoxide, hydrogen and nitrogen that is then combusted in a gas turbine.

Internal resistance
The opposition to current flow that results from the various electronic and ionic resistances within an electrochemical or photo-electrochemical cell.

Ion
An atom that has lost or gained one or more orbiting electrons and thus becomes electrically charged.

Ionization
Any process by which an atom, molecule or ion gains or loses electrons.

Isoelectronic compounds
Compounds with the same number of electrons.

Isothermal process
A process that takes place without change in temperature.

Joule–Thompson cooling
The cooling of a gas by allowing it to expand without gaining any external heat.

Latent heat
The heat absorbed or released by a substance when it changes state (*e.g.*, from solid to liquid or vice versa) at constant temperature and pressure. The term **Specific latent heat** denotes the heat absorbed or released per unit mass of a substance during the course of its change of state.

Lower heating value of a fuel (LHV)
The heat of combustion (MJ kg^{-1}) of a fuel, based on complete combustion to carbon dioxide and steam at 100 °C.

Maximum power point
The point on the current–voltage curve for a photovoltaic cell at which the cell generates maximum power.

Megawatt (MW)
Unit of energy equal to 10^6 W. MW_e denotes electrical output and MW_{th} denotes thermal heat output.

Membrane–electrode assembly
A component of a fuel cell that consists of a polymer membrane electrolyte coated with (or sandwiched between) positive and negative electrodes and then placed between bipolar plates.

Monopolar
The conventional method of constructing an electrolyzer (or battery) in which component cells are joined in parallel and have their electrodes at the same common potential.

Negative electrode
The electrode in an electrochemical cell that has the lower potential.

***n*-type semiconductor**
A semiconductor in which electrical conduction is due mainly to the movement of electrons.

Oil
In this book, oil and petroleum are used synonymously for crude (unrefined) oil.

Oil sands
A mixture of sand, clay, water and bitumen from which oil may be recovered and refined.

Oil shale
Rocks rich in organic material (kerogen) from which petroleum may be recovered by dry distillation.

Open-circuit voltage
The voltage of a power source, such as a battery, fuel cell or photo-electrochemical cell, when there is no net current flow.

Overpotential
The shift in the potential of an electrode from its equilibrium value as a result of current flow.

Peak power
The sustained pulsed power that is obtainable from a fuel cell or battery under specified conditions, usually measured in watts over a period of 30 s.

Petrol
Term used in the UK for a light hydrocarbon liquid fuel for spark-ignition internal combustion engines. Other terms for such fuel are gas(oline) and motor spirit.

Petroleum
See **Oil**.

Photo-electrochemical cell
Solar cells that extract electrical energy from light, including visible light. Each cell consists of a photo-sensitive electrode and a conducting counter electrode immersed in an electrolyte. Some photo-electrochemical cells simply produce a direct current, whereas others liberate hydrogen in a process similar to the conventional electrolysis of water.

Photolysis
A chemical reaction (often a decomposition) caused by exposure to light.

Photosynthesis
The chemical process by which green plants synthesize organic compounds from carbon dioxide and water in the presence of sunlight.

Photovoltaic (PV) cell
A semiconductor device for converting light energy into low-voltage, direct current electricity.

pH
A measure of the acidity/alkalinity (basicity) of a solution. The pH scale extends from 0 to 14 (in aqueous solutions at room temperature). A pH value of 7 indicates a neutral solution. A pH value of less than 7 indicates an acidic solution; the acidity increases with decreasing pH value. A pH value of more than 7 indicates an alkaline solution; the basicity or alkalinity increases with increasing pH value.

Physisorption
Adsorption of gases on solid surfaces whereby the bonding is by means of a weak intermolecular (van der Waals) attraction rather than by chemical bonding.

Positive electrode
The electrode in an electrochemical cell that has the higher potential.

Power density
The power output of an energy device per unit volume, usually expressed in terms of $W\ L^{-1}$ or $W\ dm^{-3}$.

Producer gas
A mixture of carbon monoxide and nitrogen made by passing air over very hot carbon. Usually some steam is added to the air and the mixture then contains hydrogen. The gas is used as a fuel in some industrial processes.

***p*-type semiconductor**
A semiconductor in which electrical conduction is due mainly to the movement of positively charged holes.

Pyrolysis
Thermal decomposition of a substance at elevated temperatures in the absence of air or oxygen.

Quantum yield
For photocells, the fractional number of electrons generated per photon incident on the cell or the ratio of the number of photon-induced reactions occurring to total number of incident photons.

Redox reaction
A chemical reaction that involves the transfer of an electron from one species (which is thereby oxidized) to another (which is thereby reduced). The species are known as redox reagents.

Regenerative braking
The recovery of some fraction of the energy normally dissipated during braking of a vehicle and its return to a battery or some other energy-storage device.

Renewable energy (Renewables)
All natural energy forms that do not derive from the combustion of fossil fuels.

Reversible potential
The potential of an electrode when there is no net current flowing through the cell.

Reversible voltage
The difference in the reversible potentials of the two electrodes that make up the cell.

Sensible heat
The heat absorbed by a substance that gives rise to an increase in temperature of the substance. See **Latent heat**.

Separator
An electronically non-conductive, but ion-permeable, component of a battery or fuel cell that prevents electrodes of opposite polarity from making contact.

Sequestration
The capture of carbon dioxide from streams of mixed gases and its subsequent indefinite storage.

Solar array
An assembly of solar panels electrically connected together.

Solar–thermal plant
A large-area assembly of mirrors to focus solar radiation to a point where a solar furnace may be used to raise steam, heat liquids or carry out chemical reactions.

Specific energy
Stored energy per unit mass, expressed in $MJ\ kg^{-1}$, $Wh\ kg^{-1}$ or $kWh\ kg^{-1}$.

Specific power
The power output of a battery or fuel cell per unit mass, usually expressed in $W\ kg^{-1}$.

Standard cell voltage
The reversible voltage of an electrochemical cell with all active materials in their standard states. See **Reversible voltage**.

Standard electrode potential
The reversible potential of an electrode with all the active materials in their standard states. Usual standard states specify a pressure of 101.325 kPa for gases and unit activity for elements, solids and $1\ mol\ dm^{-3}$ solutions – all at a temperature of 298.15 K. See **Reversible potential**.

Steam reforming
The reaction of fossil fuels with steam at high temperature to generate a mixture of hydrogen and carbon monoxide.

Stoichiometry
The branch of chemistry concerned with the relative proportions in which atoms or molecules react together to form chemical compounds; derivative: **stoichiometric**.

Sustainable energy ('sustainability')
A set of energy technologies that meets humanity's needs on an indefinite basis without producing irreversible environmental effects. (Note: various definitions exist in the literature, but they all convey the same message.)

Synthesis gas ('syngas')
A mixture of carbon monoxide and hydrogen made by reacting natural gas with steam and air or oxygen.

Synthetic natural gas
Methane produced by the catalytic reaction of carbon monoxide with hydrogen or from coal by reaction with hydrogen.

Thermocline
The depth in the ocean where the temperature changes abruptly between surface warm water and deep cold water.

Thermohaline circulation
A natural oceanic phenomenon whereby cold water in the Arctic/North Atlantic sinks to the ocean floor and then circulates at depth around the world, to surface eventually in the Indian and Pacific oceans some hundreds of years later. This circulation is sometimes called the **Ocean conveyor belt**, the **Great ocean conveyer**, the **Global conveyor belt**, or, most commonly, the **Meridional overturning circulation** (often abbreviated as **MOC**).

Thermolysis
The dissociation or decomposition of a molecule by heat.

Town gas
See **Coal gas**.

Traction battery
A battery designed to provide motive power.

Valence band
The range of energy levels of electrons that bind atoms of a crystal together. When electrons are excited from the valence band to the conduction band, the resulting electron hole is mobile and gives rise to p-type conduction in the valence band. See **Conduction band**, **Energy band**.

Voltaic efficiency
The ratio, usually expressed as a percentage, of the average voltage during discharge to the average voltage during charge.

Water gas
A mixture of carbon monoxide and hydrogen produced by passing steam over hot carbon (coke). The reaction is strongly endothermic but may be combined with the exothermic reaction for producer gas. See **Producer gas**, **Steam reforming**, **Water-gas shift reaction**.

Water-gas shift reaction
The reaction of water gas with steam to yield hydrogen and carbon dioxide.

Wind farm
A collection of wind turbines, grouped together to form a single generating unit.

Conversion Factors for Units and Useful Quantities

All factors have been rounded to the significant digits given in accordance with accepted practice.

Energy Conversion Factors[a]

To: From:	GJ Multiply by	Gcal	Quad	GWh	Mtoe[b]
GJ	1	0.239	9.479×10^{-10}	2.778×10^{-4}	2.4×10^{-8}
Gcal	4.184	1	3.968×10^{-9}	1.163×10^{-3}	1×10^{-7}
Quad	1.055×10^{9}	2.520×10^{8}	1	2.931×10^{5}	25
GWh	3.600×10^{3}	860	3.412×10^{-6}	1	8.6×10^{-5}
Mtoe[b]	4.2×10^{7}	1×10^{7}	4.0×10^{-2}	11.6×10^{3}	1

[a] 1 quad = 10^{15} BTU; 1 BTU = 1055 J; 1 calorie = 4.184 J; 1 kWh = 3600 kJ.
[b] The calorific value of crude oil varies by a few percent, as determined by its origin and composition. These values are therefore approximate.

Volume Conversion Factors

To: From:	Gal US Multiply by	Gal UK	Barrels[a]	Litres	m^3	ft^3
US gallon	1	0.8327	0.02381	3.785	0.0038	0.1336
UK gallon	1.201	1	0.02859	4.546	0.0045	0.1606
Barrel	42	34.97	1	159.0	0.159	5.616
Litre	0.2642	0.220	0.0063	1	0.001	0.0353
Cubic metre	264.2	220.0	6.289	1000	1	35.32
Cubic feet	7.480	6.227	0.1781	28.32	0.0283	1

[a] There are approximately 7.4 barrels of crude oil to a tonne, as determined by the density of the crude.

Pressure Conversion Factors

1 standard atmosphere (atm) = 1.01325 bar = 101.325 kPa = 14.696 psi.

Hydrogen Data

1 mol hydrogen = 2.016 g = 22.414 L (gas at NTP).
  (NTP is defined as: $T = 273.15$ K; $P = 1$ standard atm).

Density of gaseous hydrogen = 0.0899 g dm^{-3} (NTP).

1 kg hydrogen occupies 11.12 m^3 at NTP.

1 m^3 hydrogen at NTP weighs 89.9 g.

Density of liquid hydrogen = 70.8 g dm^{-3}.

Higher heating value (HHV) of hydrogen = 142 MJ kg^{-1} = 39.4 kWh kg^{-1}.

HHV (volumetric) = 12.77 MJ N-m^{-3} = 3.55 kWh N-m^{-3}.

Lower heating value (LHV) of hydrogen = 120 MJ kg^{-1} = 33.3 kWh kg^{-1}.

LHV (volumetric) = 10.79 MJ N-m^{-3} = 29.98 kWh kg^{-1}.

Energy content of pressurized hydrogen gas:

Pressure/MPa	Energy content/MJ dm^{-3}
20	2.53
55	6.96
70	8.86
80	10.12

Energy supplies

Total world energy supply in 2004 = 11 059 Mtoe = 440 quads.[1]

World production of crude oil in 2004 = 3793 Mtoe $\approx$ 28 000 M barrels = 76.7 M barrels per day.[1]

World production of hydrogen = 45–50 Mt per year.

Energy content of 50 Mt (using HHV) $= 7.1 \times 10^{18}$ J $= 6.7$ quads $= 169$ Mtoe.

Hydrogen production represents around 1.5% of total world energy supply.

Calorific Values of Fuels

Natural gas

Values range from 38.1 to 42.0 MJ m^{-3}, as determined by origin.
  (A notable exception is Netherlands gas, which gives only 33.3 MJ m^{-3}.)

Petrol (gasoline)

By volume:	32.80 MJ L^{-1} $= 9.11$ kWh L^{-1} (HHV)	
	31.17 MJ L^{-1} $= 8.66$ kWh L^{-1} (LHV)	
By mass:	46.70 MJ kg^{-1} $= 12.97$ kWh kg^{-1} (HHV)	
	44.38 MJ kg^{-1} $= 12.33$ kWh kg^{-1} (LHV)	

Methanol

By volume:	18.60 MJ L^{-1} $= 5.17$ kWh L^{-1} (HHV)	
	16.02 MJ L^{-1} $= 4.45$ kWh L^{-1} (LHV)	
By mass:	23.30 MJ kg^{-1} $= 6.47$ kWh kg^{-1} (HHV)	
	20.10 MJ kg^{-1} $= 5.58$ kWh kg^{-1} (LHV)	

Note: not all of these units are necessarily employed in this book, but they are commonly encountered in the energy literature.

References

1. *Key World Energy Statistics*, 2006 edition. International Energy Agency, Paris, 2006.

CHAPTER 1

Why Hydrogen Energy?

From time immemorial, mankind has burnt wood in order to keep warm and to cook food. With the discovery of coal and the development of mining engineering, a new source of fuel, of higher calorific value, became available. As populations expanded and became urbanized, wood was less readily accessible and coal assumed greater importance for heating purposes. Following the introduction of rotative steam engines in the 1780s, coal was used as the prime source of energy for the production of mechanical power. Steam engines propelled ships, railway locomotives and traction engines and also provided a universal means for generating power in factories and on farms.

Late in the 19th century, the internal combustion engine was developed. Liquid petroleum was exploited – first in North America and then across the world – and was refined to provide fuels for both petrol and diesel engines. With its greater efficiency and convenience, this new technology soon replaced steam engines for most applications. Consequently, in many countries the use of coal declined (at least in percentage terms), while that of petroleum grew rapidly.

Since the mid-20th century, natural gas fields have been found in abundance. Some of the gas is associated with oil wells, but exists on its own in other places. Where oil wells are remote from centres of population, the gas was initially seen as a by-product that had no commercial value and was therefore flared. This situation changed with the development of technology for liquefying natural gas and conveying it to market by road or by sea in cryogenic tankers. Thus, once considered to be a waste product associated with oil, natural gas is now regarded as a prime fuel. With improvements in offshore drilling technology, it became possible to seek and access reservoirs in ever-deeper waters. Often these were located conveniently close to customers, *e.g.*, in the North Sea and the Gulf of Mexico, for the gas to be delivered by pipeline. The result is that much of the developed world has now adopted gas as the preferred fuel for space heating and cooling, for use in industry and for electricity generation. Starting with wood (a form of biomass), mankind has moved to fossil fuels – first to coal, then to petroleum and latterly to natural gas – to provide the energy needed by society. Electricity also is a useful, but secondary, form of energy since it is manufactured from primary energy sources. In the mid-1950s, commercial nuclear power was added to the range of primary energy sources.

1

Fossil fuels are laid down over geological time and, once used, cannot be replaced on any realistic time-scale. These fuels represent the world's energy capital. By contrast, many renewable (sustainable[†]) forms of energy, *i.e.*, those derived from wind, solar or marine (tidal, wave, ocean) sources, must be used as they are produced; otherwise, they are wasted. Other 'renewables' may have some storage element associated with them: biomass can be stored for short periods, while hydro energy is contained in mountain lakes or reservoirs held back by dams. Geothermal energy, like fossil fuels, is retained underground until it is required. Renewables comprise the world's current account in energy. As in financial matters, where it is easier to raid the capital account than work hard to earn money for the current account, so it is easier and cheaper to dig or drill for fossil fuels than it is to extract useful energy from renewable sources. In essence, although renewable energy is widely available, the world faces major problems in harnessing this resource — many of the forms of this energy are small-scale, diffuse and, as yet, hardly cost-competitive with fossil fuels. Moreover, those that generate electricity directly have no storage component.

Most authorities attribute the rise in the mean global temperature over recent years to the combustion of fossil fuels that has grown steadily since the Industrial Revolution. The concentration of carbon dioxide in the atmosphere has risen steadily from 280–300 ppmv in the 18th century to 360–380 ppmv today, an increase of around 25%. Carbon dioxide is known to absorb infrared radiation re-emitted from the Earth and is a principal 'greenhouse gas'; see Section 1.2. The progressive move from coal to oil and then to natural gas represents 'decarbonization' of fuels and is desirable in that it results in less carbon dioxide release per unit of energy produced. Natural gas (methane) has four atoms of hydrogen per carbon atom and is the limit of decarbonization without going all the way to hydrogen, which is obviously a carbon-free fuel; see Figure 1.1. The idea of introducing hydrogen as the universal vector for conveying renewable forms of energy, and also as the ultimate non-polluting fuel, is encapsulated in idealized form in Figure 1.2. This proposition is commonly known as the 'Hydrogen Economy'. The upper part of the diagram is generally referred to as the transitional phase, during which hydrogen is produced from fossil fuels; the lower part relates to the long-term, post fossil-fuel, age when hydrogen will be manufactured from renewable energy sources and used as a storage medium and as a super-clean fuel.

There is, however, a problem with the concept of a sustainable Hydrogen Economy. Within our present span of vision, renewables alone do not afford a path to a carbon-free future because they are difficult to harvest on a large scale and, as noted above, breakthroughs in cost must be achieved if these sources are to supplant fossil fuels and become commonplace. Also, there is often local opposition to the construction of renewable facilities such as hydroelectric dams or wind generators, which may spoil areas of scenic beauty or interfere

---

[†]A 'sustainable' energy source is one that will meet the needs of the human race on an indefinite basis without causing long-term damage to the environment. This concept is termed 'sustainability'. Various definitions of 'sustainability' exist in the literature, but they all convey the same message.

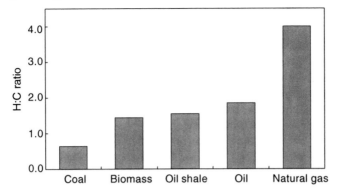

Figure 1.1   Hydrogen-to-carbon atomic ratio in carbon-based fuels. 'Oil' refers to heavy residues and other petroleum fractions.

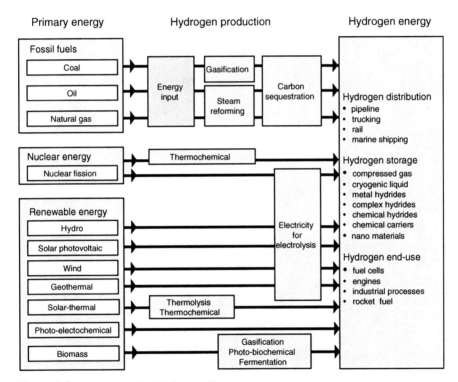

Figure 1.2   A sustainable Hydrogen Economy.

with natural habitats. The counter proposition of increasing the deployment of nuclear power, which is not usually regarded as renewable energy but at least is carbon-free, is unpopular in many quarters because of concerns over radio-active waste. Nevertheless, some countries already rely on nuclear power to provide an appreciable percentage of their domestic electricity requirements, *e.g.*, France (78%), Sweden (46%), Ukraine (45%) and Korea (36%).[1]

Hydrogen is the most abundant element in the universe. It is a major component of stars, including the Sun, whose heat and light are produced through the nuclear-fusion process that converts hydrogen into helium. Elemental hydrogen does not occur in significant amounts on Earth and energy has to be supplied in order to extract it from water or fossil fuels. Hydrogen is therefore not a primary energy source but a secondary energy vector. Energy from primary sources can be stored in hydrogen by decomposing water using chemical, thermal or electrical energy.

The various primary energy sources that can be used for the production of hydrogen and potential applications for this energy vector are summarized schematically in Figure 1.3.[2] In the near-term, it is probable that hydrogen will, as now, be derived principally from fossil fuels, since this is the most economic route.

Economic development and poverty eradication depend on secure and affordable energy supplies. Although reserves of conventional oil and gas are

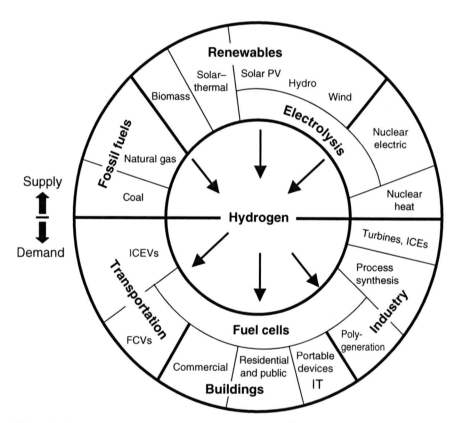

Figure 1.3   Hydrogen supply options and major uses.[2] PV, photovoltaic (cells); ICEs, internal combustion engines; IT, information technology; FCVs, fuel cell (electric) vehicles; ICEVs, internal-combustion-engined vehicles. (Courtesy of International Energy Agency Clean Coal Centre).

diminishing relatively rapidly, other fossil fuels (heavy oils, tars, coal) are widely distributed and can meet the need for energy security in the medium-term, even though they are environmentally challenging and more costly to process. Technology, driven by the right incentives, offers possible answers to the environmental problems. For example, 'clean coal technology' (see Section 2.7, Chapter 2) is designed to use coal in a more efficient and cost-effective manner while enhancing environmental protection through the capture and storage of emissions, principally carbon dioxide.

As an energy-storage medium that is manufactured from a primary energy source, hydrogen may be used to convey energy to where it may be utilized. In this respect, it is analogous to electricity, which is also a secondary form of energy. Hydrogen and electricity are complementary: electricity is used for a myriad of applications for which hydrogen is not suitable, whereas hydrogen, unlike electricity, has the attributes of being a fuel and an energy store. These two energy vectors are, in principle, inter-convertible; electricity may be used to generate hydrogen by the electrolysis of water, while hydrogen may be converted to electricity by means of a fuel cell. It should be noted, however, that such electrochemical devices are less than 100% efficient and there is a significant loss of useful energy in the inter-conversion. Nevertheless, as discussed in Chapter 6, the realization of a Hydrogen Economy is linked irrevocably with that of the fuel cell.

Many forms of renewable energy are manifest as electricity, which is used immediately it is generated. For applications that require a portable fuel (*e.g.*, road transportation), it would be necessary to convert this form of renewable energy to a chemical fuel, for example methanol or hydrogen, which could be used in an internal combustion engine or re-converted back to electricity in a fuel cell to drive an electric vehicle. In energy terms, this would be a grossly inefficient process and uneconomic under most conditions. Generally, it would be better to utilize electricity derived from renewables directly since the energy lost in local distribution of electricity is comparatively small. There may, however, be exceptions to this rule. One situation might be on islands or in isolated communities where there is plenty of renewable energy in the form of wind or solar electricity, but no means of storing it from times of surplus to times of peak demand. Hydrogen could then provide an energy store and later be reconverted to electricity, although this approach to storage would be in competition with batteries or standby diesel generators. Another exception might be in countries such as Norway or Iceland, where there is an abundance of cheap hydro or geothermal electricity, some of which could be converted to hydrogen for use in road vehicles, ships, *etc.* Ultimately, when fossil fuels are really scarce and expensive and when renewable energy technology has become economically competitive, it may prove practical on a much wider basis to convert renewable electricity to hydrogen fuel. But that day may yet be distant. The International Energy Agency (IEA) has forecast that renewables (excluding nuclear and hydro) will still account for only 10% of world energy supply by 2030; of this, more than half is expected to be derived from biomass.[1] Meanwhile, it is perfectly feasible to manufacture hydrogen from all types of fossil fuel and this is likely to be the route forward in the medium-term to a sustainable future, always provided that technologies for carbon capture and storage become available; see Chapter 3.

Before embarking on a discussion of hydrogen production and utilization (in later chapters), it is pertinent to consider the forces that drive the present international push for hydrogen energy and why it is that many people view the Hydrogen Economy as the fulfilment of an environmental dream for the long-term future. The four key 'drivers' are:

- national security of energy supplies;
- climate change (global warming);
- atmospheric pollution;
- electricity generation.

These drivers, which are examined in detail below, are inter-related in a complex fashion in an increasingly complex world. First, however, we should consider briefly the many obstacles that have to be surmounted before a Hydrogen Economy can become a reality. The obstacles fall into a number of different categories, as follows:

Institutional

- the difficulty in a free market of building and sustaining consensus on a long-term energy policy for a nation;
- the short time horizon of many politicians and much of industry;
- the inflexibility of the existing energy infrastructure and the long time-scale associated with effecting major changes to energy supply and usage;
- the lack of a hydrogen infrastructure and the huge cost of introducing one;
- the large scale of hydrogen production required to make a national impact and the inability of present-day electrolyzers even to approach this scale;
- the small-scale and present cost of generating electricity from most renewable forms of energy;
- the little near-term market demand for hydrogen as a fuel.

Technical

- technological barriers associated with the production, distribution and utilization of hydrogen – starting from coal and lower-grade fossil fuels with capture and storage of the carbon released as carbon dioxide;
- the problems of producing hydrogen efficiently and affordably using clean technology;
- the lack of a satisfactory hydrogen-storage medium, particularly for mobile applications (vehicles);

- the present limited performance, reliability and lifetime of fuel cells especially for mobile applications.

Regulatory

- concerns over the safety aspects of hydrogen installations and, in the case of vehicle refuelling, possible widespread use by the general public;
- the absence of internationally consistent codes and standards to ensure hydrogen safety and to facilitate its commercialization;
- the task of training and certifying mechanics, technicians and others involved in implementing hydrogen energy.

Financial

- the requirement for huge amounts of capital, including risk capital, to establish a hydrogen infrastructure;
- the need to reduce system costs to compete with traditional fuels and, for mobile applications, with internal combustion engines;
- the present high cost of fuel cells, particularly when compared with engines.

These barriers present a formidable challenge and it is necessary to develop a strategy for an integrated approach to overcoming them. Whilst petroleum and natural gas remain widely available and relatively inexpensive, it will prove difficult for hydrogen energy to compete on a favourable cost basis. Competition will be possible if/when premium fossil fuels fail to meet market requirements and become very much more expensive, or when concern about climate change leads to the introduction of high taxes for the liberation of carbon dioxide into the atmosphere. The various impediments to the introduction of hydrogen, as listed under the above four broad categories, must always be borne in mind when proposing hydrogen as a universal solution to present-day energy problems.

1.1 Security of Energy Supplies

The total primary energy supply throughout the world has increased from 6035 Mtoe in 1973 to 11 059 Mtoe in 2004, a rise of 83% in 31 years, and is projected by the IEA to reach 16 500 Mtoe by 2030, a further 49% on the 2004 level; see Table 1.1.[1] This increase is a result of growth in the world population and a general rise in prosperity. There is a well-established link between the gross domestic product (GDP) of a nation and its energy consumption, although nations are now trying hard to break this link.

The growth in demand for different fuels and energy sources from 1973 to 2002 and projected through to 2030 is plotted in Figure 1.4.[3] It is expected that fossil fuels will account for most of the increase in energy supply between 2002

Table 1.1   Actual and projected growth in world demand for primary energy
sources 1973–2030[1]. (Supply, based on Mtoe for each fuel and
given in percentage terms).

	Year		
Energy supply	*1973*	*2004*	*2030*
Oil	45.1	34.3	34.1
Coal	24.8	25.1	22.9
Gas	16.2	20.9	24.2
Combustible renewables: (biomass and waste)[a]	11.1	10.6	
Nuclear	0.9	6.5	4.7
Hydro	1.8	2.2	2.3
Renewables[b]	0.1	0.4	11.8
Total (%)	100.0	100.0	100.0
Total (Mtoe)[c]	6034	11 059	16 500

[a] The historical data separate out combustible renewables (mostly wood and dung) and waste,
whereas the prospective data include these figures under 'renewables'. This explains the apparent
increase attributed to renewables, which is mostly illusionary.
[b] Includes geothermal, solar, wind and tidal for 1973 and 2004 and also biomass and waste for 2030.
[c] To convert Mtoe to TJ, multiply by 4.18×10^4. See also: Conversion Factors for Units and Useful
Quantities, pages xxxvi–xxxviii.

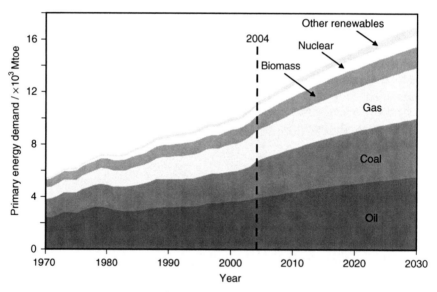

Figure 1.4   World primary energy demand per year, 1970–2030.[3] Mtoe represents 10^6
tonnes of oil equivalent, which is a convenient unit for comparing very
large quantities of energy in different forms; 1 Mtoe $= 4.18 \times 10^4$ TJ. Other
sources quote oil production in mega barrels (1 Mb $= 10^6$ barrels) or giga
barrels (1 Gb $= 10^9$ barrels); see Figure 1.5. There are 7.3–7.4 barrels per
tonne, as dictated by the density of the oil.
(Courtesy of International Energy Agency).

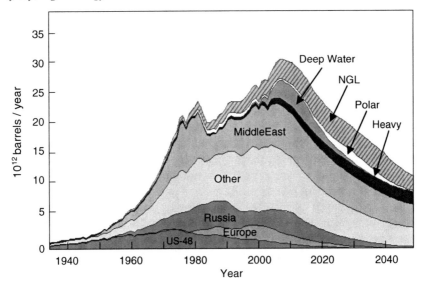

Figure 1.5 Annual production of oil from various regions.[4] Deep water: oil lying in more than 500 m of water. NGL: natural gas liquids, e.g., ethane, propane, butane. Polar: oil produced above the Arctic Circle. Heavy: oil with an American Petroleum Institute (API) specific gravity > 17.5°, i.e., viscosity > 10 cP. US-48: all USA except Alaska and Hawaii.
(Courtesy of the Association for the Study of Peak Oil and Gas).

and 2030. These data demonstrate the magnitude of the task facing mankind and the time-scales involved in substituting renewable, sustainable energy for a significant fraction of the fossil fuel consumed.

In looking ahead to 2030, it is predicted that the percentage of the world energy market supplied by oil will not change greatly with respect to that of 2002. By contrast, in percentage terms: gas will rise, coal will decline slightly and nuclear will fall significantly as old nuclear stations are retired and not replaced. In many ways, these forecasts are counter-intuitive, in the light of declining reserves of oil and gas and the economic growth of China and India with their vast reserves of coal. They are, however, the considered view of the IEA, the world's leading authority, and reinforce the competition bound to be faced by renewable forms of energy and the difficulty of introducing the Hydrogen Economy on this time-scale. Naturally, these projections to 2030 are only forecasts, albeit made by experts, and may be proved wrong by future events.

Many geologists and petroleum engineers are of the opinion that the Earth's ultimate reserves of petroleum are around 2×10^{12} barrels, of which over 40% has been used already. The concept of 'reserves' is open to debate and some authorities opt for 3×10^{12} barrels. Nevertheless, it is claimed that over 90% of all available oil has been discovered and mapped. Some important oil-producing regions (e.g., USA, North Sea) have passed their peak production rates and are in decline; others are expected to peak within 10 years; see Figure 1.5.[4]

Moreover, the rate at which oil is being pumped (28×10^9 barrels in 2003) greatly exceeds the rate at which new reserves are found.

Even when new oil fields have been identified, substantial investment of time and money is required, particularly offshore in deep water, before extraction can begin. It is important to distinguish between reserves in place and excess production capacity available at short notice. Among the major producing countries, only Saudi Arabia has excess capacity (about 3×10^6 barrels per day $\sim 4\%$ of world consumption) that could be brought into use quickly. Elsewhere, the exploitation of fresh reserves will require substantial investment and concern has been expressed that the necessary capital may not be available.

If these developments are not serious enough, an even more alarming fact is that much of the remaining 'conventional' oil (over 60%) is concentrated in just five Middle Eastern countries: Saudi Arabia, Iraq, Kuwait, United Arab Emirates and Iran. New oilfields will no doubt be discovered, for instance in the countries of the former USSR and/or off the coast of West Africa, but are unlikely to compare in size with those of the Middle East and will not significantly change the overall picture. Alberta (Canada) and Venezuela have vast reserves of 'oil sands' and heavy oil (bitumen), respectively, that can be mined and refined to petroleum. In principle, both resources are sufficiently extensive to replace much of today's oil supply, although at a considerably higher cost – not only economically, but also in terms of environmental impact. About 2 tonnes of oil sands have to be dug up, moved and processed to obtain one barrel of crude oil. This involves the input of large amounts of energy and the release of corresponding quantities of carbon dioxide. Despite these limitations, the operation in Alberta already accounts for almost half of Canada's oil output and is growing rapidly.

Between 1973 and 2004, world crude oil production increased by 40% compared with 83% for primary energy as a whole.[1] This was because natural gas and nuclear power took over many of the duties formerly assigned to oil (*e.g.*, electricity generation, heating and cooling of buildings). Accordingly, the lower growth rate in oil production has not reflected the much larger increase in demand (90%) from the transportation sector over the same period. Oil usage is now largely confined to this sector of the economy and also to the fuelling of agricultural and earth-moving machinery and to the production of industrial chemicals and fertilizers. The requirement for petroleum will doubtless intensify as the developing countries aspire to Western-style mobility. For example, given the population sizes of China and India, it is clear that if just these two countries were to become fully mechanized their petroleum requirements would constitute a large fraction of present oil production. Clearly, this could not happen in a sustainable energy future. Meanwhile, present indications suggest that there will be growing competition for oil – not in the distant future, but within the next two decades when world production peaks and starts to decline. This has been dubbed 'The Big Rollover'. From that point on there will be oil shortages and a new political dynamic in which countries compete for limited supplies. Unless there is a widespread acceptance of this disturbing prospect and an urgent response, just five Arab countries (and

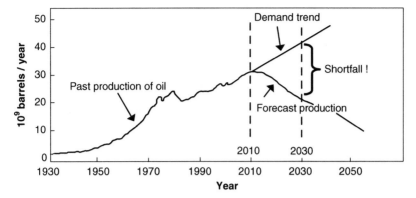

Figure 1.6 Past and projected future production of oil compared with demand.

possibly, later, Canada and Venezuela) will effectively control the supply of petroleum.

Among oil-importing nations, notably the USA, Japan and the European Union, there is real cause for concern over whether adequate supplies will be available. The IEA data predict a growth in demand for oil of 62% from 2002 to 2030.[3] This is largely a result of rising expectations for personal transportation in the developing world. The demand probably cannot be met by conventional oil and forward estimates are that the shortfall is likely to be almost 50% and will rise to over 70% by 2050; see Figure 1.6. This deficiency will have to be made good by different sources of oil, such as oil sands and shales, by bio-diesel and by emerging gas-to-liquid technology; see Chapter 2. Other contributory factors to filling the petroleum supply gap will be greater efficiency in its use (improved engines and hybrid drive systems) and reduced consumption (brought about by mode shifts in transportation and higher petroleum prices). All this supposes, of course, that concerns over climate change do not lead to sudden and dramatic political acts to curtail greenhouse gas emissions by limiting the use of fossil fuels.

What will be the response of the oil-rich countries to this developing situation? They may decide to restrict production for political reasons, to extend the life of their reserves or to force up the price of oil products. Even if there were a desire to expand production at the rate needed to meet market demand, this might not prove possible either practically or in terms of capital requirements. Whatever the outcome, it is almost bound to result in rapidly escalating prices. For a long time, the price of a barrel of oil (42 US gallons) was around US$20–25, but from 2005 onwards there has been a dramatic increase to over US$60. Rather than falling back to lower levels, it is widely forecast that, in the medium-term, oil prices will continue to spiral upwards. To bring future demand back into line with the supplies available, a five- or ten-fold increase in the price of a barrel of oil is certainly possible and this may well occur within 10–20 years, a short timeframe. It is salutary to recall that at US$20, the price until fairly recently, a barrel of oil cost about the same as a

litre of whisky – and the latter did not take geological ages to mature!
Obviously, oil was and still is, grossly undervalued in terms of its usefulness
to society and its finite availability.

The situation as regards natural gas is somewhat better. Again, there
are very large proven reserves and these are more widely scattered around
the world than those of oil, with about one-third in the Middle East and
another third in the Russian Federation.[1,5] The global demand for natural gas
is, however, expanding very rapidly because it is a convenient and clean fuel
that is relatively low in carbon. Most new power stations are based on natural
gas since electricity is produced more efficiently in combined-cycle gas turbines (CCGTs) and there are fewer gaseous pollutants than with coal. Pipelines are being laid across the world to bring gas supplies to market. Where
the distances are too great, for example to Japan and Korea, the gas is
liquefied and shipped in cryogenic tankers. With depletion of the gas fields in
the North Sea, the UK and other Western European nations will soon be
importing much more gas by pipeline from Russia and Kazakhstan or
from the nations around the Caspian Sea. The UK plans also to import
liquefied natural gas (LNG). Some authorities believe that global reserves of
natural gas have been greatly underestimated and that, despite growing
demand, there is unlikely to be any supply limitation throughout the course
of this century.[6] In this context, it is noteworthy that total world reserves
stood at $42 \times 10^{12} \, m^3$ in 1970 and are now $176 \times 10^{12} \, m^3$, despite $56 \times 10^{12} \, m^3$
having been consumed during this period. If this optimistic view is correct,
then natural gas will provide a cost base against which hydrogen as a
fuel will have to compete. In the longer term, if/when natural gas supplies
become depleted, the world can fall back on its massive reserves of coal, which
are sufficient to last for a century at least. This fuel switching will be
dependent upon clean coal technologies being well advanced to produce
hydrogen for use in gas turbines or fuel cells with almost no emissions of
carbon dioxide.

When importing oil and gas over long distances, security of supply is not just
a matter of availability and cost. It also depends on the reliability of the
delivery system. Pumps have been known to fail, gas pipelines have been
breached as a result of accident and there is always the possibility that political
action, terrorist attack or industrial dispute will cause a serious disruption of
supply. These are matters of grave concern.

Another worry, not strictly related to security but equally important, is the
impact of imported oil and gas on a nation's balance of trade. For instance, the
UK has enjoyed its own indigenous supplies from the North Sea for about
30 years and these have helped to provide economic stability. As the gas fields
become exhausted and fossil fuels have to be imported, there will be a significant impact on the country's trade balance, especially in an era of rising
energy prices. Many other nations suffer similarly.

These are all reasons why countries strive to be self-sufficient in energy
and tend to favour the use of any indigenous resources that they may have. The
USA, for example, has its own supplies of coal, gas and nuclear power, but

is now dependent upon imports for much of its oil. The attendant cost is approaching US$1 × 10¹¹ per year and is rising rapidly. Germany and Poland have appreciable deposits of coal, but little gas and no oil. France has some coal but relies heavily on nuclear and hydro power for its electricity. Australia is rich in both coal and natural gas. Those countries that have substantial coal reserves are certainly going to use them in a situation of rising prices for oil and gas. The challenge is to develop new technologies to convert coal cleanly and efficiently to electricity, to synthetic gas and possibly to oil products, without releasing pollutants to the atmosphere. This is a huge task, but one that will become more economically feasible with increases in the cost of petroleum and natural gas. Similarly, the conversion of coal to hydrogen, a longer-term goal, will become a more viable proposition.

Renewable sources of energy, in particular wind and solar power, will grow progressively and the costs of manufacture and installation will decline through the benefit of large-scale production. Despite these developments, the new energy forms will make only a modest contribution to the global energy supply during the next couple of decades, with marked variation from country to country according to the local conditions. ExxonMobil has forecast that the contribution of non-fossil fuels to the total world energy supply will be less than 20% in 2020 and that wind and solar energy will provide only 0.3% of the total; see Figure 1.7. Indeed, in the foreseeable future, few major countries are likely to derive more than 10−25% of their electricity from renewables (excluding hydro). To put the problem in context (and leaving aside the issue of whether electricity can substitute for petroleum as a transportation fuel), the projected shortfall of about 21 × 10⁹ barrels of oil per year by 2030 is equivalent, in energy terms, to 33 000 TWh of electricity. This is almost twice the total world production of electricity in 2004.[1] To produce so much electricity from renewables by 2030 would be quite a task! Nevertheless, both hydro power and wind power are now economically viable

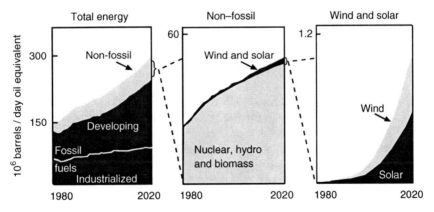

Figure 1.7 Contribution of non-fossil energy to total world energy supply, extrapolated to 2020.
(Courtesy of ExxonMobil).

and it is expected that both will grow rapidly in the years ahead in those regions where the resource is plentiful. The growth of hydro power will be constrained by geography, topography and environmental concerns, although there is plenty of scope for further development in some countries. As mentioned earlier, so long as renewables constitute a relatively minor component of the electricity supply, it will generally be more efficient in energy terms to utilize the electricity directly rather than to convert it to hydrogen.

Diversity of energy supply is now recognized as the key to energy security for any country that has to import much of its fuel. In the case of the UK, for example, it has been suggested that an electricity scenario based upon 30% coal, 30% gas, 30% nuclear and 10% renewables would be a good strategic mix; although whether the electricity utilities and the investing community would agree with this proposal is another matter. With de-regulation of electricity generation, it is difficult for governments to develop and implement an overall national generating strategy, although this may be possible with a suitable blend of incentives and taxes.

Superimposed on all these strategic considerations is a political issue. By and large, politicians and the general public are concerned more with the affairs of the moment. When there are immediate threats of war, terrorist attacks, famine, drought, earthquakes, hurricanes and general political discontent, who can blame politicians for focusing on these and the prospects for winning the next election as opposed to more distant issues? Similarly, the 'man in the street' is occupied with short-term matters that involve family, food, job, housing, finance, recreational activities, *etc.*, and has little time left to worry about distant threats to civilization. The mantle therefore devolves to a minority of concerned citizens and energy analysts to alert the public to the developing energy situation when supplies of conventional oil and gas show signs of becoming inadequate. In particular, a heavy responsibility falls on them to educate teachers so that they, in turn, will encourage the next generation to take seriously the issues of energy supply and security that loom ahead. Such an approach has worked well in the areas of environmental science and bio-diversity and it is now time for educationalists to address the world's energy future. Already this is happening through the widespread anxiety over climate change and 'carbon footprints' – topics that are now widely appreciated and debated.

Much of the discussion in this book revolves around a period of decades or, at longest, the rest of the 21st century. What of the centuries to come? By then, fossil fuels will mostly have been utilized or deemed unacceptable because of the greenhouse gas problem, while the costs of the remaining oil and gas supplies will have risen inexorably. At this point in the future, it is difficult to envisage any option other than nuclear fusion — still speculative — or renewables, which are a more certain but diffuse, small-scale and costly source of energy. Evidently, on this time-scale, today's economics are irrelevant; energy would become a more expensive commodity and people would simply have to adjust their lifestyles accordingly.

1.2 Climate Change (Global Warming)

In all the debate over climate change and its origin, three facts seem incontrovertible: (i) the Earth's climate is changing and generally warming; (ii) the concentration of carbon dioxide in the atmosphere has risen steadily since the Industrial Revolution when fossil fuels started to be burnt in large quantities; (iii) carbon dioxide is a greenhouse gas that absorbs infrared radiation reflected from the ground and prevents its escape into space. Most authorities link these three facts and conclude that carbon dioxide derived from fossil fuels is largely responsible for the observed change in climate. Is this, however, the complete story?

The factors that control our climate are very complex and only partly understood. Carbon dioxide is released in vast quantities from natural processes, notably by the respiration and decay of plants and animals on land and plankton in the oceans. Releases from respiration and decay are approximately balanced by a similar quantity taken up by photosynthesis (the process whereby living organisms use light to produce complex substances from carbon dioxide and water) and by re-absorption in the oceans. Other natural sources of carbon dioxide include volcanoes, forest fires and evaporation from oceans, lakes and rivers. By comparison, anthropogenic releases of carbon dioxide from combustion processes in, for example, power stations and motor vehicles are small. With extensive de-forestation taking place, however, one of the principal sinks for carbon dioxide is being removed and this is compounding the effect of burning fossil fuels. The results of two independent analyses[7,8] of the global flux of carbon dioxide (expressed as Gt of carbon) are summarized in Table 1.2. Aside from the apparent discrepancy between the amount of carbon exchanged between the land and the atmosphere (see footnote to Table 1.2), the data are in

Table 1.2 Global carbon cycle.

	Gt carbon/year[a]	
Carbon dioxide flux	*Ref. 7*	*Ref. 8*
Loss by terrestrial plant respiration and decay	60.0[b]	119[b]
Uptake by photosynthesis	61.7[b]	120[b]
Net gain by land mass from the atmosphere	1.7	1
Loss from the oceans by evaporation	90.0	88
Dissolution in the oceans from the atmosphere	92.2	90
Net gain by oceans from the atmosphere	2.2	2
Loss through combustion of fossil fuels	6.0	6.3
Loss through changing land use (deforestation)	1.4	−0.2
Net increase in atmospheric carbon	**3.5**	**3.1**

[a] 1 tonne of carbon equates to 3.67 tonnes of carbon dioxide.

[b] The apparent discrepancy between the two columns is thought to be due to Ref. 8 quoting gross values for carbon photosynthesized and lost by all mechanisms, whereas Ref. 7 gives only the net amounts respired by living plants (autotropic respiration) and ignores that stemming from decayed plants by microbial breakdown (heterotropic respiration). We are indebted to the Hadley Centre, UK Meteorological Office, for this explanation.

fair agreement, particularly as regards the net effects. If the analyses are at all accurate, fossil fuels contribute only ~6 Gt to the 150–200 Gt of carbon taken up by the troposphere each year (*i.e.*, 3–4%). In terms of the annual balance, fossil fuels and to a lesser extent deforestation are held to be entirely responsible for the growing concentration of carbon dioxide.

Carbon dioxide is not the only greenhouse gas. Others are as follows.

(i) *Nitrous oxide*, which is a further product of combustion (from nitrogen and oxygen in the air) and also emanates from agricultural practices (soil management, fertilizers, waste burning), from some chemical manufacturing processes and from a diversity of natural biological processes in soil and water.

(ii) *Methane*, which arises from a variety of both human-related activities (coal mining, natural gas extraction and distribution, oil recovery, fossil fuel combustion, biomass burning, wet rice cultivation, animal husbandry, waste management) and natural sources (anaerobic decay of vegetation in wetlands, enteric fermentation in ruminant animals and some insects, anaerobic digestion in marine zooplankton, release from methane hydrates). Some recent research indicates that living trees may liberate methane directly. This finding, if confirmed, throws into doubt the precise benefit of forests as a sink for carbon dioxide.

(iii) *Hydrofluorocarbons* (HFCs), which were introduced as a replacement for chlorofluorocarbons (CFCs) and hydrochlorofluorocarbons (HCFCs) in the 1970s for use in aerosol cans, refrigerators and air conditioners (because CFCs and HCFCs were found to contribute to breakdown of the Earth's ozone layer, which offers protection from harmful ultraviolet rays).

(iv) *Perfluorocarbons* (PFCs), which are used principally in the electronics and semiconductor industries, and also as a refrigerant blend and a fire suppressant.

(v) *Sulfur hexafluoride*, which is used as an electrical insulator in transformers and switchgear, a cover gas during both aluminium and magnesium production and processing to prevent excessive oxidation and other degradation of the resulting metal, and also as a feed gas for plasma etching of semiconductor devices.

The ability of these compounds to absorb infrared radiation varies widely from compound to compound, as does their life in the atmosphere before they undergo photochemical reactions or are absorbed in the oceans or on land. Methane has a concentration of only 1.7 ppmv in the troposphere, which is much less than that of carbon dioxide. On the other hand, each molecule of methane has a global warming potential (GWP) value that is 21 times that of carbon dioxide over the course of 100 years. (Note: the GWP value has been developed to compare the ability of each greenhouse gas to trap infrared radiation over 100 years relative to another gas; by convention, carbon dioxide has a GWP of 1.) Although methane has a relatively short lifetime (a few years) in the atmosphere before being oxidized to carbon dioxide, it is thought to contribute about half as much to global warming as carbon dioxide. By contrast, nitrous oxide is much more stable and has a GWP of 310. Hydrochlorofluorocarbons have very high GWPs (100–3000) and fairly long lifetimes (tens to hundreds of years). Perfluorocarbons have extremely high GWPs (5000–10 000) and are stable for thousands of years, but are released in relatively small amounts, as is sulfur hexafluoride, which has by far the highest GWP (23 000) and is extremely stable. It should be noted, however, that as yet there is no *unequivocal* proof that global warming results from the observed increase in anthropogenic greenhouse gases because the observed effect is small (less than about 0.8 °C over the past 100 years). Nevertheless, despite complications arising from natural phenomena, the evidence that greenhouse gases are largely responsible for climate change is growing ever stronger. The change in global temperature from 1860 to 2000 is shown in Figure 1.8(a),[9] while data that demonstrate that the concentration of carbon dioxide in air has increased from ~270 ppmv before the Industrial Revolution to almost 380 ppmv today are given in Figure 1.8(b) along with the annual emissions of carbon that have grown from negligible amounts in 1860 to a present level of almost 6.5 Gt.[10] Much of the rise in atmospheric carbon has occurred since 1950.

A recent conference of climate scientists concluded that the world should be aiming to keep the concentration of carbon dioxide below 400 ppmv to avoid serious climatic consequences. Potential long-term problems include melting of the Western Siberian permafrost and the Arctic ice cap; see Figure 1.8(c); collapse of the Western Antarctic ice-sheet into the ocean; melting of the Greenland ice sheet; shutting down of the Gulf Stream‡; enhanced liberation of carbon dioxide from soils as the Earth warms up; uncontrolled release of methane locked away as crystalline hydrate in deep ocean sediments and polar ice-fields; and acidification of the oceans that would result in a reduced ability to take up carbon dioxide.[11] Most of these possible outcomes are interactive and accumulative and their consequences would be devastating. If the climate scientists are correct, then there is cause for grave concern. Calculation shows that the ice in the Greenland ice sheet, if fully melted, would lead to a rise in ocean levels of 7 m. Similarly, the West Antarctic ice sheet would equate to a 6 m rise in global sea level.

Likewise, it should be recognized that water vapour, which is present in the atmosphere in far greater amounts than carbon dioxide, is also a potent greenhouse gas – it accounts for at least 60–70% of the natural greenhouse effect. One has only to experience the difference in temperature on a frosty, starlit night in winter from that on a cloudy, overcast night to appreciate the

---

‡ It should be noted that the Gulf Stream is a limb of the global ocean circulation system, which is known by a variety of names; see Glossary of Terms. Driven by water temperature and salinity, the system takes warm surface water from the Pacific Ocean and returns colder, denser water from the North Atlantic at depth. This process significantly moderates average winter temperatures in North Atlantic regions. Increased melting of ice would put more fresh water into the seas and, in addition to curtailing the transfer of heat to the atmosphere, would weaken the salinity driving force with consequent impacts on climate. Progressive shutting down of the ocean circulation system could result in Western Europe becoming as cold as Canada.

(a)

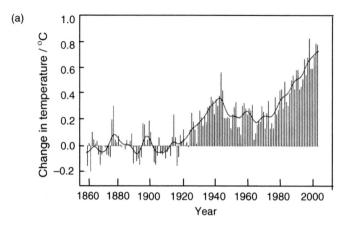

(b)

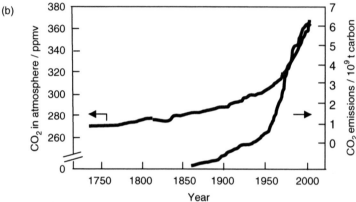

(c)

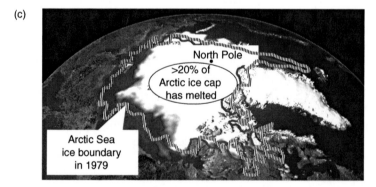

Figure 1.8 (a) Global near-surface temperature averaged over land and ocean, 1861–2004.[9] Differences are expressed relative to the end of the 19th century.
(Courtesy of UK Meteorological Office).
(b) Atmospheric concentration of carbon dioxide from 1750 to 2000 (left axis) and global annual releases of carbon (right axis).[10]
(Courtesy of Oak Ridge National Laboratory).
(c) Shrinkage of summer Arctic ice cap since 1979. The frozen peninsula to the right of the photograph is Greenland.
(Courtesy of National Aeronautics and Space Administration).

role of water vapour in absorbing infrared radiation emitted from the Earth's surface. Indeed, without this intervention of water, it has been estimated that the average surface temperature of the Earth would be about 33 °C lower. One of the consequences of enhanced global warming will be a higher overall atmospheric concentration of water vapour due to increases in the evaporation rates from oceans, lakes and rivers, as well as from plants. This is a positive feedback that will exacerbate the greenhouse effect. On the other hand, any increase in cloud cover will result in greater reflection of sunlight from the upper surface of the clouds so that less insolation will reach the Earth's surface. At present, there is no exact understanding of the balance between these two opposing effects of water vapour, although it is believed that clouds play an important, but ill-defined, role in determining climate.

Recent research appears to show that some pollutants, in particular ash and sulfur dioxide, cause the nucleation of smaller droplets of water in clouds. These smaller droplets are thought to reflect more sunlight from the upper surface of clouds back into space and thereby act in opposition to carbon dioxide and so *reduce* the warming effect. This is a negative feedback effect and, since less sunlight is observed to reach the Earth's surface, the phenomenon is referred to as 'global dimming'. The corollary is that the more one reduces pollution by removing particulates and sulfur dioxide, the worse the warming becomes. The fact that this effect has been discovered only recently illustrates the great complexity of climate science and how much remains to be elucidated. Nevertheless, forecasts of global warming may now have to be drastically revised upwards. Thus, increasing measures to curb pollutants, unless enacted in concert with reductions in greenhouse gas discharges, could lead to even greater changes in the climate system and so pose a cataclysmic threat to society.

This discussion highlights the complexity of atmospheric science that originates from a variety of poorly-understood feedback loops, both positive and negative. Nevertheless, there is growing concern internationally that if the problem of build-up of greenhouse gases is not addressed soon, human beings will be sowing the seeds of their own demise. Unless vigorous action is taken now to move towards renewable (or carbon-free) energy, the concentration of carbon dioxide is predicted to rise to 700 ppmv or even higher by the end of this century. The consequences would clearly be catastrophic, especially given the expressed belief that even at the present level of 360–380 ppmv further small increases will have a disproportionate effect on the climate.

The first tentative steps to combat global warming were taken at the United Nations Framework Convention on Climate Change that was held in Kyoto in December 1997. The resulting 'Kyoto Protocol' called for the industrialized nations to reduce the average of their individual emissions by at least 5% below baseline 1990 levels by 2008–2012. The specific reductions proposed for the European Union, Japan and the USA were 8, 6 and 7%, respectively. Targets were not set for developing countries. The great majority of nations have ratified the Kyoto agreement, but notably not the USA and Australia, whose administrations consider it to be counter-productive. For the other nations, the Kyoto Protocol entered into force on 16 February 2005.

The target for the USA is, in practice, totally unrealistic in view of the marked growth in emissions that has already occurred (*e.g.*, 13% between 1990 and 2001) and the fact that most of the power stations that will operate in 2012 have already been commissioned. Through the burning of fossil fuels, the USA, with only 4.6% of the world population, produces about 25% of the global output of carbon dioxide. When, however, this is expressed in terms of GDP, *i.e.*, the amount of carbon liberated per unit of economic activity (the 'carbon intensity'), the USA is somewhat below the global average, although above the average for industrialized countries.[1]

The European Union has been more proactive in seeking to meet the Kyoto targets. Subsidiary goals for carbon releases have been set for individual nations who, in turn, have made allocations to large industrial emitters such as power stations, blast furnaces and cement manufacturers. These companies are issued with permits that grant them permission to emit a certain quantity of carbon dioxide in a particular year. Those that do not hold sufficient permits will have to purchase them on the open market. A trading scheme has been initiated across Europe with a new European Carbon Exchange in London. In the UK alone, the government has awarded more than 1000 annual permits to cover the period 2005–2008. The country had some prospect of reaching its goal of 20% reduction by 2010, thanks largely to the construction since 1990 of gas-fired power stations in preference to coal-fired alternatives, together with an ambitious programme to introduce wind turbines. The latest figures, however, make it less likely that the target will be met. Looking further ahead, the UK government now aims for a 60% reduction in carbon dioxide emissions by 2050, and this is much more challenging in an era of possible reductions in gas supplies.

The market for carbon permits between non-European nations may prove more difficult to establish. It has been estimated that, internationally, permits worth a trillion dollars or more per year would be required and even then, there is no guarantee that the Kyoto targets would be met. Moreover, there is no provision for how future targets are to be allocated or for enforcement of the scheme. Ultimately, it will be necessary to have a judicial body to enforce compliance, but international law is too weak at present to take on this task. Inviolate targets and time-scales are simply not practicable in the light of changing circumstances.

The consumption of energy in large countries that are in the process of becoming industrialized and that are not included in the Kyoto Protocol, notably China and India, is growing rapidly, as is the world population. Any move towards low-carbon fuels (natural gas and renewables) will be offset by the greater use of high-carbon coal, and there is every expectation that global discharges of carbon dioxide will increase rather than decrease unless dramatically new technology and legislation is adopted widely. This is largely a matter of economics and politics. At present, the IEA predicts that global release of carbon dioxide will rise from 26.6 Gt (7.3 Gt carbon) in 2004 to 40 Gt (10.9 Gt carbon) in 2030, in line with the projected consumption of fossil fuels.[3]

Obviously, it is too early to gauge the extent to which the Kyoto Protocol may prove to be successful, but it should be acknowledged that the initiative is only a first, small step along the path towards carbon-free energy. Far greater international commitment will be required post-2012 if a significant global impact is to be made in reducing anthropogenic emissions of carbon dioxide and other greenhouse gases. A start was made at the United Nations Climate Change Conference, which was held in Montreal in December 2005. At this meeting, 150 of the world's nations agreed to hold discussions to define a programme of carbon reductions for the post-2012 era. Agreement was also reached on a mechanism to monitor compliance to the Kyoto Protocol in the run-up to this date.

In recent years, there has been a growing awareness that solid waste materials should not be simply discarded to landfill, but recycled or reprocessed. Similarly, the atmosphere should not be regarded as a free dumping ground for waste gases. This is gradually becoming accepted wisdom, but practical means for dealing with the large volumes of carbon dioxide remain elusive. A significant shift towards hydrogen as a fuel would go a long way towards solving the problem, provided that carbon dioxide is not released in large quantities during the manufacture of the hydrogen itself. If the carbon dioxide originates from large centralized plants, it should be possible to separate and capture ('sequester') the gas and then dispose of it underground (or even in the ocean). This is far more feasible than treating the exhaust from a myriad of domestic heating boilers, industrial furnaces and motor vehicles. Scientists are now beginning to address the best way to sequester carbon dioxide; see Chapter 3.

1.3 Atmospheric Pollution

In many parts of the world, considerable success has been achieved in the abatement of atmospheric pollution. Action taken in the UK provides a good example. In the 1950s, cities were badly polluted with smoke from coal burnt on open domestic fires. It was customary to burn soft, bituminous coal and in winter the air became heavy with aromatic hydrocarbons and soot particles. This led to pernicious 'smogs' (smoke-laden fogs) and many elderly people died from chest infections. In 1956, the Clean Air Act banned the burning of soft coal on open fires; smokeless fuel (*e.g.*, anthracite) and closed stoves were introduced. This, together with widespread use of gas for fires and central heating, had a dramatic remedial effect on urban air quality.

Since the 1960s, there has been a marked increase in road traffic, world-wide. At first, this gave rise to serious air pollution of a different type. In sunny climates, photochemical reactions among the automotive exhaust gases led to the formation of ozone and other lachrymators. The resulting 'photochemical smog' was intensified when the climatic and geographic conditions were such that there was an inversion layer over a city situated in a basin surrounded by hills. Los Angeles experienced the problem first, but soon, with increase in traffic everywhere, it became a more general phenomenon, with severe pollution in Athens, Rome and Tokyo, for example. Under the stimulus of stringent legislation emanating from the US Environmental Protection Agency, catalysts were developed for fitting to vehicle exhausts. These catalysts removed hydrocarbons and carbon monoxide by oxidation to carbon dioxide, as well as nitrogen oxides (NO_x) by reduction to nitrogen. Catalytic converters were first introduced to vehicles in the USA and Japan around 1975, but have since become ubiquitous. Over the years, the legislation concerning the permitted concentration of pollutants in vehicle exhausts has become progressively stricter in many countries and now, for new cars fitted with petrol engines, the problem is almost solved. Diesel-engine vehicles emit more particulates (soot) and this is technically a more difficult problem, but steady progress is being made towards its solution. A ceramic matrix through which the exhaust gases flow and which supports the catalyst is shown in Figure 1.9, together with a schematic of a catalytic converter in action.

Other undesirable components of both petrol and diesel fuel are sulfur-containing compounds that may be present in small quantities. These give rise to sulfur dioxide in the exhaust gas. In addition to being another acid pollutant, it is also detrimental to the life of the catalytic converter. Consequently, the principal petroleum companies have developed and brought to the market ultra-low sulfur ('green') petrol and diesel. The action to reduce dramatically sulfur levels in motor fuel has also been made in response to regulations in the USA and the European Union.

Road vehicles are by no means the only source of pollution. Ships and trains are subject to less legislation and frequently emit visible plumes of particulate pollution along with gases. Industries that are heavy users of energy, and therefore potential polluters, are iron and steel, cement manufacture, paper and board and, especially, power stations. Coal-burning power stations emit sulfur dioxide, which arises from the sulfur contained in the coal, and also nitrogen oxides produced during combustion. It is these gases that give rise to 'acid rain', which is detrimental to the natural habitat. Technologies exist for the suppression of these acid gases, but are not yet universally employed. In some countries, power stations are licensed as to the quantity of sulfur dioxide that they may discharge, which encourages them to burn low-sulfur coal. The alternative is to fit devices to the exhaust stack to absorb sulfur dioxide in limestone (calcium carbonate), but this is a costly solution that may not be economically viable, although it is widely used in some countries (*e.g.*, Germany).

In summary, the march of technology has made major inroads into solving the problems of air pollution in the more advanced countries. By contrast, the legislation is often less strict and the power station technology less sophisticated in developing nations, with the result that pollution is severe. Nevertheless, as these countries become more affluent and their present generating plant reaches the end of its useful life, more modern technology will be employed. Although air pollution is not nearly as severe a problem as it once was, a transition to hydrogen energy would improve matters further.

There is, however, a new pollution scare associated with hydrogen energy itself. In a global Hydrogen Economy, the quantities handled would be

(a)

(b)

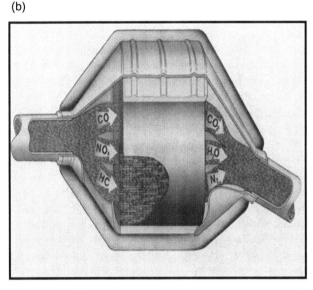

Figure 1.9 Catalytic exhaust conversion for petrol engines: (a) ceramic catalytic converter; (b) schematic of a catalytic converter in action. (Courtesy of Johnson Matthey plc).

enormous and, inevitably, a small proportion would escape into the atmosphere. Both the troposphere and the stratosphere contain hydroxyl radicals that play important roles in climate chemistry. Concern has been raised that escaped hydrogen will react with these to form water (clouds). In the stratosphere, it is considered that this might deplete the ozone layer that protects the Earth, while in the troposphere it could result in a longer residence time for methane and so enhance the greenhouse effect.[11] The Hydrogen Economy would then be simply a matter of replacing one greenhouse gas (carbon dioxide) by a more potent one (methane). There are, as yet, few real data to justify or refute these theoretical concepts.

1.4   Electricity Generation

Electricity is highly valued as an extremely flexible, versatile and clean form of energy. So many tasks of everyday life can be accomplished only by electricity. Electric lighting, electric motors and electronic devices in general, notably televisions and computers, are all integral to modern society. In 2004, 16.2% of the world's final consumption of energy was supplied as electricity.[1] A further merit of electricity is that it can be generated in so many different ways: steam cycles based on any of the fossil fuels (coal, gas, oil) or biomass; gas turbines operating in combined-cycle mode; nuclear reactors; hydro electricity or any of the other renewable forms of energy, *i.e.*, wind, solar photovoltaic, solar thermal, waves, tidal, marine currents, geothermal. As an energy vector, electricity is versatile in production and versatile in application. The contribution of different energy sources to world electricity generation in 1973 and 2004 are listed in Table 1.3.[1] The data show that generation has increased 2.8-fold over this period of 31 years. Moreover, the rate of change is accelerating; from 2002 to 2004 the increase in world electricity generation was 8.7%. Approximately 41% of the electricity generated world-wide is consumed in industry, less than 2% in transport and the remainder in all other sectors of the economy.

The downside of electricity generated from fossil fuels is the vast quantity of greenhouse gas (carbon dioxide) that is liberated, especially when burning coal, which is the most widely available and cheapest of the fossil fuels. The efficiencies of the three leading types of electricity generation plant are shown in Figure 1.10(a),[12] with and without capture of carbon dioxide. The three plants are: (i) coal-fired using pulverized fuel (PF); (ii) coal-fired integrated gasification combined-cycle (IGCC; see Chapter 2); and (iii) natural gas-fired, combined-cycle gas turbine (CCGT; see Chapter 2). The coal IGCC plant uses pre-combustion capture, whereas the coal PF and natural gas CCGT plant employ post-combustion capture. There is a significant drop in overall plant efficiency (8–13%, depending on the plant and fuel employed), associated with the energy consumed in the capture process. The corresponding quantities of carbon dioxide released, expressed as kg per MWh generated, are shown in

Table 1.3   Percentage of fuel shares of world electricity generation in 1973 and 2004.[1]

Energy source	1973	2004
Coal	38.2	39.8
Oil	24.7	6.7
Gas	12.1	19.6
Hydro	21.0	16.1
Nuclear	3.4	15.7
Other[a]	0.6	2.1
Total (%)	100.0	100.0
Total (TWh)	6111	17450

[a] 'Other' includes combustible renewables and waste, geothermal, solar, wind.

(a)

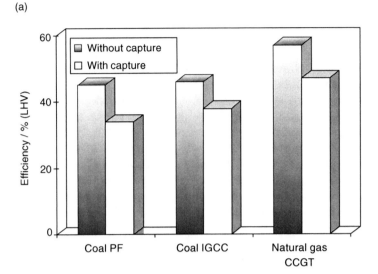

(b)

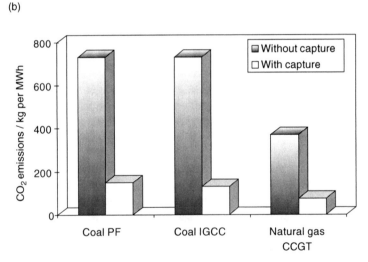

Figure 1.10 (a) Power generation efficiencies and (b) emissions per MWh of electricity generated for major electricity plant, without capture and anticipated with capture of carbon dioxide.[12]
(Courtesy of International Energy Agency).

Figure 1.10(b).[12] The capture is not perfect, but emissions of carbon dioxide are greatly reduced.

Great strides have been made in bringing mains electricity to the masses. In the past 15 years, China is said to have connected 700 million people to the electricity grid so that the country is now 98% electrified;[13] it is still building 30–40 GW of new coal-fired plant each year. India also relies heavily on coal for its electricity production, which is also expanding rapidly. Nevertheless, about 25% of the world population, many living in sub-Saharan Africa, still do not have access to mains electricity. Even more (around 40%) rely on biomass fuels for cooking and heating. This is largely a consequence of poverty, although a sparse population spread out over a wide area is not conducive to centralized generation and distribution by a grid.

A worrying feature of the electricity utilities in developed countries is that many of the large coal-fired and nuclear electricity plants were built in the 1960s and 1970s and are due to be de-commissioned in the 2010–2020 period. Thus, a gap between demand for electricity and supply will doubtless open up and this will have to be filled. In the UK, for example, it is estimated that a further 2.5 GW will be required annually for at least 10 years; this presents a major challenge. In general, the date for the introduction of new generating capacity is too close for radical new technology to be introduced. Consequently, shortfalls in supply will have to be met by gas-fired combined-cycle plant (despite concerns over security of gas supply) or by conventional coal-fired stations (despite the large liberation of carbon dioxide), with the possibility of new nuclear-build always present in the background.

A future trend in the electricity supply industry, much discussed, is a move towards distributed (local) generation.[14] This might be based on combined heat and power (CHP) schemes that would employ gas microturbines or gas engines, on solar energy (photovoltaic, solar–thermal) or on wind turbines. Most of the renewables are best utilized on a small-scale, local basis. Distributed generation is applicable in both semi-urban areas and in remote locations and may be the answer for many of those who currently are isolated from mains supplies. Sadly, often these people cannot afford such technology and their economic position is unlikely to improve until they do have access to electricity and fuels.

One problem with localized electricity generation from wind or solar sources is the need for a back-up system to match a fluctuating electricity supply with a fluctuating demand. This may be (i) mains electricity, where available, (ii) another local source, e.g., a diesel generating set, or (iii) an energy store. The prime candidates for local electricity storage are batteries and hydrogen; the latter requires fuel cells to convert the hydrogen back to electricity.

Fuel cells supplied with hydrogen hold many attractions for distributed electricity generation. They are silent in operation, flexible in energy output and, compared with gas-turbine/generator sets, much cleaner sources of power. Like gas turbine or diesel generating sets, they can also supply heat as well as electricity. The practical value of this heat depends on the type of fuel cell that is being used and its temperature of operation; see Chapter 6. At first, fuel cells are expected to find service in individual buildings (factories, hospitals, departmental stores, apartment blocks, etc.) where both electricity and heat are required. When operating in this mode, their energy efficiency is particularly high. In urban locations, fuel cells could be supplied with hydrogen by pipeline from a central source, but in remote areas it would be necessary to generate the hydrogen on site.

There are also substantial efforts to develop hydrogen fuel cells for the propulsion of electric vehicles. To date, attention has focused on private cars, as they represent the mass market, and on buses, which have more space to accommodate the fuel cell and hydrogen store. These activities are discussed further in Chapter 7. The replacement of conventional vehicles by electric vehicles powered by hydrogen fuel cells is an ambitious goal of the automotive companies. If these vehicles are to be widely available by, say, 2025, it is vital that development work on this challenging technology is pursued vigorously. Projects are under way in the European Union, Japan and the USA. Although these activities are a step in the right direction, it should be remembered that until renewable energy becomes widely and cheaply available – probably not for several decades – the hydrogen will be derived mainly from fossil fuels and so a switch to fuel cell vehicles will not contribute significantly, if at all, to the abatement of greenhouse gases unless the carbon dioxide formed during the manufacture of the hydrogen is captured and disposed of safely. Meanwhile, society must face the more immediate requirement to reduce its petroleum consumption and refocus its lifestyle in anticipation of the price rises soon to come.

Altogether, there are many reasons why hydrogen is stimulating world-wide interest. As noted above, a particularly desirable feature is its complementarity to electricity as an energy vector and also its potential use as an energy store. Much depends upon future political and economic factors that are difficult to forecast and, for this reason, hydrogen is seen to be a speculative venture. If oil and gas supplies are restricted or even disrupted, if climate change is as severe as many fear, if there is a breakthrough in lowering the cost of renewable energy, then, suddenly, hydrogen could become highly desirable. In any event, growing reliance upon coal and other carbon-rich fossil fuels for world energy, as oil and gas supplies decline, is almost a foregone conclusion. Coal is best utilized, from an environmental standpoint, through the media of hydrogen and electricity, *i.e.*, through gasification (see Section 2.7, Chapter 2) with sequestration of carbon dioxide. Both the development of hydrogen technology and the establishment of an infrastructure will, however, be neither quick nor easy, and this provides the justification for undertaking the basic science and engineering sooner rather than later.

Early in the 20th century, the invention of the internal combustion engine had a revolutionary impact on the transportation sector, with a shift from horses and carts to railway locomotives, motorized vehicles and aircraft. It has taken at least a century to develop this technology and the associated markets, and the process is not finished yet. The realization is now dawning that, for all the reasons mentioned above, the time is approaching for another Industrial Revolution, based on new forms of energy, which will involve massive disruptions in developed societies. Central to this revolution will be the growing use of natural gas in gas turbines, the development of clean coal technology that involves gasification and sequestration of carbon dioxide, the distributed (local) generation of electricity based (in part) on renewables, and the replacement of petrol and diesel engines by other propulsion units or fuels. The precise role (if any) for hydrogen in this complex scenario has yet to be established. History shows that periods of industrial upheaval also provide opportunities for the generation of new forms of employment, stronger economic growth and greater scope for nations to work together to evolve a sustainable global community. A change to hydrogen as the ultimate solution to energy security is now widely recognized, but equally many acknowledge that it will probably take at least a century to bring to realization.

1.5 Hydrogen as a Fuel

If hydrogen is to be employed as an energy vector and a non-polluting fuel, then it is necessary to take account of its basic physical properties. Hydrogen is a colourless, odourless, tasteless and non-toxic gas. It is the lightest of all molecules (molecular weight $= 2.016$) and, consequently, has a density of only $0.0899 \, \mathrm{kg \, m^{-3}}$ at normal temperature (273.15 K) and pressure (101.325 kPa), *i.e.*, 7% of the density of air. Liquid hydrogen also has a low density of $70.8 \, \mathrm{kg \, m^{-3}}$ (7% of that of water)[§]. The liquid has a very low boiling point (20.3 K) and therefore requires fairly sophisticated equipment to prepare and maintain it in this state.

By virtue of its exceptionally low density, hydrogen has the best energy-to-weight ratio ('heating value') of any fuel, but its energy-to-volume ratio is poor (Table 1.4). The thermodynamic heat of combustion of hydrogen (its heating value) equates to the standard heat of formation (ΔH_f°) of the product water, *i.e.*,

$$\mathrm{H_2} \, (gas) + \tfrac{1}{2}\mathrm{O_2} \, (gas) \rightarrow \mathrm{H_2O} \, (liquid) \qquad \Delta H_f^\circ = -285.83 \ \mathrm{kJ \ mol^{-1}} \qquad (1.1)$$

In alternative units, the standard heat of formation is 141.78 MJ per kg of hydrogen. This is the maximum amount of heat that can be derived from the combustion of hydrogen when the product water is condensed to 298.15 K; it is known as the 'higher heating value' (HHV) of hydrogen. In most engineering practice, the product water is released as steam whose calorific value (including the latent heat of condensation) is lost. If the steam is released at about 150 °C, then the effective (or engineering) heat of combustion will be around $120 \, \mathrm{MJ \, kg^{-1}}$. This is a practical value known as the 'lower heating value' (LHV). Hydrogen is unique among fuels in having such a high difference (18%) between the HHV and the LHV. For calculating the *absolute* efficiency of energy-conversion devices (*e.g.*, fuel cells) it is important to use the HHV, but when comparing the *practical* or *relative* efficiencies of engineering devices (*e.g.*, boilers) it is often convenient to use the LHVs. Modern condensing boilers do, however, recover a proportion of the lost heat and have efficiencies that lie between the two heating values.

[§]It is interesting to note that a given volume of water or petrol contains more hydrogen than liquid hydrogen itself.

Table 1.4 Technical comparison of hydrogen with other fuels.

	Hydrogen	Petroleum	Methanol	Methane	Propane	Ammonia
Boiling point/K	20.3	350–400	337	111.7	230.8	240
Liquid density/kg m^{-3}, NTP^a	70.8	702	797	425	507	771
Gas density/kg m^{-3}, NTP^a	0.0899	–	–	0.718	2.01	0.77
Heat of vaporization/kJ kg^{-1}	444	302	1168	577	388	1377
Higher heating value^b (mass)/MJ kg^{-1}	141.9	46.7	23.3	55.5	48.9	22.5
Lower heating value^b (mass)/MJ kg^{-1}	120.0	44.38	20.1	50.0	46.4	18.6
Lower heating value (liquid)^b (volume)/MJ m^{-3}	8520	31 170	16 020	21 250	23 520	14 350
Diffusivity in air/cm^2 s^{-1}	0.63	0.08	0.16	0.20	0.10	0.20
Lower flammability limit/vol.% (in air)	4	1	7	5	2	15
Upper flammability limit/vol.% (in air)	75	6	36	15	10	28
Ignition temperature in air/°C	585	222	385	534	466	651
Ignition energy/mJ	0.02	0.25	–	0.30	0.25	–
Flame velocity/cm s^{-1}	270	30	–	34	38	–

^{a} NTP=normal temperature (273.15 K) and pressure (101.325 kPa).
^{b} Different authors give slightly different figures for the heating values.

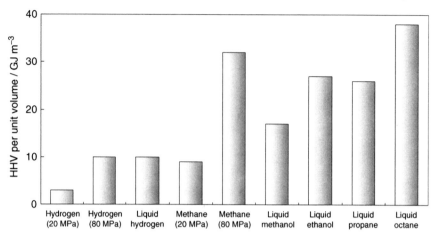

Figure 1.11   Higher heating value per unit volume for various fuels.[15]

A comparison of the volumetric energy densities of compressed hydrogen gas, liquid hydrogen and other fuels is presented in Figure 1.11;[15] methane represents natural gas and octane represents petrol. The data show that highly compressed methane and liquid fuels are much superior to hydrogen.

Hydrogen burns cleanly in air; water is the only product apart from traces of nitrogen oxides at high combustion temperatures. It has a wide range of flammability (lower flammability limit 4% by volume, higher flammability limit 75% by volume). Since the lower explosive limit of hydrogen in air (13% by volume) is higher than the lower flammability limit, hydrogen generally burns rather than explodes. The speed of propagation of the wave front (the flame velocity) is determined by diffusion through the nitrogen of the air. Because hydrogen has a high diffusivity in air, its flame velocity is much higher than that of conventional gaseous fuels. When mixed with pure oxygen in a 2:1 molecular ratio of hydrogen to oxygen and ignited, hydrogen detonates violently since there is no inert nitrogen to slow down the wave front. The energy required to ignite a hydrogen–air mixture is exceptionally low, 0.02 mJ (Table 1.4), and is only one-fourteenth of the energy needed to ignite natural gas.

The combination of physical properties exhibited by hydrogen (low density, low boiling point, wide range of flammability, low ignition energy, high diffusivity in air, high flame velocity) is unique among fuels (see Table 1.4), and this combination has safety implications for the use of hydrogen in bulk. On the positive side, the low density of the gas and its high diffusivity in air mean that outdoors leakage of hydrogen soon disperses safely. By contrast, liquid fuels remain on the ground and evaporate slowly, which poses a fire hazard. Even liquid hydrogen, if spilled, evaporates almost instantaneously on account of its low boiling point and then diffuses rapidly away. The situation is different in an enclosed space; the wide flammability and explosive ranges of hydrogen in air mean that a leakage is very likely to give rise to a fire or explosion. Since hydrogen contains no carbon, it burns with a non-luminous flame that does not radiate heat. Consequently, bystanders are not subject to radiation heat burns. The non-luminous flame does, however, have a negative feature: because it is almost invisible and does not radiate heat, there is always the possibility of inadvertently straying into the flame and being seriously burnt. Also, given the low ignition energy of hydrogen compared with other fuels, extreme precautions should be taken to avoid static electricity when working with hydrogen in bulk. These may include wearing garments of cotton or wool rather than synthetic materials and earthing (grounding) all tools that might give rise to a spark.

Clearly, the distinctive properties of hydrogen make it a unique fuel to handle and operatives need to be well trained and experienced. For example, those skilled in the bulk handling of hydrogen have expressed reservations about allowing untrained personnel to participate in the refuelling of hydrogen-powered vehicles. Escape of hydrogen in enclosed areas (*e.g.*, garages and tunnels) would constitute a particular risk. The safety of hydrogen from both a technological and societal perspective will be a key issue if the Hydrogen Economy is to be taken forward.

Liquid hydrogen as a fuel presents its own handling problems. Because of its exceptionally low boiling point, all containment vessels and transfer lines have to be specially designed. This normally implies a high-vacuum enclosure with multi-layer insulation and heat reflection surfaces arranged alternately. Often there is a surrounding vessel containing liquid nitrogen (boiling point 79 K) as an additional barrier between the hydrogen at 20 K and the air at ambient temperature. Care must also be taken to ensure that ice condensation does not lead to blockages in the storage and dispensing system. Custom-designed connectors are needed when pumping liquid hydrogen from cryostat to cryostat.

A further complication with liquid hydrogen arises from the fact that the hydrogen molecule exists in two forms, namely, ortho-hydrogen (the ortho isomer) in which the nuclear spins of the two atoms are aligned parallel and para-hydrogen (the para isomer) in which the nuclear spins are anti-parallel. At ambient temperatures, the equilibrium mixture contains around 75% ortho and 25% para. On cooling, the ortho isomer converts progressively to the para counterpart; practically all the hydrogen has undergone transformation at 20 K. The ortho–para conversion releases heat ($703 \, \mathrm{J \, g^{-1}}$ of ortho-hydrogen) in excess of the heat of evaporation of hydrogen ($446 \, \mathrm{J \, g^{-1}}$). The consequence is loss of liquid by boil-off and the need for greater energy to be expended in the liquefaction process.

Extensive experience has been gained in the safe handling of liquid hydrogen, both in physics laboratories and, on a tonnage scale, for use in the space industry as a rocket fuel. No insuperable technical problems are encountered. The specialized equipment is, however, very costly and is one reason why liquid hydrogen has not been seriously considered as a fuel outside the space industry, where its low density is a particularly valuable property. Experimentally, liquid hydrogen has been employed as a fuel in automotive applications and there has been some preliminary consideration of using it as an aircraft fuel; see Section 7.3, Chapter 7.

1.6   A Note of Caution

By virtue of its physical and chemical properties, it is clear that hydrogen has the potential to occupy a unique position in the future world energy scene. Not only could it become ultimately a universal means of conveying and storing energy – especially if renewable energy is to become dominant – but also an entirely novel fuel with properties that are distinct from those of other fuels. Hydrogen is the obvious choice for a low-carbon economy in that it would liberate no pollutants to the atmosphere[†] and, when coupled to carbon sequestration or derived from non-fossil primary energy sources, little or no carbon dioxide to contribute to climate change.

Before becoming too enchanted by this vision, it is important to realize that the Hydrogen Economy is a complex concept that embraces a range of possibilities. Among the more important categories, it is therefore important to distinguish between:

- hydrogen as a chemical and hydrogen as an energy vector, where the economics of these two applications are different;
- hydrogen derived from fossil sources and that from non-fossil sources (hydro, nuclear or renewable electricity);
- hydrogen combusted in internal combustion engines and that utilized in fuel cells;
- hydrogen for stationary applications and for portable or mobile applications.

Within these broad categories, there are numerous detailed production routes from primary energy sources (both fossil fuels and renewables) and manifold applications to be evaluated, each of which will impose its own individual technical and economic specifications. The role that hydrogen will play in future energy scenarios will only be known when all the remaining research and development has been completed, when demonstrations projects have been conducted and when full technical and economic evaluations have been made within the context of the prevailing energy scene at that time.

References

1. *Key World Energy Statistics*, 2004 and 2006 Editions, International Energy Agency, Paris, 2004 and 2006.
2. A.-G. Collot, *Prospects for Hydrogen from Coal*, Report CCC/78, IEA Clean Coal Centre, London, December 2003.
3. *World Energy Outlook 2006*, International Energy Agency, Paris, 2006.
4. The Association for the Study of Peak Oil and Gas, *Newsletter No. 55*, July 2005; see www.peakoil.ie/newsletters/aspo55.

[†] Unless hydrogen itself proves to be a pollutant, as discussed above.

5. R.M. Dell and D.A.J. Rand, *Clean Energy*, Royal Society of Chemistry, Cambridge, 2004.
6. M.A. Adelman and M.C. Lynch, *Natural Gas Supply to 2100*, International Gas Union, Hoersholm, Denmark, October 2002; see: www.igu.org.
7. *Technology Opportunities to Reduce US Greenhouse Gas Emissions*, Prepared by the National Laboratory Directors for the US Department of Energy, October 1997; see www.ornl.gov (quoted by C.F. Edwards in Stanford University Report: *Carbon-free Production of Hydrogen from Fossil Fuels*).
8. *Greenhouse Gases, Climate Change and Energy*, Brochure from the US Energy Information Administration, which quotes *Climate Change 2001: The Scientific Basis*, a publication from the Intergovernmental Panel on Climate Change (IPCC), Cambridge University Press, Cambridge, 2001.
9. *Climate Change and the Greenhouse Effect. A Briefing from the Hadley Centre*, Hadley Centre for Climate Prediction and Research, UK Meteorological Office, Exeter, December 2005 (see also *New Scientist*, 12 February 2005, 41).
10. Oak Ridge National Laboratory, Carbon Dioxide Information Analysis Center, http://cdiac.esd.ornl.gov/.
11. *New Scientist*, 15 November 2003, 6–7.
12. J. Davison, P. Freund and A Smith, *Putting Carbon Back into the Ground*, IEA Greenhouse Gas R&D Programme, International Energy Agency, Cheltenham, February 2001; see www.ieagreen.org.uk.
13. *The Role of Coal as an Energy Source*, World Coal Institute, London, 2003; see www.wci-coal.com.
14. W. Patterson, *Transforming Electricity*, Royal Institute of International Affairs, Earthscan Publications, London, 1999.
15. U. Bossel, B. Eliasson and G. Taylor, *The Future of the Hydrogen Economy: Bright or Bleak?* Updated version distributed at the Lucerne Fuel Cell Forum, 30 June–4 July 2003.

CHAPTER 2

Hydrogen from Fossil Fuels and Biomass

Hydrogen was first identified as a distinct entity in 1766 by the British scientist Henry Cavendish (1731–1810) after he produced the gas by reacting zinc metal with hydrochloric acid. This gas he called 'inflammable air'. In a paper to the Royal Society in London, Cavendish provided exact measurements of the weight and density of the gas and thereby revealed its inherent lightness. At that point, Cavendish turned his attention to other areas of research and did not return to the study of gases until the early 1780s. He found that water was formed when 'inflammable air' was ignited with a spark in ordinary air, but continued to hold the belief that, in its production, the gas was released from the metal itself rather than from the acid.

In June 1783, news of Cavendish's latest investigations reached Antoine-Laurent Lavoisier (1743–1794) in France. Lavoisier published the results of his own research into the composition of water and thus became the first to appreciate that the liquid is a compound substance formed from a combination of inflammable air and oxygen. Indeed, Lavoisier gave hydrogen its name by coupling together the Greek words '*hydro*' meaning 'water' and '*genes*' meaning 'genesis', hence the 'maker of water'. Although Lavoisier himself did not give due credit to the work that had been conducted by Cavendish, the discovery of hydrogen is now accredited to the latter in 1766. In fact, other experimenters produced hydrogen well before the work of Cavendish and Lavoisier; as far back as the early 16th century, the Swiss alchemist Paracelsus (1493–1541) reported a gaseous product that was generated when iron was dissolved in sulfuric acid. Following the identification of hydrogen as an element of water, large amounts of the gas were first manufactured in France in 1794 for filling captive balloons that were used as military observation platforms by the French Republican Army. The hydrogen generator consisted of a furnace with a cast iron tube, which contained iron filings, with steam piped in at one end and hydrogen emerging at the other. Sadly, Lavoisier went to the guillotine in May 1794, after being charged with an unrelated matter. At just 50 years of age, his untimely death was a great loss to contemporary chemical science.

The splitting of water to generate hydrogen is conveniently done by electrolysis, although in most countries this is not the favoured economic route.

Rather, most hydrogen today is obtained from natural gas or naphtha by steam reforming or from oil by a partial oxidation process. There is growing interest in the gasification of coal to make hydrogen, on account of the widespread availability of this fossil fuel, and also in the use of renewable energy sources (solar, wind, hydro and wave power) as primary energy input for hydrogen production. Which processes will be favoured in the future will depend upon many factors, *e.g.*, the primary energy sources that are available and their costs, the required purity of the hydrogen, the scale of operation, and other technical, political and commercial considerations. All of these factors will vary from country to country and from time to time. It should be noted that hydrogen is only as clean as the method by which it is produced in the first place, even though there may be local environmental benefits at the point of utilization.

2.1 Present and Projected Uses for Hydrogen

To date, the major consumption of hydrogen has been in the petroleum industry for the refining and upgrading of crude petroleum and in the chemical industry for the manufacture of ammonia (*e.g.*, for fertilizers), methanol and a variety of organic chemicals. Other important uses are found in the food industry for the hydrogenation of edible plant oils to fats (margarine) and in the plastics industry for making various polymers. Lesser applications occur in the metals, electronics, glass, electric power and space industries. A summary is given in Table 2.1.

The world production of hydrogen is around $45-50$ Mt per year, some 20% of which takes place in the USA. Most hydrogen is derived from natural gas by steam reforming. The remainder is obtained principally from oil and coal by the partial oxidation process (see below). The transformation of natural gas and liquid hydrocarbon feedstocks into hydrogen is straightforward, but the transformation of coal requires an initial step of high-temperature gasification. Only 4% of the hydrogen world-wide is generated by electrolysis, invariably when there are special reasons that make this route economic, *e.g.*, where there is a surfeit of cheap hydroelectricity or when the hydrogen is a by-product of the chlor-alkali process for the manufacture of chlorine and caustic soda.

By far the bulk of the hydrogen produced in petrochemical complexes is consumed on site. For example, an ammonia or methanol plant is likely to be located next to a hydrocarbon cracker or steam reformer. Distribution by pipeline is therefore confined to the same or an adjacent plant, in short runs from one unit to another. Very little of the hydrogen is supplied to other consumers, for example for use in the electronics or the nuclear industries, and this is termed 'merchant hydrogen'. As the quantities tend to be relatively small, merchant hydrogen is generally distributed in the form of compressed gas in steel cylinders. A small proportion of hydrogen is liquefied and transported in cryogenic vessels. This is mostly for specialized applications where liquid hydrogen (LH_2) is required, *e.g.*, in bubble chambers to detect interactions between charged particles and as fuel for space rockets. Sometimes hydrogen is

Table 2.1 Principal non-energy uses of hydrogen.

Industry	Uses
Oil	• Removal of sulfur and other impurities ('hydrodesulfurization') • Conversion of large oil molecules to fuel distillates for blending with petroleum ('hydrocracking')
Chemical	• Production of ammonia (and then fertilizers), methanol, hydrogen peroxide, acetic acid, oxo alcohols, dyestuffs, hydrochloric acid
Food	• Conversion of sugars to polyols (bulk sweeteners) • Conversion of edible oils (from coconuts, cotton seed, fish, peanuts, soybeans, *etc.*) to fats, *e.g.*, margarine • Conversion of tallow and grease to animal feed (and soap)
Plastics	• Production of nylons, polyurethanes, polyesters, polyolefins • Cracking of used plastics to produce lighter molecules that can be recycled in new polymers
Metals	• Reductive atmosphere for production of iron, magnesium, molybdenum, nickel, tungsten • Heat treatment of ferrous metals to improve ductility and machining quality, to relieve stress, to harden, to increase tensile strength, to change magnetic or electrical characteristics • Oxygen scavenger in metalworking • Welding torches
Electronics	• 'Epitaxial' growth of polysilicon • Manufacture of vacuum tubes, light bulbs • Heat bonding of materials ('brazing')
Glass	• High-temperature cutting torches • Reductive atmosphere for float-glass process • Glass polishing • Heat treatment of optical fibres
Electric power	• Coolant for large generators and motors • Nuclear fuel processing

liquefied just for transportation purposes, particularly where substantial quantities are ordered and the distances between production and end-users are considerable.

The calorific value of the world output of hydrogen (45–50 Mt) is equivalent to about 1.5% of the global energy supply. When account is taken of losses in processing, about 2% of the total energy usage is employed in producing hydrogen. Very little of this is yet utilized as a fuel. It is clear, however, that if hydrogen is to become a significant new player on the international energy scene, then the expected demand for the gas will be one to two orders of magnitude greater than that of today. The US Department of Energy considers that, by 2040, the USA alone will have 300 million light-duty vehicles and these would require 65 Mt of hydrogen annually if powered by fuel cells. This is a massive increase compared with the present annual domestic production of 9–10 Mt. Such an expansion in activity would present a formidable challenge to the fossil-fuel industry and may indeed be wishful thinking. Typically, a large natural-gas steam reformer produces up to 3.3×10^6 N-m^3 (300 t) of hydrogen per day[†] or, say, 0.1 Mt per year when allowance is made for downtime. Some indication of the size and complexity of the operation may be gained from the photograph of one such reformer shown in Figure 2.1. To meet the US annual target of 65 Mt would necessitate the construction of 600–700 large steam-reformer plants. The capital investment and the time needed to construct facilities on this scale are daunting prospects and the target date of 2040 seems unrealistic. The establishment of a totally new energy industry based on water-splitting reactions powered by renewable forms of energy (see Chapter 4) would be an even more formidable task.

The centralized generation of hydrogen is, however, beset by a fundamental problem – there are no existing networks of pipes for the bulk distribution of the gas. To install a system with nation-wide coverage would be a massive and expensive operation and is unlikely to happen unless and until the Hydrogen Economy becomes well established. In a possible transitional phase, hydrogen might be generated locally by the small-scale steam reforming of natural gas, with the existing gas grid being used to bring the feedstock to the reformer. This scheme has merit in allowing time for the market to develop before too much fixed investment is set in place. Localized generation of hydrogen, using either small-scale gas-reforming or electrolyzer technology, is already being adopted in several countries for the refuelling of fuel cell vehicles; see Chapter 7. Gas reforming at such a level would not, however, offer an effective means of sequestrating carbon dioxide (as discussed in Chapter 3).

It may be possible to generate some hydrogen centrally, with sequestration, and then transport it together with natural gas in the existing distribution network. The mixture would be similar to the old-style coal gas. The percentage of hydrogen that could be added satisfactorily and the specific uses to which the

---

[†] A normal cubic metre (N-m^3) of gas is that measured at 273.15 K and at 101.325 kPa pressure. The hyphen is introduced to avoid confusion with N, the international symbol for the newton.

Figure 2.1   A large natural-gas steam reformer.

gas could be put would have to be determined, but clearly it would be unsuitable as feed for low-temperature fuel cells; see Section 6.4, Chapter 6.

2.2   Natural Gas

Natural gas is found in association with petroleum in oil fields and also on its own in gas fields. The composition of natural gas varies according to the field, formation or reservoir from which it is extracted. It consists predominantly of methane (CH_4), with lesser quantities of ethane (C_2H_6), propane (C_3H_8) and butane (C_4H_{10}). Generally, it is a fairly pure hydrocarbon, but may contain significant amounts of carbon dioxide and minor quantities of helium. Gas fields are in fact a commercial source of helium, which is formed by the radioactive decay of uranium or thorium present in the rocks of the reservoir. Often, there is little or no sulfur in natural gas, although where present it has to be removed before steam reforming in order to avoid poisoning of the catalyst. For pipeline distribution or commercial use, natural gas may have to be processed to remove water vapour, carbon dioxide, solids or other contaminants.

When natural gas occurs in oil fields that are remote from centres of population, as in much of the Middle East or off the north-west coast of Australia, it is common practice to flare this gas to waste for lack of a suitable market. The flaring of natural gas is exceedingly wasteful, with huge quantities of energy being lost and carbon dioxide formed. In recent years, plants have been constructed to harness this 'stranded' gas by liquefaction (liquefied natural gas, LNG; the boiling point of methane is $-164\,°C$) for export to centres of population in ships equipped with insulated, cryogenic tanks. This has necessitated the construction of port facilities close to the liquefaction plant, the installation of pipelines to convey the gas from the fields to the port, and the building of a fleet of cryogenic tankers. New liquefaction plants and port terminals in Nigeria and Oman have now come on-stream. A major undertaking is being built on Sakhalin Island, off the Russian Siberian coast, and will be capable of producing 4.8 Mt per year to supply the growing markets in Asia and the west coast of the USA. There is also scope for increased imports into Europe. These developments and opportunities will considerably increase world trade in LNG and end the routine flaring of gas from the associated oil wells.

All the above activity stems from the expanding availability of natural gas, its clean-burning properties as a fuel and the world's insatiable demand for ever-increasing amounts of energy, both for electricity generation and for the heating and air-conditioning of buildings. Given the expected growth in the market, it is difficult to see where the extra natural gas that is needed for hydrogen-fuelled road transport is to be found; it may be necessary to utilize other fossil fuels; see Sections 2.4 and 2.7 below.

2.3 Reforming of Natural Gas

Compared with other fossil fuels, natural gas is the most desirable feed for making hydrogen. It is widely available, easy to handle and relatively cheap (at present). Moreover, the gas has the highest hydrogen-to-carbon ratio of all fuels (see Figure 1.1, Chapter 1) and this serves to minimize the quantity of by-product carbon dioxide.

The steam reforming of natural gas to manufacture hydrogen is a mature industry. In this process, methane is reacted with steam over a nickel-based catalyst at around $900\,°C$ and at elevated pressure (several MPa):

$$CH_4 + H_2O\,(gas) \rightarrow CO + 3H_2 \tag{2.1}$$

The resulting mixture is known as 'synthesis gas' (or 'syngas') because it may be used for the preparation of a range of commodities that include various organic chemicals such as methanol, formaldehyde, oxo alcohols and poly-carbonates. (It should be noted that 'synthesis gas' is in fact a generic term that is used to describe the combined products – hydrogen and carbon monoxide – from the gasification of any carbonaceous fuel.) The steam reaction is endothermic, *i.e.*, heat absorbing, and it is therefore essential to provide a large quantity of heat, namely, 252 kJ per mole of methane when all the reactants are under 'standard' conditions of temperature (298.15 K) and pressure (101.325 kPa) – so-called STP conditions – or 206 kJ per mole of methane when the input water is already in gaseous form as steam[‡]. This reaction heat may either be supplied externally (allothermal reformer) or internally by adding oxygen or air to the reaction mixture (autothermal reformer). In the former case, the catalyst is suspended in an array of long (12 m) tubes that are mounted in a hot box, which is heated externally by burning a fraction of the incoming methane (up to 25%). Steam reforming is reversible and the forward reaction is thermodynamically favoured by high temperature and modest pressures. The use of high temperatures, however, introduces the possibility of carbon formation, which can lead to deactivation of the catalyst and blockages in the reactor. This reaction can be minimized by adjusting the composition of the catalysts and by careful control of the operating conditions, especially the steam-to-carbon ratio (typically, ratios of 2−3 are required to suppress carbon build-up). When the natural gas contains sulfur, this is removed in a preliminary step by catalytic hydrogenation to hydrogen sulfide, which is then removed by chemical reaction with an amine solvent, *e.g.*, triethanolamine. The spent solvent is pumped to a desorber where it is heated to reverse the reaction. The recovered amine is recycled.

'Partial oxidation' is an alternative procedure for producing hydrogen. Air or oxygen is added to the gaseous mixture so that some of the methane is oxidized internally via an exothermic (heat evolving) reaction:

$$2CH_4 + O_2 \rightarrow 2CO + 4H_2 \tag{2.2}$$

Here, all the hydrogen comes from the methane and, consequently, only two hydrogen molecules are produced per methane molecule. Generally, sufficient air or oxygen is added so that the heat effects in the reactor balance out and the reaction is self-sustaining. This gives the advantage of a more compact reactor design since there is no need for a heat-exchanger. The disadvantage of employing air in partial oxidation is that the product gases are diluted by nitrogen. On the other hand, if pure oxygen is used, it has first to be obtained by the cryogenic distillation of liquid air and this adds a further, costly, process step. For these reasons, the heat required for reaction (2.1) is usually supplied externally, rather than generated internally by adding air or oxygen. The provision of a cryogenic air-separation plant is economic only for large-scale reformers.

At present, syngas is used largely for the manufacture of organic chemicals, as mentioned above, or is converted to hydrogen by the 'water-gas shift (WGS) reaction', as represented by reaction (2.5) below. There is, however, a further option that has sometimes been employed in the past and is likely to prove of

---

[‡] In thermodynamic terms, the quantity of heat absorbed under STP conditions is referred to as the standard enthalpy of the reaction, $\Delta H°$, and this function has a positive sign when the reaction is endothermic.

growing importance in the future. This is the conversion of syngas to liquid fuels by the so-called 'Fischer-Tropsch process' [reaction (2.3)], which is carried out over a nickel-, cobalt- or thorium-based catalyst at 200 °C.

$$nCO + (2n + 1)H_2 \rightarrow C_nH_{2n+2} + nH_2O \qquad (2.3)$$

Nowadays, this is more commonly known as the 'gas-to-liquids' (GTL) process and is likely to become an important source of transport fuels when native petroleum becomes scarce. Synthesis gas can also be converted to methanol (CH_3OH) and this, too, is a potential motor fuel and a means of immobilizing, storing and conveying hydrogen; see Chapter 5. Methanol is usually prepared by reacting synthesis gas of appropriate composition over a catalyst of copper and zinc oxides at 300–450 °C; see reaction (2.4). Various designs of reactor have been described for operation at both high pressure (25–30 MPa) and low pressure (5–10 MPa).

$$CO + 2H_2 \rightarrow CH_3OH \qquad (2.4)$$

When the object of reforming natural gas is to produce hydrogen and not to prepare organic chemicals or carbonaceous fuels, it is necessary to complete the conversion by subjecting the syngas to the WGS reaction. This converts the carbon monoxide to carbon dioxide by further reaction with steam over a catalyst at a much lower temperature, *i.e.*,

$$CO + H_2O \, (gas) \rightarrow CO_2 + H_2 \qquad (2.5)$$

The process may be undertaken in two steps by which the carbon monoxide content is first reduced to <2 vol.% at 400 °C and then to <0.2 vol.% at 200 °C. If ultra-pure hydrogen is required for use in fuel cells, the remaining small quantity of carbon monoxide is selectively oxidized to <0.002 vol.% by admitting air at 100 °C. As in the case of reaction (2.1), the use of excess steam assists the WGS reaction and enhances the yield of hydrogen, although at the expense of a reduction in overall thermal efficiency.

Summing reactions (2.1) and (2.5) yields:

$$CH_4 + 2H_2O \, (gas) \rightarrow CO_2 + 4H_2 \qquad (2.6)$$

Thus, each molecule of methane gives rise to four molecules of hydrogen, of which two molecules originate from the steam. In reality, the practical yield of hydrogen is less than this because, as already mentioned, some of the methane is burnt externally to provide the heat needed for reaction (2.1). A schematic representation of the overall steam–methane reforming process is shown in Figure 2.2. To improve overall efficiency, unreacted fuel and steam are recycled to the reformer.

It is also possible to reform methane by reaction with carbon dioxide, rather than with steam, so-called 'dry reforming':

$$CH_4 + CO_2 \rightarrow 2CO + 2H_2 \qquad (2.7)$$

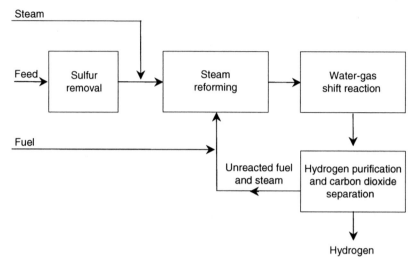

Figure 2.2   Schematic of process for steam reforming of methane.

This is a well-documented catalytic reaction and is a potential route for 'fixing' carbon dioxide. Little is known about the reaction mechanism, or the parameters that determine the stability of the catalyst. From the viewpoint of carbon dioxide disposal, there would be little point in using dry reforming for the manufacture of hydrogen, since all the carbon dioxide would appear again in the subsequent WGS reaction.

2.3.1   Gas Separation Processes

Finally, in the preparation of hydrogen, it is necessary to separate the product gas from the carbon dioxide and any other gases that may be present. This is termed 'pre-combustion capture'[§] and is generally accomplished by 'pressure swing adsorption' (PSA), which involves isolating the carbon dioxide and impurity gases by adsorbing them at high pressure (up to 4 MPa) on a suitable adsorbent (*e.g.*, a molecular sieve or activated carbon) in a packed bed. Impurities are selectively adsorbed, while pure hydrogen is withdrawn at high pressure. To allow continuous operation, multi-beds connected in parallel are often used. Once a bed has become saturated with impurities, the feed is switched automatically to another fresh bed to maintain a continuous flow of hydrogen. Reducing the pressure in a number of discrete steps, which releases the adsorbed gases, regenerates spent adsorbent. Usually, for existing hydrogen plants, all the desorbed carbon dioxide is vented to the atmosphere, though in principle it is possible to separate it out for storage. Indeed, there are

---

[§]'Pre-combustion capture' is when the carbon dioxide is separated from the hydrogen before it is combusted in a gas turbine. 'Post-combustion capture' is the more usual process where the fuel is burnt first and the carbon dioxide is extracted from the exhaust gas. See Section 3.2, Chapter 3.

Figure 2.3 Industrial pressure swing adsorption unit. (Courtesy of Universal Oil Products).

instances where it is pumped underground as a means of facilitating enhanced oil recovery; see Section 3.3, Chapter 3.

Pressure swing adsorption yields high-purity hydrogen (up to 99.999%) and recoveries of up to 90% are routinely achieved in industrial plants. The process is highly reliable, flexible and easy to operate with a modest energy input. On the other hand, PSA suffers from limited capacity and a requirement to cool the gases to low temperature to effect separation. The product hydrogen is therefore now at low pressure and low temperature, which is not ideal as feedstock for a gas turbine or a high-temperature fuel cell if either system is to operate at maximum efficiency. Due to the modular nature of the PSA process, scaling-up can be readily achieved with no loss of efficiency. Industrial units have production capacities that range from 500 to $100\,000\,\text{N-m}^3\,\text{h}^{-1}$ and typically require up to 12 beds for maximum hydrogen recovery. A photograph of an industrial PSA unit is shown in Figure 2.3.

There are numerous ways, other than pressure swing adsorption, of separating hydrogen from carbon dioxide. These include temperature swing adsorption, both physical and chemical *absorption* processes, and cryogenic separation. In the last-mentioned process, carbon dioxide is separated from hydrogen and other gases by cooling and condensation. The technology is used commercially but is energy intensive and, when water is present, may experience operational problems through the formation of carbon dioxide clathrates and ice that can result in serious blockages of the system. Nevertheless, there is the advantage that, after re-warming under pressure, carbon dioxide is produced in a liquid state that can be easily transported to a disposal site by tanker.

Selective membrane diffusion processes also offer a promising approach to separating hot gases and these are discussed below in Section 2.6.

2.3.2 Characteristics of Steam Reforming of Methane

The overall thermal efficiencies of steam-reformer plants are, at best, only 75–80% and thus, in the absence of carbon dioxide capture and storage ('sequestration'), the conversion of methane to hydrogen releases significantly more carbon dioxide to the atmosphere (per unit of energy output) than burning the methane directly. Clearly, for steam-reforming activity to expand substantially and sustainably, the emissions of carbon dioxide must be captured and stored in perpetuity. This would add materially to the cost of the hydrogen. Without sequestration, however, there would be little incentive to move towards a Hydrogen Economy, except possibly as a transitional phase whilst awaiting the advent of widespread, low-cost renewable energy.

Steam reformers may be scaled down to smaller sizes for local generation of hydrogen, but in so doing the efficiency declines and the capital cost per unit of hydrogen delivered rises considerably. At small sizes, a more cost-effective approach is to use a reformer that operates at a lower temperature (700 °C) and pressure (0.3 MPa) and hence allows construction from less-expensive materials. 'Micro' steam reformers operating in the range 50–3000 m^3 of hydrogen per day, running on natural gas, have been developed for integration into fuel cell systems. So also have micro reformers to produce hydrogen from methanol and diesel fuels. Medium-scale steam reformers with outputs of 3000–30 000 m^3 hydrogen per day have been demonstrated for distributed generation and it has been proposed that these might be located at vehicle service stations to provide hydrogen for fuel cell vehicles. This would then take advantage of natural gas networks that already exist and thus would avoid the installation of hydrogen pipelines from a central facility. There would remain, however, the problem and costs of training sufficient technical personnel to operate steam reformers safely and efficiently at refuelling stations, given that present petrol stations largely rely on self-service by drivers. The initial capital cost of manufacturing and installing such reformer units must also be taken into account.

Another limitation of local, dispersed generation of hydrogen is that, in most instances, the carbon dioxide by-product would have to be vented to the atmosphere and not stored. In that case, it would certainly be easier and cheaper to run the vehicles on compressed natural gas (CNG) directly, so long as supplies are available. A more likely application for distributed hydrogen generation is to fuel stationary fuel cells to produce combined heat and power (CHP) for large building complexes (hotels, hospitals, airports, apartment blocks, *etc.*). At this scale, solar–thermal energy (*v.i.*) could conceivably be employed to provide the heat needed for the reforming reaction, but even then the economics are uncertain given the added capital costs of the solar installation. Such a reformer–fuel cell combination would be in direct competition with an engine or micro-turbine fuelled by natural gas for CHP applications.

2.3.3 Solar–Thermal Reforming

In Australia, the Commonwealth Scientific and Industrial Research Organisation (CSIRO) is investigating the possible use of solar energy, in the form of a solar furnace, to provide the heat required for the steam reforming of natural gas and other methane-containing gases, *e.g.*, landfill and coal-bed methane. As the resulting syngas would contain a substantial amount of embodied solar energy (up to 25%), solar–thermal reforming offers the prospects of high thermal efficiencies and greatly reduced emissions of carbon dioxide. Moreover, the emissions would be in concentrated form and thus more amenable to gas separation.

There are three basic designs of solar furnace: (i) a simple parabolic dish that focuses the sun's rays on to a thermal receiver mounted above the dish at its focal point; (ii) parabolic trough mirrors that track the sun as it crosses the sky and that have receivers located at their foci; (iii) an array of thousands of individual mirrors ('heliostats') around a central receiver set on top of a tall tower. To demonstrate the feasibility of solar–thermal reforming, CSIRO erected a solar dish of 48 curved mirrors, but is now exploring decentralized generation through the development of more practical 'mini' versions of the solar-tower approach [design (iii)] coupled to a small steam reformer; see Figure 2.4. Apart from the benefit of permitting the generation of hydrogen close to where it is needed, this modular technology is less expensive than a dish and is more flexible in that it allows easier integration of additional units to

Figure 2.4 CSIRO solar tower array.

meet any growth in demand. Obviously, the solar furnace cannot function at night or during periods in the day when there is no sunshine. Therefore, to maintain continuity of hydrogen supply to the customer, this calls for an adequate storage system or back-up from a conventional steam reformer.

2.4   Partial Oxidation of Hydrocarbons

As mentioned above in Section 2.3, partial oxidation is an alternative process to steam reforming for producing hydrogen from methane. It is also applicable to a wide range of liquid hydrocarbons that include heavy oils found in low-value refinery residues (refinery 'bottoms') and to the gasification of coal; see Section 2.7. In general terms, the process may be expressed as:

$$2C_mH_n + mO_2 \rightarrow 2mCO + nH_2 \tag{2.8}$$

Just sufficient oxygen is employed to oxidize the carbon content of the hydrocarbon to carbon monoxide while leaving the hydrogen in elemental form; hence the name 'partial oxidation'. The reaction is exothermic and no indirect heat-exchanger is necessary. The partial oxidation unit is more compact than a steam reformer, in which heat must be added indirectly via a heat-exchanger. The plant consists of an oxidation reactor, followed by a WGS stage and a hydrogen purification system. A cryogenic oxygen facility is often incorporated, since operation with pure oxygen reduces both the size and the cost of the reactors and also removes the need to separate nitrogen from the product gas. One disadvantage of partial oxidation compared with steam reforming, as mentioned earlier, is that only two as opposed to three molecules of hydrogen are produced per methane molecule before the WGS reaction; *cf.*, reactions (2.1) and (2.2). Compensating advantages of partial oxidation are the flexibility of the feedstocks that can be used and the fact that external heat is not required. Depending on the composition of the feed and the type of fossil fuel used, partial oxidation is carried out either catalytically or non-catalytically. The latter approach operates at high temperatures (1100–1500 °C) and can be applied to any possible feedstock, including heavy residual oils and coal. By contrast, the catalytic process is performed at a significantly lower range of temperatures (600–900 °C) and, in general, uses light hydrocarbon fuels as feedstocks, *e.g.*, natural gas and naphtha.

2.5   Other Processes

2.5.1   Autothermal Reforming

Autothermal reformers combine some of the best features of steam reforming and partial oxidation systems. A hydrocarbon feedstock (methane or a liquid fuel) is reacted with both steam and air (or oxygen) to produce a hydrogen-rich gas, *i.e.*,

$$4C_mH_n + 2mH_2O\,(gas) + mO_2 \rightarrow 4mCO + 2\,(m+n)H_2 \tag{2.9}$$

The reaction takes place at high temperature (950–1100 °C) and at pressures up to 10 MPa. During this process, both the steam reforming and partial oxidation reactions run concurrently. With the right mixture of input fuel, air/oxygen and steam, the latter process supplies all the heat required to drive the former.

The autothermal reformer requires no external heat source and no indirect heat-exchangers. Consequently, the technology is both simpler and more compact and will probably have a lower capital cost. Moreover, since all the heat generated by partial oxidation is fully utilized to drive steam reforming, autothermal reformers typically offer higher system efficiency (80–90% is possible) than partial oxidation units, in which the excess heat is not easily recovered. They also require a WGS reactor and a hydrogen purification stage, just as with the other systems. Combustion conditions have to be carefully controlled to avoid carbon formation but, overall, the unit has the advantage of being able to operate at very low steam-to-carbon ratios (as low as 0.6).

2.5.2 Sorbent-enhanced Reforming

By mixing a solid *absorbent* for carbon dioxide with the reforming catalyst, it is possible to combine the steam reforming of methane [reaction (2.1)] and its subsequent shift reaction [reaction (2.5)] into a single step and, at the same time, reduce the temperature of the former from about 900 to 400–500 °C. Removal of the carbon dioxide from the reaction zone causes a shift in the equilibrium of the combined reaction so that the production of hydrogen is enhanced, while the carbon monoxide is oxidized to carbon dioxide. A typical product gas is composed of 90% hydrogen and about 10% unreacted methane, with a small percentage of carbon dioxide and a trace of carbon monoxide.

When the sorbent has become saturated with carbon dioxide, it is regenerated by purging with steam. The exit steam is condensed and the released carbon dioxide can be captured ready for compression and conveyance to underground storage. For continuous operation, two parallel beds are required – one bed undergoes reaction while the other is regenerated. The sorbent-enhanced reaction process offers an elegant approach in that it removes the need for a second (shift) reactor and for additional gas-separation steps. Also, the lower temperature of operation results in reduced heat loss and permits the use of lower-cost materials of construction. It should be cautioned, however, that the process is still in the early stages of development and technical problems may yet be encountered in scaling to commercial size. This process has been under development at The Netherlands Energy Research Centre.

2.5.3 Plasma Reforming

Hydrogen can also be produced by the direct thermolysis or thermocatalytic decomposition ('cracking') of methane or other hydrocarbons. The energy requirement per mole of methane is in fact less than that for steam reforming (although only half as much hydrogen is produced) and the process is simpler.

In addition, a useful by-product – clean solid carbon in the form of soot – is produced. Obviously, this can be captured and stored more easily than gaseous carbon dioxide. Whereas thermocatalytic cracking offers the benefit of operating at a much lower temperature than direct thermolysis, it does suffer from progressive catalyst deactivation through carbon build-up. Moreover, reactivation would result in unwanted emissions of carbon dioxide.

The carbon build-up problem can be avoided by the application of thermal plasma technology. This is because a thermal plasma, which is created by an electric arc, is characterized by temperatures of 3000–10 000 °C and under such conditions no catalyst is required. Plasma reforming can operate in pyrolytic mode on all forms of hydrocarbons, including heavy oil fractions, so that again the carbon is turned into soot, but now there is no catalyst to deactivate. Moreover, because the reaction occurs much faster, the technology allows for a more compact and lighter design of reactor compared with those used in conventional operations. Of course, the process is a heavy consumer of electricity (high-intensity and low-voltage current is used) and this detracts from its merits. In addition, conversion levels are generally low, but may be improved to some degree by the addition of steam and oxygen. Thus, although potentially a route to producing hydrogen without liberating carbon to the atmosphere, plasma systems are unlikely to be justified on energy efficiency grounds.

2.6 Membrane Developments for Gas Separation

As an alternative to pressure swing adsorption and the other methods mentioned in Section 2.3, research is in progress to find an effective means to separate hydrogen from carbon dioxide subsequent to the WGS reaction. In particular, it is desirable to separate the gases while hot and so conserve heat energy. The work is directed primarily towards the use of membranes that are selective to the diffusion of hydrogen (a small molecule) while excluding carbon dioxide. By using ceramic membranes, it should be possible to effect the separation at close to the temperature of the shift reaction. Although this approach is already practised to some extent, there remains great scope for improving the performance and lowering the cost of all types of membrane.

2.6.1 Membrane Types

Membrane separators offer the possibility of compact systems that can achieve fuel conversions in excess of equilibrium values by continuously removing the product hydrogen. Many different types of membrane material are available and a choice between them has to be made on the basis of their compatibility with the operational environment, their performance and their cost. Separators may be classified as (i) non-porous membranes, *e.g.*, membranes based on metals, alloys, metal oxides or metal–ceramic composites, and (ii) ordered microporous membranes, *e.g.*, dense silica, zeolites and polymers. For the separation of hot gases, the most promising are ceramic membranes.

Metal-based, non-porous membranes can produce a hydrogen stream of very high purity that can be utilized directly in a fuel cell. The separation process relies on the ability of the metal to allow only the diffusion of hydrogen. The most widely known metallic, hydrogen-permeable, membrane materials are palladium and its alloys. The permeation of hydrogen is thought to proceed via several steps, namely, adsorption of molecular hydrogen, dissociation to the monatomic form, ionization, diffusion, re-association and, finally, desorption. The hydrogen flux density (and hence the membrane performance) is a function of the inherent diffusion characteristics of the material, which in turn are a function of temperature, the thickness of the membrane and the pressure drop across it. Although materials based on palladium and its alloys are commercially available, the cost of the metal is a major barrier to application in large-scale hydrogen separation. Two approaches to reducing the cost would be:

- the use of very thin films ($< 10\,\mu m$) of palladium or its alloys supported on a metal or ceramic substrate that has controlled pore size and distribution;
- the adoption of cheaper amorphous alloys as alternatives to palladium.

Other key issues confronting the development of metal membranes include maintenance of a defect-free structure during manufacture, degradation in membrane strength from hydrogen embrittlement and decreased performance over time due to hydrogen entrapment within the structure.

A further type of non-porous membrane is based on the use of dense, proton-conducting metal oxides from the perovskite family. These ceramic materials have the general formula ABO_3 or $A_{1-x}A'_xB_{1-y}B'_yO_{3-\delta}$, where x and y are fractions of dopants in the A and B sites, respectively, and δ is the number of oxygen vacancies. Considerable research has been undertaken on $SrCeO_3$ and $BaCeO_3$ that have been variously doped with trivalent cations such as those of yttrium, ytterbium or gadolinium. Although these oxides can operate at much higher temperatures (up to 800 °C) than metal membranes, they are difficult to make and also suffer from low mechanical strength, poor hydrogen flux and low chemical stability in the presence of both carbon dioxide and water.

Ceramic–metallic ('cermet') membranes are also under development for hydrogen treatment. Here, the finely-divided ceramic and metal are sintered together to form a uniform body. The ceramic functions as both an electronic and a proton-conducting phase, whereas the metal component serves to enhance the hydrogen permeability of the ceramic by increasing the electronic conductivity. Depending on its hydrogen permeability, the metal may also provide an additional transport path for the hydrogen. In an alternative design of composite membrane, a permeable ceramic matrix acts solely as a structural support and the hydrogen is transported almost exclusively by a thin metal film that is deposited on the surface of the ceramic. Both of these types of membrane offer good stability in the presence of trace contaminants such as sulfur compounds and mercury, but further research is required to optimize the amounts, microstructure and compatibility of the ceramic and metal components.

A porous membrane – the second category of hydrogen separation membrane – usually consists of a thin layer of a microporous sieve material such as silica, carbon or zeolite on a thicker and highly porous support. The hydrogen is transported through the pores of the membrane predominantly by molecular diffusion, which is a purely physical process with a characteristic that is determined by the pore diameter of the membrane. To separate hydrogen efficiently, the pores must be less than 1 nm in diameter. A variety of established manufacturing techniques could be used to fabricate microporous membranes on either ceramic or metallic macroporous supports. To maximize the flux, the membranes are usually made as thin as practically possible; films with thicknesses below 10 µm are routinely made. Generally, 1–3 thin layers are applied to a porous support. Membranes are usually produced in 'tube-and-shell' configurations that are assembled in multi-tube modules for efficient distribution of feed and product gases; see Figure 2.5. The feed gas is introduced on the shell side and hydrogen product is withdrawn from the tube side (or *vice versa*). Metal support tubes are easier to install into modules and are more robust and not as prone to catastrophic failure as ceramic supports. On the other hand, metal supports are subject to the problem of thermal expansion mismatch between the support and the membrane. Because a perfect porous membrane with a discrete pore size is impossible to fabricate, an infinite separation factor is unachievable, compared with non-porous membranes that can give perfect separation. Nevertheless, separation factors of up to 100 have been achieved and these would give a single-stage hydrogen purity of 99%. Higher purities would require a number of stages to be used, for example, two stages would result in 99.99% hydrogen. Hybrid systems using porous membranes in combination with pressure swing adsorption may be required to reach the high purities demanded by some fuel cells.

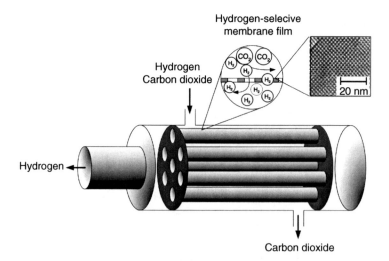

Figure 2.5   Schematic of a molecular sieve membrane.

Polymer membranes are widely used for the separation of carbon dioxide from natural gas and the recovery of hydrogen from oil refinery streams. Separation is achieved by selective permeation of one or more gases from the feed side to the permeate side of the membrane, *i.e.*, gas transport takes place under the influence of a concentration gradient. Membranes with high diffusivity and selectivity are usually made from rigid glassy polymers, which are also generally less expensive to produce than inorganic membranes and can achieve very high separation factors. On the other hand, they cannot withstand high temperatures such as those experienced with integrated gasification combined-cycle (IGCC) systems; see Section 2.7. To overcome this problem, research is in progress to develop more robust polymer–ceramic composites and other types of hybrid material.

2.6.2 Membrane Reactors

Efforts are being made to combine the WGS reaction [reaction (2.5)] and the removal of carbon dioxide into a single step by using a 'membrane reactor', in which a hydrogen-transport membrane (in either tubular or planar form) is placed inside a WGS reactor. The catalyst for the shift reaction either takes the form of a packed bed (as shown schematically in Figure 2.6), so that the reaction zone is distinct from the separation zone, or is incorporated in the membrane itself so that reaction and separation occur simultaneously. The hydrogen in the syngas, and also that produced via the shift reaction, permeates through the membrane and thus allows the equilibrium to move towards the desired product, hydrogen. The reject stream consists primarily of carbon dioxide, non-recovered hydrogen and water. The end result is a stream rich in hydrogen separated from a stream rich in carbon dioxide. Depending on the specifications placed on the carbon dioxide for effective sequestration, it may be necessary to include a second conversion stage in which unconverted

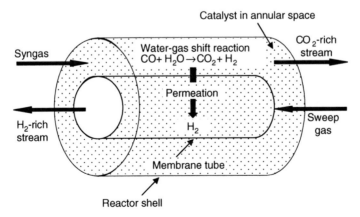

Figure 2.6 Schematic of a packed-bed membrane reactor.

fuel and combustion intermediates (carbon monoxide and hydrogen) are converted to carbon dioxide and water.

Present research is being directed towards the development of catalysts that can maintain sufficient activity at high partial pressures of hydrogen and carbon dioxide and can operate at temperatures above the present limit of 500 °C. Process-intensification steps such as these should, if successful, improve efficiency and reduce both capital and operating costs.

A further application for membranes is the separation of oxygen from air, as an alternative to low-temperature (cryogenic) distillation, and integrating this with the steam-reforming reaction in a single step. A ceramic membrane is employed which, at 900 °C, is both an ionic conductor (for oxygen ions) and an electronic conductor. Hot air passes over the outer surface of a tubular membrane where the oxygen is ionized. This membrane is integral with the steam-reformer tube. Oxide ions pass through the membrane in one direction and electrons in the opposite direction. The oxide ions are reconverted to atoms when exiting the membrane on the inside of the tube and there meet the incoming natural gas where partial oxidation takes place. Nitrogen is rejected as it is unable to pass through the membrane. The ceramic material must show long-term stability in both reducing and oxidizing atmospheres, and also long-term compatibility with the catalysts for oxygen reduction and reforming that are in contact with it.

In summary, there are several potential designs and applications for gas separation membranes in the manufacture of hydrogen from fossil fuels. Much scope exists for further research and development. A particular goal is the perfection of membranes that operate at high temperatures, as this would obviate the need to cool and reheat large volumes of gases. Similarly, membranes that function at high pressure are desirable in order to improve the efficiency of the process and to deliver the separated gases in a pressurized state. The latter feature would save the considerable energy (and cost) involved in recompressing the gases for storage and transport.

2.7   Coal and Other Fuels

Gasification is defined as the transformation of solids into combustible gases in the presence of an oxygen carrier (*e.g.*, air, oxygen, steam, carbon dioxide) at high temperatures (*e.g.*, > 700 °C). The energy required for thermochemical conversion is provided by partial combustion of either the solid or the resulting syngas. Almost any fossil fuel can be treated in this way to produce hydrogen. The most widely-distributed fossil fuels are the various types of coal, but other possibilities exist such as tar sands (now renamed 'oil sands'), asphalts, heavy oils extracted from shale or refinery residues, and coke. Biomass, which includes agricultural waste, forestry waste, energy crops and municipal solid waste, may also be processed by gasification technology to produce hydrogen; this is discussed separately in Section 2.8.

2.7.1 Gasification Technology

The world production of coal is projected to rise from 2772 Mtoe in 2004 to 2818 Mtoe in 2010 and 3779 Mtoe in 2030.[1] Much of this coal will be used for electricity generation. In future, rather than employing conventional boilers, coal will probably be converted to a gaseous fuel that will power gas turbines as part of an IGCC scheme. Compared with conventional power plants that are fired with pulverized coal, IGCC offers the following advantages:

- it can achieve greater thermal efficiency through the use of gas turbines (up to 55% efficient) and thus there will be lower emissions of carbon dioxide and solid waste (ash) per unit of electricity produced;
- the process uses significantly less cooling water;
- it can operate on a variety of fuels, *e.g.*, heavy oils, petroleum cokes and coals;
- pre-combustion capture of carbon is easier than post-combustion capture (where the exhaust gases are heavily diluted with nitrogen from the air) and hence costs of sequestration will be lower;
- there is greater removal of sulfur (95% minimum);
- the exhaust gas contains lower levels of nitrogen oxides (NO_x), *i.e.*, less than 50 ppmv.

As supplies of natural gas become depleted, gasification of coal is seen as the preferred technology of the future for electricity production, along with the renewable forms of energy. This is because coal is plentiful and widely available geographically, the gasification process is more efficient than conventional coal combustion, and the carbon dioxide can be extracted from the syngas before combustion and then stored permanently underground. All these benefits, however, have yet to be realized practically on a commercial scale. Major developments are needed to improve the performance and reduce the cost of the technologies and their integration with gas and steam turbines to give IGCC power plants.

The syngas made from coal, like that produced by the steam reforming of natural gas, may be further processed to hydrogen and carbon dioxide by the WGS reaction or used to produce a variety of liquid fuels and chemicals via the Fischer-Tropsch process [see reaction (2.3)]. The first large-scale production of liquid hydrocarbons derived from coal for use in internal combustion engines took place in Germany during World War II, when there was an acute shortage of petroleum. Later, South Africa, which has coal deposits but little oil, built plants to produce its own petroleum during the period of international trade sanctions. Thus, there are three major and distinct applications for gasification technology:

- more efficient generation of electricity from coal;
- synthesis of liquid fuels and chemicals from coal, *i.e.*, 'gas-to-liquids' (GTL) process;
- production of hydrogen from solid fuels and residual oils.

Gasification technology is being driven primarily by the first two of these applications.

Coal gasification facilities are being constructed around the world, particularly in China – primarily for the manufacture of chemicals and fertilizers, but also for more efficient generation of electricity. A world survey in 2004[2] identified 117 operating plants with 385 gasifiers and a total capacity of 45 000 MW_{th}. Of these, approximately half were coal-fired and the remainder were operating with petroleum residues and other fuels. The applications for these gasifiers were 37% chemicals manufacture, 36% Fischer–Tropsch synthesis, 19% power and 8% gaseous fuels. The growth forecast for total capacity was 5% per year.

It is only now that the production of pure hydrogen from solid fuels, for use as an energy vector, is seen as a long-term goal. Here, the driving force is to utilize the world's extensive coal deposits in an environmentally friendly fashion to meet future energy demands. The US Department of Energy has taken a lead, together with industry, in developing coal gasification for electricity generation as part of its Clean Coal Technology programme. This initiative is motivated by the huge reserves of coal in the USA, in addition to concerns over atmospheric pollution.

There are three basic designs of coal gasifier:

- entrained-flow;
- moving-bed (sometimes also referred to as fixed-bed!);
- fluidized-bed.

These technologies are shown schematically in Figure 2.7 and their main operating features are as follows.

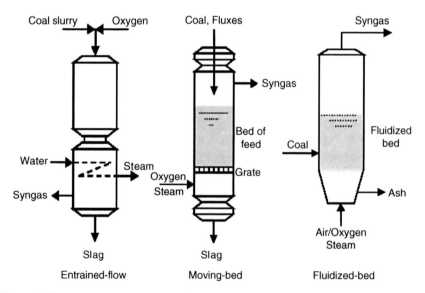

Figure 2.7   Schematics of entrained-flow, moving-bed and fluidized-bed gasifiers.

2.7.1.1 Entrained-flow Gasifier

Entrained flow is the most aggressive form of gasification, in which pulverized coal and oxidizing gas flow co-currently. Optionally, the pulverized coal may be fed to the gasifier in the form of a slurry with water that then provides the source of the steam required for the reaction. Under operating conditions of high pressure (2–3 MPa) and high temperature (> 1400 °C), almost complete gasification is achieved with little formation of tars or of char. Extremely turbulent flow results in the coal particles experiencing significant back-mixing and residence times are measured in seconds.

Entrained-flow gasification is specifically designed for low-reactivity coals and high coal throughput. Single-pass carbon conversions are in the range 95–99%. To experience smooth operation, with the removal of the ash as a molten slag, the temperature of the gasifier must lie above the coal–ash fusion temperature. Alternatively, it is necessary to add fluxes to lower the melting temperature of the mineral matter in the coal.

2.7.1.2 Moving-bed Gasifier

Moving-bed gasifiers operate at about 3 MPa and closely resemble a blast furnace. Coal and fluxes are placed on the top of a descending bed in a refractory lined vessel. On moving downwards, the coal is gradually heated and contacted with an oxygen-enriched gas flowing upwards countercurrently. Pyrolysis, char gasification, combustion and ash melting occur sequentially. The oldest and best known gasifier of this type is the Lurgi moving-bed gasifier. The temperature at the top of the bed, where the syngas off-take is located, is typically 450 °C and at the bottom approaches 2000 °C. Mineral matter in the coal melts and is tapped as an inert slag. Ash melt characteristics influence bed permeability and fluxes may need to be added to modify slag flow characteristics. The off-gas contains tars that must be condensed and recycled. The production of tars makes downstream gas cleaning more complicated than with other gasification processes. The residence time is between 30 min and 1 h, which places stringent restrictions on the physical and chemical properties of the coal. Long residence times mean that moving-bed gasifiers have a low throughput and hence have limited application in large-scale IGCC plants.

2.7.1.3 Fluidized-bed Gasifier

A fluidized-bed reactor is a vessel in which fine solids are kept in suspension by an upward-flowing gas such that the whole bed exhibits fluid-like behaviour. Finely-divided coal is injected into a bed of inert particles that is fluidized by steam and air (or oxygen) at high pressure. Rising oxygen-enriched gas reacts with suspended coal at a temperature of 950–1100 °C and a pressure of 2–3 MPa. High levels of back-mixing give rise to a uniform temperature distribution within the gasifier. This type of reacting system is characterized by high rates of heat and mass transfer (*i.e.*, increased reaction rates) between the solid and the gas. The unusual characteristic of fluidized-bed gasifiers is that the majority of the bed material is not coal but accumulated mineral matter and sorbent (for *in situ* desulfurization). Operating with a high inventory of inert bed material has two advantages: (i) the coal experiences a high rate of heat transfer on entry and (ii) the gasifier can operate at variable load, that is, the rate of syngas production can be varied at will within wide limits.

The comparatively low temperature of operation limits the use of fluidized-bed gasifiers to reactive and predominantly low-rank coals such as lignite or brown coal. Most such units require recycling of entrained fines to achieve 95–98% carbon conversion. To reduce the extent of fines recycling, it has been proposed that the gasifier be linked with a fluidized-bed combustor (*i.e.*, an 'air-blown gasification cycle'). In this process, the coal is first gasified to yield a carbon conversion of 70–80%. The unreacted char is then fed to the combustor where generated heat is used for steam production. The gasifier−combustor combination would permit the use of low-reactivity coals in an IGCC system. In general, however, the type of gasifier should be matched to the properties of the coal available, especially with respect to its gasification characteristics and mineral content (ash melting temperature, chemical composition). The three major classes of gasifier in their various modifications can, between them, cope with most types of coal.

Gas cleaning is an essential part of the overall gasification technology, both to protect catalysts from poisoning and to meet the end specification for the syngas (*e.g.*, for use in gas turbines or chemical manufacture) or the pure hydrogen required by low-temperature fuel cells. Compared with the steam reforming of natural gas, the gasification of coal yields a syngas that has (i) higher levels of carbon monoxide, which have to be shifted to carbon dioxide and hydrogen, and (ii) generally higher levels of impurities, which have to be cleaned from the gas before use. The main contaminants in syngas produced from coal are particulates, sulfur oxides (SO_x), nitrogen oxides (NO_x), the alkali metals sodium and potassium, and mercury. Minor or trace amounts of ammonia, arsenic, beryllium, cadmium, chromium, hydrogen chloride, hydrogen fluoride, lead, nickel and selenium can also be present. All these species have to be removed to acceptable limits in order to protect downstream process equipment (in particular, the gas turbines) against fouling, erosion and/ or corrosion, to prevent poisoning of the catalyst used for the shift reaction, and to meet environmental regulations. At present, barrier 'candle' filters are used to remove solid contaminants. The filters have a porous tubular structure and can be classified into two broad designs: (i) 'ceramic' filters, which are made from materials such as alumina, silica and zeolites; and (ii) 'metallic' filters, which are made from nanofibres or wires of alloys, *e.g.*, iron aluminide, Incoloy, Monel, Hastelloy, and are protected from corrosion by a ceramic coating.

To avoid degradation of the construction materials used for the treatment unit, it has been customary to undertake cleaning after the gas has been cooled to below 100 °C, but this reduces the overall energy efficiency of the plant and increases both capital and operating costs. When using IGCC technology, the syngas is produced at much higher temperatures and pressures. Hence there are obvious operational advantages to be gained from the development of 'hot-gas cleaning' systems. Indeed, such an advance is considered essential if the comparative economics of IGCC plants are to become attractive. As mentioned above with respect to membrane reactors, hot-gas cleaning technologies are in the early demonstration stage with units currently being evaluated at temperatures above 600 °C (which still necessitates considerable cooling of the gas) and at pressures up to 2.5 MPa. Calcium-based sorbents (limestone and dolomite) and zinc titanate are the leading candidates for use in the desulfurization of high-temperature fuel gas, but even these have serious limitations that will have to be overcome before they may be regarded as satisfactory.

2.7.2 Combined-cycle Processes

Combined-cycle gas turbines (CCGTs) are widely used for electricity generation from natural gas. The gas is burnt in a high-temperature gas turbine that is coupled to a generator. The exhaust gas from the turbine is used to raise steam, which is then fed to a conventional steam turbine and a further generator. An IGCC plant for generating electricity from coal is based on a similar process. Cleaned syngas replaces natural gas as fuel for a large gas turbine coupled to a generator. Again, the gas turbine exhaust passes through a heat-exchanger (the 'heat-recovery steam generator') to raise high-pressure steam. This is mixed with the steam from the gas cooler and used to operate a conventional steam turbine that generates more electricity. The overall process is an example of so-called 'clean coal' technology.

The scale of the IGCC operation is impressive. Techno-economic studies have indicated[3] that a large gasifier will have a coal feed in the range 2000–3000 t per day and will produce syngas at an energy efficiency of 75–80%. On combusting this syngas in an IGCC scheme, the net power generated will lie in the range 270–420 MW$_e$, with an overall thermal efficiency (coal-to-electricity) of 38–45%, although eventual efficiencies of 50% are widely anticipated. This performance is significantly better than that of a conventional coal-fired electricity utility. A 400 MW$_e$ system will incur a capital expenditure of the order of US\$500 × 10^6 and will emit around 6500 t of carbon dioxide per day.

In the early 1990s, a coal-based IGCC plant was built at Wabash River in west-central Indiana, USA. This was a demonstration project in which the combustion turbine generated 192 MW$_e$ and the steam turbine 104 MW$_e$. With a parasitic load of 34 MW$_e$, the net power production was 262 MW$_e$. Between 1995 and 2000, 1.5 Mt of coal was processed and 4 TWh of electricity produced. Other IGCC plants are under construction in Kentucky and Ohio; the latter is a 580 MW station for the co-generation of hydrogen and electricity from petroleum coke. Based on the successful demonstration of this technology, it should be possible to build a power station of 1–2 GW output by combining several of the gasification units. Whether or not this will happen depends on both the

(a)

(b)

Figure 2.8   (a) Wabash River IGCC facility; (b) gas-cleaning unit for removal of
hydrogen sulfide and ammonia from the syngas.[4]
(Courtesy of the US Department of Energy).

economics and environmental considerations. A photograph that illustrates
the size and complexity of the overall Wabash River facility is shown in
Figure 2.8(a) and a close-up of the gas-cleaning unit alone in Figure 2.8(b).[4]

To keep a perspective, as noted above, only about one-fifth of the world's
gasifiers are employed in power generation. Most of the rest are dedicated to

the manufacture of chemicals. In 2003, 1800 MW_e of IGCC plant was operating for electricity generation and a further 3150 MW_e was planned.[5]

Improvements in performance and cost may be expected on the basis of the operational experience now being gained. As the technology matures and the sizes of the power plants increase, costs are expected to decrease accordingly. The question of gas purity also needs to be addressed because it is likely that environmental regulations will set restrictions on the impurities that may be discharged.

The next key steps in the utilization of IGCC technology to further the Hydrogen Economy are:

- to separate syngas that has been upgraded by the WGS reaction into its principal components – hydrogen and carbon dioxide;
- to purify the hydrogen so that it is suitable for use in fuel cells;
- to develop a practical process for the sequestration of the carbon dioxide in order to prevent its escape to the atmosphere.

All these advances have to be achieved within a framework of costs that are commercially acceptable. The US Department of Energy is addressing these issues in its FutureGen programme.

2.7.3 FutureGen Project

FutureGen is a major IGCC project, backed by the US government and intended to last 10 years, that has three key objectives:

- to demonstrate the production from coal of pure hydrogen for use in fuel cells;
- to co-generate electricity at the 275 MW_e level with zero emissions of carbon dioxide;
- to meet an electricity cost target not more than 10% greater than that of a plant in which the carbon dioxide is released.

The project, which is a joint industry and government venture, has a budget of US$1 billion over its 10 year life. At its launch in February 2003, the US Secretary of Energy described FutureGen as: 'one of the boldest steps our nation has taken toward a pollution-free energy future'. The initiative will serve as a platform to demonstrate hydrogen production and carbon dioxide sequestration under realistic conditions at a commercial scale and also to test new components and innovative technologies that are aimed at raising the performance and lowering the cost of IGCC systems. Construction of FutureGen is expected to be completed by 2012.

The overall process involves seven steps:

- separation of oxygen from air;
- gasification of coal to raw syngas;

- sulfur removal and further purification;
- water-gas shift reaction;
- separation of hydrogen from carbon dioxide and its further compression and storage;
- sequestration of the by-product carbon dioxide;
- combined-cycle generation of electricity using syngas or hydrogen as fuel.

All of these steps require considerable research and engineering development to be carried out if the cost target is to be met or even approached. One of the features of the project is that, as an alternative to making pure hydrogen, the syngas could be converted to methanol or other organic chemicals. Thus, FutureGen will be a versatile facility that is capable of producing pure hydrogen (or organic chemicals) and electricity from coal in any desired ratio – a so-called 'poly-generation plant'. The hydrogen would be available for use in fuel cells. A block diagram of an integrated gasification combined-cycle process is shown in Figure 2.9.

In Europe, the German company RWE is planning a similar project for IGCC power generation with the capture and storage of carbon dioxide. The proposed unit would operate at the $450\,MW_e$ level and be commissioned in 2014. It is estimated that between 25 and 50% more fuel will be required as a result of fitting carbon-capture equipment and that the cost of producing electricity will rise by 30–100%. This is not a project to be undertaken lightly.

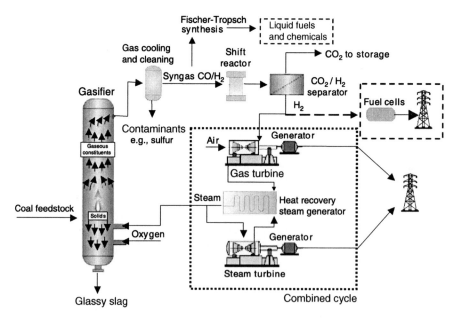

Figure 2.9 Concept of a poly-generation plant based on the IGCC process.

2.8 Biomass

Biomass is clearly not of fossil origin. It is, however, a carbon-based solid fuel and is therefore included in this chapter. As a renewable form of energy, biomass is not deemed to be a net contributor to the build-up of greenhouse gases. Strictly, however, biomass is not 'carbon neutral' on account of the external energy consumed in growing (ploughing, fertilizer, *etc.*), harvesting (reapers, power saws, *etc.*) and transporting the crop to the combustion site.

The contention that biomass is 'carbon neutral' merits further examination. Irrefutably, biomass derives from atmospheric carbon dioxide through photosynthesis. Less obvious, however, is the argument that the carbon dioxide arising from the burning of biomass, once released, will speed up subsequent photosynthesis so as to restore 'neutrality'. The main question is: 'does the combustion of biomass contribute to the atmospheric burden of carbon dioxide, *i.e.*, to the global greenhouse effect?' This is a matter of the equilibrium between the carbon dioxide in the air and that on the Earth (both terrestrial and oceanic) and the kinetics of exchange between the two – a slow process and a complex issue. Of course, if a biomass crop is harvested and replaced by a new growth then (leaving aside the energy input mentioned above) it is, to a first approximation, carbon neutral. When left untreated, all biomass eventually decays with the liberation of carbon dioxide and methane (or burns through grass or forest fires) and therefore much of the carbon inevitably returns to the atmosphere. Hence it makes sense to extract the useful energy rather than allow it to go to waste.

2.8.1 Dry Biomass

Dry biomass, in the form of straw and wood chips, has been extensively employed, particularly in Denmark, to raise steam for district heating schemes. In rural communities, individual farmers have installed boilers to provide heat on a smaller scale for their immediate locality ('neighbourhood heating schemes'). Dry biomass is also employed as fuel for combined heat and power (CHP) schemes. Most such ventures based on waste agricultural projects are restricted in size to about 10 MW_{th} because of the limited local availability of the raw material.

One practical means of utilizing some forms of dry biomass on a larger scale is through co-combustion with coal. Already some coal-fired power stations are conducting co-combustion trials by adding sawdust, wood chippings, bagasse (residue of sugar cane) or pine kernels to the coal feed. Up to 10% of biomass (depending upon which type is employed) may be mixed in with the coal. For a large (2000 MW_e) power station, this represents a substantial amount of biomass to collect and handle. The co-combustion of coal and dry biomass is a relatively recent concept, introduced to lower the consumption of coal so as to keep within imposed emission limits for carbon dioxide and sulfur dioxide (biomass contains much less sulfur than most coals). Some power stations have, however, experienced operational problems with co-combustion. For example, biomass additions to some varieties of coal can change the ash characteristics, which are important when using the ash in established markets such as building block manufacture. Biomass ash can also cause increased fouling and reduce the fusion temperature of coal ash. Other coal-fired power stations have successfully added a few percent of biomass to the feed with no ensuing operational problems. Nevertheless, it is clear that blending biomass with pulverized coal for electricity generation is not entirely straightforward and the control parameters of the plant have to be adjusted to suit each coal and biomass combination.

The second option for dealing with dry biomass is to gasify it, either alone or by addition to the feed of a coal gasification plant. In principle, hydrogen may be produced by the gasification of any form of biomass. The most suitable materials are those with water contents in the range 5–30 wt.%, including fast-growing energy crops such as *Miscanthus* ('silver') grass, hemp, poplars and willows, or any forestry or agricultural waste such as wood chippings, sawdust, bagasse, coconut or nut shells, cotton stalks, jute sticks, maize and jowar cobs, rice husks, and wheat and rice straw. Waste paper and municipal waste (after pre-sorting) may also be converted to crude syngas. The gasification is generally conducted at lower pressures and temperatures than those used for coal and is limited to mid-size operation due to the heterogeneity of biomass, its localized production, and the relatively high costs incurred in its gathering and transportation. The largest such plants are capable of processing just a few hundred tonnes of material per day.

Modern biomass gasification operations use moving-bed or fluidized-bed technology. The combined thermal efficiency of heat and power generation of a biomass gasifier coupled to a gas turbine can be in the range 22–37%; lower figures are reported for gas engines. Similarly, direct biomass combustion technologies that involve steam generation and steam turbines have efficiencies of only 15–18%. For CHP applications that employ biomass, there is a distinct trend away from conventional boilers towards gasifiers.

Properties such as moisture content, mineral fraction, density and chemical composition (*e.g.*, chlorine and nitrogen contents) differ widely depending on the source of the biomass. This may necessitate the development of a specific design of gasifier to accommodate a particular type of biomass. The wide diversity in the characteristics of biomass is a challenge for gasification processes and in practice a range of technical designs of gasifier will be employed to convert different feedstocks into biogas and into hydrogen, as dictated by the scale of operation and the end market. For localized production of hydrogen, the gasifier will need to be married to a gas cleanser and shift reactor. One of the major attractions of preparing hydrogen from biomass is that this fuel is rated as a renewable (since it is generally regarded as carbon neutral) and provides the only direct route for the bulk production of hydrogen from a renewable source without first going via electricity.

In addition to the nature and composition of the biomass, the feasibility of gasification is influenced by the entire supply system. Depending on the choice of gasification technology, various preparatory steps such as sizing (shredding, crushing, chipping) and drying have to be included to meet process

(a)

(b)

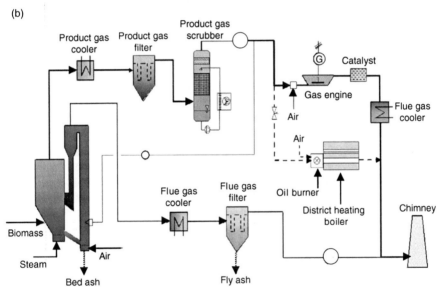

Figure 2.10 (a) Biomass CHP plant at Güssing, Austria; (b) flow diagram of the operation.
(Courtesy of REPOTEC – Renewable Power Technologies).

requirements. Storage, processing (*e.g.*, densification) before transport and pre-treatment all affect moisture content, ash content, particle size and sometimes even the degree of contamination (*e.g.*, biomass obtained from salt marshes must be washed thoroughly).

A number of demonstration gasifers are being operated in Europe under the auspices of the Bioenergy Programme of the International Energy Agency. An example of this is the 8 MW$_{th}$ facility that has been built to supply heat and power to the town of Güssing in Austria; see Figure 2.10(a). This is based on a fluidized-bed reactor that consists of two zones; see Figure 2.10(b). The gasifier zone is fluidized purely with steam, so as to produce syngas that is free from nitrogen. The combustion zone is fluidized with air and the resulting flue gas (mostly nitrogen and carbon dioxide) is discharged up the chimney. The reactor design relies on the fluidized bed bringing about thorough mixing of the particles from the two zones and thus transferring heat from the combustion to the gasification process. The product syngas has a high ratio of hydrogen to carbon monoxide, *viz.*, 1:6–1:8 by mass (equivalent to 2.3:1–3.4:1 in terms of moles).

A US report[6] dismisses biomass gasification as a national source of hydrogen for fuel cells, particularly fuel cell vehicles, on the grounds of the huge areas of land that would be required, the uncertainty over the environmental impact on the land itself, and the unacceptably high costs associated with a multitude of mid-size generators. As discussed above, most biomass has been used for the heating of buildings, for cooking in rural areas and, more recently, for electricity production, either in conventional plant or in gasifiers. Where, in future, there may be a local requirement for hydrogen in modest amounts and where a suitable source of biomass is available, gasification may be an acceptable means for disposing of waste with the added benefit of obtaining a useful product. It should be noted, however, that the overall energy efficiency of converting solar energy to hydrogen via the growing and harvesting of energy crops, gasification and hydrogen separation is very poor, namely, less than 1%. Moreover, the cost of the resulting hydrogen is several times that of hydrogen produced by the large-scale steam reforming of natural gas. Account must also be taken of the environmental impact associated with the growing, production and transport of biomass, together with any potential degradation in land quality that might arise from the intensive farming that is required. Also, energy inputs in the form of fertilizers and transport fuel are not inconsiderable. Finally, in a world that is both food-limited and carbon-constrained, there remains the question as to whether agricultural land is not better utilized in growing food, rather than in raising energy crops.

The use of dry biomass for energy production in the world has been, and continues to be, dominated by combustion. In terms of net energy produced, gasification, pyrolysis, fermentation (*e.g.*, to produce ethanol), extraction (*e.g.*, bio-diesel) and other options play only a marginal, but growing, role. Nevertheless, new and improved technologies in the field of gasification, such as biomass integrated gasification combined-cycle technology, and also in the production of biofuels and hydrogen may eventually make biomass more attractive.

2.8.2 Wet Biomass

Wet forms of biomass, such as sewage, liquid manure and silage, would normally be treated by anaerobic digestion at ambient temperature. The resulting 'biogas' is a mixture of methane, carbon monoxide and carbon dioxide, with rather little hydrogen. It is similar to the biogas produced in sealed landfill sites used for the disposal of municipal waste. Small bio-digesters have been commercialized in some developing countries to produce biogas for cooking purposes. In rural areas of developed countries, farmers sometimes employ biogas to heat their buildings or to fuel gas engines for combined heat and power schemes. Biogas is unsuitable for use in low-temperature fuel cells, but should be acceptable for high-temperature molten carbonate or solid oxide fuel cells; see Chapter 6. This is a future possible path to local electricity generation in rural communities.

Bacterial digestion typically involves the breakdown of biological material by many and varied micro-organisms, *i.e.*, cryophiles, mesophiles and thermophiles that thrive at cold ($<15\,°C$), medium ($25–40\,°C$) and high ($>45\,°C$) temperatures, respectively. The resulting product is biogas of the composition mentioned above. There are, however, some specific bacteria that give a much higher yield of hydrogen in the gas. For example, bacteria in the *Clostridia* family, which occur in cow dung and sewage sludge, are particularly effective in yielding hydrogen – biogas containing over 60 vol.% hydrogen has been demonstrated under laboratory conditions. Research is continuing to determine the best strains of bacteria to use for specific feeds and the optimum conditions. This is a potential route for the direct conversion of wet biomass to hydrogen.

2.9 Basic Research Needs

A large number of requirements for basic research have been identified to improve the performance and lower the cost of converting fossil fuels to hydrogen.[7] This is particularly the case when coal is the feedstock and when pure hydrogen for use in low-temperature fuel cells is the desired product. The challenge identified by the US Department of Energy is to reduce the cost of pure hydrogen, produced from coal with the sequestration of carbon dioxide, to about one-quarter of that of hydrogen obtained by the steam reforming of natural gas with discharge of carbon dioxide to the atmosphere. This is a rather arbitrary target and has been set to enable fuel cell vehicles to compete economically with current operating costs for conventional vehicles in USA. Such a target presents a formidable challenge to chemical engineers and at present seems unrealistic, but it may become less onerous as supplies of petroleum and natural gas are depleted and their market prices rise.

Much of the enthusiasm for hydrogen energy does indeed revolve around the use of hydrogen in fuel cells to generate electricity, especially to drive electric vehicles. For this application, low-temperature fuel cells are the best choice (see Chapters 6 and 7), but they are susceptible to poisoning by impurities in the gas. Both carbon monoxide and sulfur have to be removed to very low levels, particularly for use in low-temperature fuel cells. For instance, the WGS reaction, as now practised, leaves sufficient carbon monoxide in the product stream to deactivate the negative electrode in these fuel cells. Existing methods for removing this impurity, which include pressure swing adsorption and preferential oxidation, add cost and complexity to the fuel-processing system. Accordingly, more active catalysts are required for the WGS reaction, and also better processes for the separation of hot gases.

A more radical approach is to develop membrane processes that combine hydrogen generation and separation in a single reactor. Further research on membranes is also needed to yield a new and cheaper process for the separation of oxygen from air. The present process of cryogenic distillation is too costly for the bulk production of hydrogen as a fuel.

The removal of sulfur from the feedstock can present difficulties. Natural gas is generally fairly low in sulfur, but improved methods are required to decrease the sulfur content to the desired ultra-low level. Coal feedstocks are often high in sulfur and this impurity is extracted from the syngas by catalytic conversion to hydrogen sulfide and subsequent absorption in an amine solvent. This, too, is expensive and may not yield hydrogen at the very high level of purity that is demanded by many types of fuel cell. It is therefore essential to find better electrocatalysts with greater activity, specificity and stability, in addition to being less susceptible to poisoning and fouling.

In summary, it is clear that vigorous basic research programmes on catalysts and on membrane separation processes should be instigated if hydrogen from fossil sources is to reach its full potential and become cost-effective. Similarly, further investigation is necessary to define the optimum conditions for the direct conversion of wet biomass to hydrogen. Sequestration of carbon dioxide, which is another important topic for research, is discussed in the following chapter.

References

1. *Key World Energy Statistics*, 2006 Edition, International Energy Agency, Paris, 2006.
2. J. Childress and R. Childress, *2004 World Gasification Survey: a Preliminary Evaluation*, Gasification Technologies 2004, 4–6 October 2004, Washington, DC.
3. A.-G. Collot, *Prospects for Hydrogen from Coal*, Report CCC/78, International Energy Agency Clean Coal Centre, London, December 2003.
4. *Clean Coal Technology; Topical Report No 20: The Wabash River Coal Gasification Repowering Project*, An Update, US Department of Energy, Washington, DC, September 2000.
5. *The Role of Coal as an Energy Source*, World Coal Institute, London, 2004.
6. *The Hydrogen Economy: Opportunities, Costs, Barriers and R & D Needs*, National Research Council and National Academy of Engineering of the National Academies, National Academies Press, Washington, DC, 2004.
7. *Basic Research Needs for the Hydrogen Economy*, Office of Science, US Department of Energy, Washington, DC, January 2004.

CHAPTER 3
Carbon Sequestration

With growing evidence that increasing anthropogenic emissions of greenhouse gases – principally, carbon dioxide – are inducing abnormal changes in climate (specifically, global warming) comes the challenge of how to continue to utilize fossil fuels without causing irreversible damage to the Earth's environment. Slowing the build-up of these gases in the atmosphere implies the need to extract carbon dioxide from exhaust gases (capture technology) and then immobilize it for the indefinite future (storage). Together, carbon capture and storage (CCS) are termed 'sequestration'. There are two main types of CCS: (i) direct sequestration, where the carbon dioxide is captured and stored permanently[†], often in geological formations, and (ii) indirect sequestration, where the carbon dioxide is removed from the atmosphere by enhancing natural processes such as the growing of trees.

Most of the candidate sequestration schemes that have been proposed to date will only be technically and economically viable when applied to large industrial plants (*e.g.*, power stations, oil refineries, cement kilns), which are commonly referred to as 'point sources'. About 40% of all anthropogenic greenhouse gases originate from electricity-generating stations. World-wide, more than 7000 point sources have been identified and these are responsible for over half of all carbon dioxide emissions from fossil fuels. There is almost no possibility of capturing and storing the carbon dioxide ejected from millions of individual internal combustion engines. Therefore, strictly from the standpoint of minimizing releases of carbon dioxide, the following conclusions are drawn.

- Fossil fuels should be processed in large, centralized plants that have facilities for the capture and storage of carbon. Such plants would ideally generate electricity and/or hydrogen although, as an interim measure, some facilities may be built to produce low-carbon fuels (methanol, methane) from high-carbon feedstocks (heavy oils, coal).
- Distributed power units, such as road vehicles and trains, ought to be electrically propelled or hydrogen fuelled. The hydrogen should be produced centrally and distributed widely, either for burning directly in

---

[†] Permanence requires definition in terms of the acceptable amount of leakage and also of the rate at which 'stored' carbon dioxide may be slowly transformed into a non-gaseous state, *e.g.*, by mineralization.

internal combustion engines or for use in fuel cells coupled to electric motors. In the short term, liquid petroleum gas (LPG), compressed natural gas (CNG) or methanol might be employed more extensively as transitional low-carbon fuels in internal combustion engines. Already, bio-ethanol and bio-diesel find application as 'carbon-neutral' fuels.

- The distributed generation of electricity using fossil fuels, *e.g.*, combined heat and power (CHP) plants or micro-turbines fuelled by liquid hydrocarbons, should be seen as a near-term practice only. Whereas this concept is attractive from the viewpoint of reducing the load on the transmission system and gives rise to the possibility of improved energy efficiency,[1] there will still be emissions of carbon dioxide. One approach, yet to be evaluated, is to collect the carbon dioxide from distributed CHP plants into a new network of pipes that lead to a regional sequestration facility. Ultimately, in a post-fossil fuel Hydrogen Economy, neighbourhood electricity must be based on renewable energy sources or on fuel cells that operate on centrally generated hydrogen.

As the world population rises and with it the demand for energy, carbon sequestration will almost certainly become a key issue. This is because coal – a high-carbon fuel in plentiful supply, but one that yields more carbon dioxide than either oil or natural gas – is expected to play an increasingly important role as a primary source of energy.

3.1   The Scale of Carbon Sequestration

Globally, around 24 Gt of carbon dioxide (6.5 Gt carbon) derived from fossil fuels are currently being discharged into the atmosphere each year. Close to 60% of this emanates from point sources and the remainder from dispersed sources (central heating boilers, transportation). If projections of the usage of fossil fuels are correct and no actions are taken to change the present trend, a 50% increase in emissions is expected by 2020–2030. This would constitute an enormous quantity of carbon dioxide to store, even if it could indeed be captured. In fact, only a proportion of the gas, namely that produced centrally in large-scale plants, is at all likely to be captured. Consequently, carbon sequestration is, at best, only a partial solution to the greenhouse gas issue. A breakdown of the carbon dioxide that is produced from these stationary sources, by geographic region and by industry sector, is given in Table 3.1.[2]

The estimated ratios (by mass) of carbon released to hydrogen produced in natural gas reforming and in coal gasification (with and without carbon sequestration) are listed in Table 3.2, both for present-day operations and also for projected future technology.[3] These ratios are useful for estimating the size of the sequestration problem. For instance, in the context solely of supplying hydrogen to fuel cell vehicles (FCVs) in the USA, a demand for 100 Mt H_2 per year by 2050 has been forecast,[3] *i.e.*, about twice the current world production

Table 3.1 Carbon dioxide emissions from large stationary (point) sources in 2000 by geographic region and by industry (given in % terms).[2]

Region	%	Industry	%
China	25	Power	54
North America	20	Cement	15
OECD Europe[a]	13	Gas processing	12
Former Soviet Union	8	Iron and steel	6
Japan	6	Refineries	5
Asia	6	Ammonia	3
India	5	Ethylene	2
Middle East	4	Hydrogen	2
Eastern Europe	4	Ethylene oxide	1
Latin America	4		
Africa	3		
Oceania	2		

[a] OECD = Organization for Economic Cooperation and Development.

Table 3.2 Estimated ratio of carbon emissions to hydrogen produced (C:H, by mass) in a large central station that generates hydrogen from natural gas or coal.[3]

State of technology development	Technology	C:H ratio	
		Without CCS[a]	With CCS[a]
Present	Natural gas reforming	2.51	0.42
Present	Coal gasification	5.12	0.82
Possible future	Natural gas reforming	2.39	0.35
Possible future	Coal gasification	4.56	0.60

[a] CCS = carbon capture and storage.

of hydrogen; see Chapter 2. With the adoption of possible future technologies, without sequestration (Table 3.2), the associated annual emissions of carbon would be 239 Mt from natural gas and 456 Mt from coal plants. Carbon capture and storage would reduce these levels to 35 and 60 Mt, respectively. Thus, the scale of the task involved in sequestering most of the carbon dioxide emitted by an all-hydrogen fleet of light-duty vehicles in the USA in 2050 would equate to about 300–400 Mt of carbon per year (assuming that a mixture of gas and coal feedstocks would be used and given that there is insignificant hydrogen production from nuclear or renewable energy sources). This corresponds to 1100–1470 Mt CO_2[‡].

[‡] 44 t of carbon dioxide equate to 12 t of carbon.

By comparison, one of the largest sequestration projects to date, namely that in the Norwegian Sector of the North Sea (see Section 3.3 below), injects around 1 Mt CO_2 (0.27 Mt of carbon) annually. Clearly, the USA alone would face a massive problem in terms of carbon disposal – a capability of at least 1000 times that of the North Sea facility would be required – if all its light vehicles were to be fuelled by hydrogen generated centrally from natural gas and/or coal.

The above forecast does not consider other (non-vehicular) applications for hydrogen energy or the situation in other countries. Fortunately, however, there are a number of different sequestration possibilities. Many of these involve underground storage and potential geological sites are to be found throughout the world.

3.2   Capture of Carbon Dioxide

The overall sequestration process involves four steps:

- capture of the carbon dioxide from the emission source;
- dehydration and compression of the gas;
- transport to the storage site;
- injection into a geological reservoir.

The capture of carbon dioxide from exhaust gases is taken to be the most difficult part of the overall sequestration activity and also the most costly. The three main processes (post-combustion gas scrubbing, pre-combustion decarbonization, oxy-fuel combustion) for capturing carbon dioxide from power plants and compressing and drying it ready for underground storage are shown schematically in Figure 3.1. Note that oxy-fuel combustion is essentially a variant of post-combustion capture in which oxygen rather than air is the oxidant and the need to separate carbon dioxide from nitrogen is therefore eliminated. Carbon dioxide is also generated from many large industrial operations (petroleum refineries, ammonia plants, *etc.*) and is sometimes separated and used for chemical processes or in the food industry.

3.2.1   Post-combustion Capture

In a conventional combustion process, such as a power station boiler, the carbon dioxide is contained in the exit flue gas. The concentration is generally fairly low and ranges from about 4 vol.% for a combined-cycle system operating on natural gas to 12–14 vol.% for a traditional boiler fired by pulverized coal. The exhaust can be 'scrubbed' with an amine solution, typically monoethanolamine, which is then heated to release the absorbed carbon dioxide. The low concentration of carbon dioxide in the exit gas means that a huge volume would have to be handled and this would entail the installation

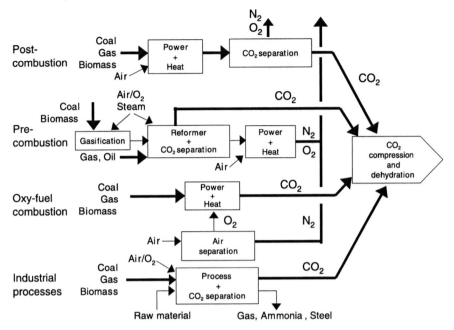

Figure 3.1 Principal routes for managing carbon dioxide emissions from power stations and industrial sources.[4]
(Courtesy of Intergovernmental Panel on Climate Change).

of large and expensive equipment. Also, considerable energy is required to desorb carbon dioxide from the amine solution and subsequently to repressurize it. The amine, which is a valuable chemical, can be recycled but soon degrades through the action of high temperature, oxygen, and impurities in the gas; it therefore has to be replaced. This is particularly true for exhaust gases from coal-fired stations that contain sulfur dioxide and other acid impurities. An alternative process, now being developed, is that of chilled ammonia scrubbing. The flue gas is cooled to $0-10\,^{\circ}$C, at which temperature the carbon dioxide reacts with ammonia to form a precipitate of solid ammonium carbonate (or urea) that may then be removed. The process consumes less energy than the amine route and is said to be more economical. A small pilot plant for testing has been built in the USA and a larger unit is planned.

Given the drawbacks of the amine route, post-combustion capture has not been demonstrated to be cost-effective in most power stations. It has been estimated that the process might well double the capital cost of a combined-cycle operation running on natural gas. In principle, retro-fitting to existing utilities is possible but there may be practical considerations such as the availability of land. By contrast, post-combustion capture has been employed in some chemical manufacturing plants; an example is given in Figure 3.2(a), and Figure 3.2(b) shows a unit for pre-combustion capture from a coal

(a)   (b)

Figure 3.2   (a) Post-combustion capture at a plant in Malaysia that employs a
chemical absorption process to separate 0.2 Mt CO_2 per year from the
flue-gas stream of a gas-fired power plant for urea production.[4]
(Courtesy of Mitsubishi Heavy Industries).
(b) Pre-combustion capture at a coal gasification plant in North Dakota,
USA, that employs a physical solvent process to separate 3.3 Mt CO_2 per
year from the gas stream.[4]
(Courtesy of Intergovernmental Panel on Climate Change).

gasification plant (see below). These photographs provide some idea of the
scale of chemical engineering involved.

In recent years, the efficiency of electricity plants that operate with pulverized
coal has improved steadily from about 35% to close to 40%, since it has been
possible to raise the temperature and pressure of the steam, consequent upon
the development of new alloys for turbine blades. The latest state-of-the-art,
coal-fired plant, which uses supercritical pressure steam conditions, has an
efficiency as high as 46%. Some new coal stations of this type are equipped with
carbon capture, while others are being built 'capture ready' (*i.e.*, with appropriate land and other facilities for the required technology to be installed later).
The reason for the latter option is that carbon dioxide capture adds to the
capital and operating costs and will not lightly be undertaken until demanded
by legislation or the economics become favourable.

3.2.2   Oxy-fuel Combustion

The concentration of carbon dioxide in flue gas can be increased greatly by
using oxygen instead of air for combustion, either in a boiler or a gas turbine –
a process known as 'oxy-fuel combustion'. The oxygen would be produced by
cryogenic air separation, which is already employed on a large scale, for
example in the steel industry. When fuel is burnt in pure oxygen, the flame

temperature is excessively high, so some exhaust gas is recycled to the combustor in order to hold the flame temperature similar to that in a normal air-blown boiler. The advantage of oxygen-blown combustion is that the flue gas has a carbon dioxide concentration of ~80% compared with 4–14% for air-blown combustion and therefore separation of the carbon dioxide is relatively simple or even unnecessary. After combustion, the gas stream is first cooled and compressed to remove the water vapour. It may prove possible to omit some of the gas-cleaning equipment that is used in modern coal-burning power stations (such as scrubbers for desulfurization of flue gas) and this would reduce the net cost of carbon dioxide capture. Some sulfur compounds and other impurities would then remain in the carbon dioxide fed to storage, which may be acceptable. A possible downside of omitting the desulfurization stage, however, is that recycled, wet flue gas will contain sulfuric acid that is highly corrosive to the plant.

The disadvantage of oxy-fuel combustion is that a large quantity of oxygen is required and this is expensive to produce in terms of both capital cost and energy consumption. Advances in oxygen production processes, such as new and improved membranes that can operate at high temperatures (see Section 2.6, Chapter 2), could improve overall plant efficiency and economics. To date, oxy-fuel combustion aimed at power generation has only been demonstrated in test rigs. Nevertheless, there is active interest in the technology within the electricity industry because of its potential for retro-fitting to pulverized coal plants that constitute the majority of the world's generating capacity. This would have the advantage of improving the generating efficiency and so reducing the carbon emissions.

3.2.3 Chemical Looping Combustion

This process is a highly speculative alternative to oxy-fuel combustion that has been proposed for the separation of carbon dioxide from nitrogen and excess air. A metal oxide (e.g., copper, cobalt, nickel, iron or manganese oxides) serves as an oxygen carrier that transfers oxygen from the air to the fuel. Two chemical reactors in the form of interconnected fluidized beds are employed: (i) a fuel reactor, in which the metal oxide is chemically reduced by the fuel, and (ii) an air reactor, in which the reduced metal oxide from the fuel reactor is re-oxidized with air. A schematic of the process is given in Figure 3.3. The outlet stream from the fuel reactor consists of carbon dioxide and water, whereas that from the air reactor contains only nitrogen and some unused oxygen. The overall chemical reaction is the same as for normal combustion with the same amount of heat released, but with the important difference that carbon dioxide is inherently separated from nitrogen, so that no extra energy and costly external equipment are required for gas separation. Although the technology is still very much in its infancy, particularly with respect to chemical engineering issues, chemical looping combustion offers the same advantages as

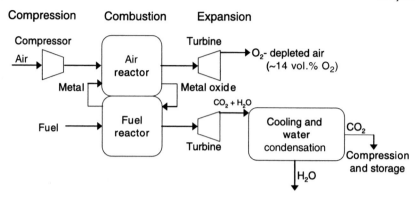

Figure 3.3   Principle of chemical looping combustion.

oxy-fuel combustion, but with the added prospects of higher thermal efficiency and no requirement to extract oxygen from air.

3.2.4   Pre-combustion Capture

Pre-combustion capture is suited to power plants that employ integrated gasification combined-cycle (IGCC) technology. As discussed in Section 2.7, Chapter 2, a solid fuel such as coal or biomass is first converted to syngas, which may be augmented by that produced by the steam reforming of natural gas or oil. After cooling to 400 °C, the syngas is subjected to the water-gas shift reaction to produce a mixture that is composed mostly of hydrogen and carbon dioxide. These gases are then separated by the pressure swing adsorption (PSA) process or its alternatives (see Section 2.3, Chapter 2) and the hydrogen is fed to a gas turbine. Pre-combustion capture involves new technology for the electricity industry, but the separation of hydrogen from carbon dioxide by PSA is well established in the manufacture of ammonia and synthetic fuels. A coal gasification plant in North Dakota, USA, with pre-combustion capture is shown in Figure 3.2(b). Although the initial fuel conversion steps are more elaborate and costly than in conventional combustion systems, the high concentrations of carbon dioxide produced by the shift reactor (typically 15–60 vol.% on a dry basis) and the high pressures often encountered in these applications are more favourable for separation. The FutureGen project in the USA, as described in Section 2.7, Chapter 2, will be a serious attempt to couple the IGCC process for electricity generation with hydrogen production and carbon dioxide sequestration.

All three major processes – post-combustion capture, oxy-fuel combustion, pre-combustion capture – require a step that, variously, involves the separation of carbon dioxide, oxygen or hydrogen from a bulk gas stream (flue gas, air or syngas, respectively). These separations can be accomplished by means of physical/chemical solvents, membranes, solid sorbents or cryogenic processes.

The choice of a specific capture technology is determined largely by the conditions under which the separation must operate. Present post-combustion and pre-combustion systems for power plants could capture 85–95% of the carbon dioxide that is produced. The net amount captured is rather less (approximately 80–90%), however, when allowance is made for the additional carbon dioxide that arises from the energy expended in the separation stages. Higher capture efficiencies are possible, but the separation devices become considerably larger, more energy intensive and more costly. Although oxy-fuel combustion systems are, in principle, able to recover nearly all of the carbon dioxide produced, the possible need for additional gas-treatment systems to remove pollutants such as sulfur and nitrogen oxides would lower the practical level to slightly more than 90%.

Power stations fitted with the above systems demand significantly more energy for their operation than conventional stations. This reduces their net efficiency, so more fuel is needed to generate each kilowatt-hour of electricity produced. For utilities that employ best current technology to collect 90% of the liberated carbon dioxide, the increase in fuel consumption per kWh generated ranges from 11–22% for natural gas combined-cycle operations, through 14–25% for coal-based IGCC systems, to 25–50% for various types of pulverized coal plants (in each case, the comparison is made with respect to similar technology that is operating without control of emissions). This extra fuel consumption not only reduces the efficiency of generation, but also gives rise to yet more gas to be sequestered. Moreover, the greater demand for fuel causes an increase in most other environmentally harmful emissions (*e.g.*, NO_x) per kWh produced relative to new state-of-the-art technology without carbon dioxide capture and, in the case of coal, proportionally larger amounts of solid wastes (ash). In addition, there is an increase in the consumption of chemicals, such as ammonia and limestone, which are used by coal-fired power stations to restrict the release of nitrogen oxides and sulfur dioxide. Nevertheless, there is optimism in some quarters that equipment will be developed to overcome the energy inefficiency problem and yield net reductions in the discharge of pollutants.

Because of the growing importance of carbon dioxide sequestration, there is currently a lively debate as to whether future coal-fired power stations should be conventional pulverized fuel, oxy-fuel or gasification designs. This is by no means a straightforward choice and involves considerations of overall fuel efficiency, engineering complexity and capital and operating costs. In addition, there are many types of coal (anthracite, bituminous coal, brown coal) with possibly dissimilar impurity contents, each of which may dictate a different plant design. The jury is still out on whether future coal-fired power stations will employ post-combustion or pre-combustion capture of carbon dioxide; this is a crucial issue to decide as the plants have a life of 40–50 years.

At present, the carbon dioxide formed during the manufacture of hydrogen by the steam reforming of natural gas is vented to the atmosphere. With a world-wide hydrogen production today of 45–50 Mt per year, this implies that about 120 Mt of carbon (440 Mt CO_2) are released annually from this source alone. This must be seen as one of the first targets for sequestration, since the hydrogen is already being separated out to leave a waste stream that is essentially concentrated carbon dioxide. Even so, disposal by storage (assuming that a suitable site is available) will impose a further cost penalty that has to be acceptable to the hydrogen producers and their customers. This situation will arise when carbon emission permits (see Section 1.2, Chapter 1) prove to be more costly than the necessary remedial action.

Once the necessity to sequester carbon dioxide from existing hydrogen plants is conceded, it should not be difficult to extend the concept to new production facilities that may be built to produce hydrogen solely for use as an energy vector or fuel. Admittedly, the economics are not the same, given that fuel hydrogen is of less value than chemical hydrogen. Also, the specification for the purity of the hydrogen will be different and is especially exacting when the gas is intended for use in low-temperature fuel cells. These factors will undoubtedly be reflected by variations in the detailed engineering of the process and in its costing. Even if emission permits for carbon dioxide were to be traded as high as US$50 per tonne of carbon, preliminary estimates[3] suggest that it would still be cheaper for steam reformers to buy these and discharge the gas rather than to undertake sequestration. With coal-derived syngas, however, the costs are said to be about equal. (These conclusions relate to the USA in 2004.) The economics for both routes, *i.e.*, natural gas and coal, will almost certainly vary from country to country and will change with time, particularly with movement in the price of emission permits.

In addition to solvent scrubbing and pressure swing adsorption (see above), at least two further potential processes exist for the separation of carbon dioxide from hydrogen. One is by cooling and condensation (cryogenics). On re-warming under high pressure, carbon dioxide forms as a liquid at near ambient temperature. This process is convenient when applied to gas streams that contain high concentrations of carbon dioxide because the liquefied gas may be more readily transported to a disposal site by road tanker or by ship. The disadvantages of the cryogenic process, as pointed out in Section 2.3.1, Chapter 2, are the requirement for low temperature and high pressure and the need to dry the gas first to prevent the formation of ice blockages in the transfer lines. The second possibility for gas separation is by diffusion of the hydrogen through a membrane that is impermeable to carbon dioxide, as discussed in Section 2.6, Chapter 2. The design of such membranes is a promising avenue for future research, particularly for mixtures of hot, pressurized gases. Ultimately, the economics of carbon capture and storage will depend critically on the cost of carbon dioxide emission permits in the open market. The higher the rate at which these permits trade, the greater will be the financial incentive to sequester the gas.

3.3   Storage Options

There are a number of possibilities for the bulk storage of carbon dioxide, the most important of which are reviewed below.

3.3.1 Geological Storage

The geological storage of carbon dioxide has been a natural process in the Earth's upper crust for hundreds of millions of years. Carbon dioxide derived from various sources (biological species, igneous activity, chemical reactions between rocks and fluids) accumulates in the sub-surface environment in the form of carbonate minerals, as a dissolved species and as a gas or supercritical liquid. The use of geological formations for the storage of anthropogenic carbon dioxide was first proposed in the 1970s, but attracted little attention until the early 1990s when international concern over greenhouse gas emissions gained momentum following the release by the Intergovernmental Panel on Climate Change (IPCC) of its First Assessment Report.[5]

Studies have shown that the storage of carbon dioxide is possible in various geological settings. The main candidates are sedimentary basins, *e.g.*, oil and gas reservoirs (working or abandoned), deep unmineable coal-seams and saline formations (aquifers). Sub-surface storage can take place at both on-shore and off-shore locations; access to the latter is via pipelines from the shore or from off-shore platforms. Other prospective sites for storage include salt caverns, basalts, oil/gas shales and disused mines. The various options are shown schematically in Figure 3.4.

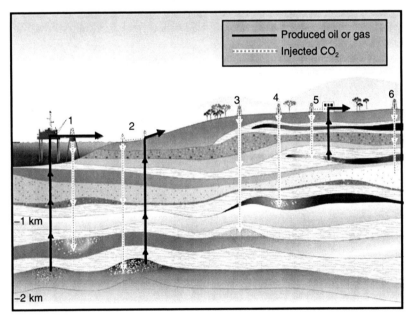

Figure 3.4   Options for the geological storage of carbon dioxide. 1. Deep unused saline formations (aquifers). 2. Use of carbon dioxide for enhanced oil recovery. 3. Depleted oil and gas reservoirs. 4. Deep unmineable coal seams. 5. Use of carbon dioxide in enhanced coal-bed methane recovery. 6. Other suggested options (e.g., salt caverns, basalts, oil/gas shales, disused mines).[4]
(Courtesy of Intergovernmental Panel on Climate Change).

Carbon dioxide is by no means the first substance to be deposited deep in the Earth's crust on an industrial scale. Fluids have been injected for many years to dispose of unwanted chemicals, pollutants and by-products of petroleum refining, or to enhance the production of oil and gas. The principles involved in such activities are well established and in many countries there are governing regulations. Large amounts of natural gas have also been pumped into depleted oil and gas wells in order to manage fluctuations in demand for this fuel or for seasonal storage between summer and winter. If such reservoirs are to be used for the permanent disposal of carbon dioxide, procedures will have to be devised for transferring ownership from the licensed field operator to a storage operator. In this context, it should be recalled that abandoned fields generally contain residual oil or gas that, eventually, may become economically recoverable given the expected rise in hydrocarbon prices. Placement of carbon dioxide in these reservoirs has, to date, been done on a relatively small scale, but if this route were to be adopted to deal with the emissions from power stations, then the extent of the operations would have to increase substantially.

The key questions to be addressed are: how large is the world's geological storage capacity and does it occur where it is needed? Not all sedimentary basins are suitable for the containment of carbon dioxide. Some are too shallow and others are dominated by rocks with low permeability or poor confining characteristics. Favourable basins have permeable rock formations saturated with saline water (saline formations), extensive covers of low-porosity rocks ('caprock', which serves as a seal to retain the gas) and structural simplicity.

The US Department of Energy has estimated that the storage capacity of depleted reservoirs in the USA is about $80-100$ Gt CO_2, which is sufficient to accommodate the waste gas from all the present power stations in the country for 50 years or more. The IPCC reports a storage capacity of $675-900$ Gt CO_2 for the whole world – a massive storage potential that is sufficient to accept the present global annual production of the gas some 30 times over.

Depleted oil and gas fields have a number of attractive features as geological repositories:

- the fields have been mapped and exploration costs are therefore low;
- the reservoirs are proven traps for gases over geological eras;
- in some cases, it may be possible to use existing oil-field equipment to transport and inject the carbon dioxide.

These are significant advantages of oil and gas wells over other geological formations. In addition, there is the possibility of enhanced oil recovery.

3.3.1.1 Enhanced Oil Recovery

The petroleum industry currently uses carbon dioxide for enhanced oil recovery (EOR). The gas may derive from a chemical plant, from an aquifer, or from a

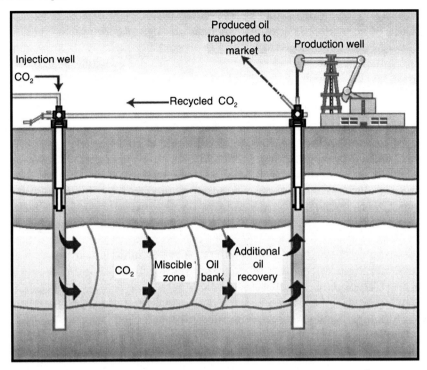

Figure 3.5 Enhanced oil recovery through injection of carbon dioxide.[4] (Courtesy of Intergovernmental Panel on Climate Change).

natural gas well where it is present as a minor component. The carbon dioxide is injected into the oil field to boost the flow of oil towards the production wells when the rate of extraction is slowing. The pressurized gas serves to force the oil through the porous sandstone structure and so facilitates its recovery, as illustrated in Figure 3.5. In North America, about 40 Mt CO_2 are used annually for EOR operations and mostly come from natural reservoirs that are abundantly available. The gas is transported over distances of up to 800 km in pipelines. Around two barrels of extra oil are produced for each tonne of gas injected. When carbon dioxide ultimately breaks through at the production well, it is normally separated from the oil stream and re-circulated back into the system. This recycling decreases the amount that must be extracted or purchased and avoids emissions to the atmosphere. Typically, half to one-third of the carbon dioxide is not recycled, of which a significant fraction is thought to be trapped permanently. The extra oil recovered is a credit that serves to offset the cost of sequestration. Enhanced oil recovery is the most mature of the geological storage technologies for carbon dioxide, although until now its success has been judged on the additional quantity of oil recovered rather than on its feasibility for the permanent containment of the gas with no leakage.

New technologies are being developed to obtain carbon dioxide as a by-product of industrial operations (such as natural gas processing and the manufacture of fertilizers, ethanol and hydrogen) for use in EOR at locations where naturally occurring reservoirs of the gas are not available. One demonstration at the Beulah, North Dakota, plant of the Dakota Gasification Company is collecting carbon dioxide as waste gas from a synthetic fuels process [Figure 3.2(b)] and delivering it *via* a new 325 km pipeline to the Weyburn oil field in Saskatchewan, Canada. The facility commenced operation in October 2000 with an initial injection rate of 5000 t per day. Before the project ends, it is expected that 20 Mt CO_2, which otherwise would have been released to the atmosphere, will have been permanently stored. On this scale, EOR is a practical and long-term solution to the problem of carbon dioxide disposal. The economics depend critically on the price of crude oil. When oil traded at US$20–25 per barrel, EOR may have been marginally viable, but with the price rising to around US$60 the process becomes extremely attractive on account of the extra oil recovered. If, at the same time, carbon emission permits are also costly, then the financial returns are even more favourable. EnCana Corporation, the operator of the Weyburn oil field, is hoping to add another 25 years to the productive life of the resource and to reclaim as much as 130 million barrels of oil that might otherwise have been abandoned as uneconomic.

In 2007, BP and Rio Tinto announced the formation of a new, jointly-owned company, Hydrogen Energy, to develop decarbonized energy projects around the world. The venture will initially focus on hydrogen-fuelled power generation, in which fossil fuels and CCS technology will be used to deliver new large-scale supplies of clean electricity. The first of these planned plants was to have been at Peterhead in the north east of Scotland. Natural gas from the North Sea would be steam reformed and the hydrogen produced would fuel a 475 MW power station. The carbon dioxide (1.8 Mt per year) would be injected into the Miller oil field, 240 km off-shore, to enhance oil recovery and to avoid the cost of buying emission credits. This was expected to extend the life of the field by 15–20 years and to produce up to 40 million additional barrels of oil. This initial project has now been set aside by BP, allegedly because the UK government's time-scale was incompatible with that of the company. The opportunity is still there, provided that the economics can be made favourable.

The second project is at Carson in California where the feed material for conversion to hydrogen will be petroleum coke (5000 t per day) from a BP refinery. After reforming, the hydrogen will supply a 500 MW power station and the carbon dioxide (4 Mt per year) will be pumped into an on-shore oil field. This plant should be on-stream early in the next decade. It has since been announced that General Electric is to link with BP with a view to building at least five such clean-coal electricity plants in the USA, with Carson being the first.

The latest joint venture to be announced is a feasibility study for a large, clean-coal power plant in Western Australia. This also would generate 500 MW and capture up to 4 Mt CO_2 annually for underground storage in an off-shore deep, saline reservoir. A final investment decision is expected in 2011 and, if favourable, the plant could be operational by 2014 or shortly thereafter. Much may depend on the encouragement and level of support provided by the Australian government.

The above initiatives are illustrative of what may be expected in the near future where opportunities exist and the economics are right. Projects such as these are likely to be among the first major manifestations of hydrogen energy coupled to carbon dioxide sequestration.

3.3.1.2   Depleted Oil and Gas Fields

Less commercially attractive, but still potentially important, is injection into oil and gas fields that are either close to exhaustion (so-called 'depleted' fields) or have been abandoned as unprofitable. Once the oil or gas has been extracted, there remains a large volume of porous rock in which carbon dioxide can be retained. In this situation, there is no credit for enhanced fuel recovery but, because the formation has safely held oil/gas for geological periods of time, there is again confidence that the carbon dioxide will remain locked up indefinitely. It is necessary, however, to ensure that any holes in the caprock due to drilling and oil extraction have been permanently sealed before introducing the gas.

3.3.1.3   Saline Aquifers

Aquifers are underground layers of water-bearing sedimentary rock or mixtures of unconsolidated materials such as gravel, sand or clay. To be useful for storing gas, the aquifers need to be sealed with impermeable shale or clays. The occluded water is usually saline or brackish, and sometimes warm or even hot. Throughout Europe and in other parts of the world, hot springs have been developed as health spas. Paris, for example, sits above an aquifer of warm water that has been used to heat apartment blocks. Indeed, hot geothermal aquifers constitute a modest form of renewable energy. Where the water is warm, the heat stems from the radioactive decay of the isotopes uranium-238, thorium-232 or potassium-40 in the rock.

There are many other saline aquifers in non-radioactive strata, where the water is cold. These are far more common and are considered to be suitable for the storage of carbon dioxide; the high salt content precludes the use of such aquifers for agricultural purposes or for drinking water (except in arid regions where desalination might be attempted). There can be no guarantee in advance that the gas will be held indefinitely, since there is no history to offer reassurance. Nevertheless, the prospects often appear to be good. Due to their lack of commercial importance, deep saline aquifers are generally not as well characterized as oil and gas reservoirs. Careful attention must be paid to investigating the overall geological formation. In particular, the incidence of fractures in the rock strata is a major concern as these may provide pathways through which the carbon dioxide may leak to the surface. Equally, possible lateral transport of the gas must be assessed. Obviously, there must also be good overlying ('capping') materials to prevent gas escaping back to the atmosphere. Finally, consideration must be given to any potential impact on the local environment; for example, biota are known to exist at depth underground. Much of the pressurized carbon dioxide is likely to dissolve in the water present in the aquifer. Under favourable conditions and depending on the chemical nature of the rock, the dissolved gas may react with the minerals and be locked away permanently as basic carbonates. This is a promising route for permanent disposal in situations where the aquifer resides within alkaline rock formations, but much more research has to be undertaken to validate the proposition. A contrary view stems from some recent research in Texas. Underground injection of 1600 t of liquid carbon dioxide appeared to dissolve away minerals (presumably basic carbonates) and thereby introduced the possibility that pathways might open up for the gas to escape to groundwater or back to the environment. The key point is that the chemistry of carbon dioxide is very different from that of natural gas and it cannot be assumed that a formation that has held methane for geological periods will necessarily do the same for carbon dioxide.

Statoil, a Norwegian petroleum company, is conducting a major project in the North Sea at its Sleipner gas field, about 250 km west of Stavanger. The natural gas from the Heimdal sandstone contains about 9 vol.% CO_2, which is more than is permissible for supply to customers and which therefore has to be removed to an acceptable level of 2.5 vol.%. Nearby, is a brine-saturated, sandstone aquifer – the Utsira Formation – at a depth of 800–1000 m below the sea floor and with a mass storage capacity of 1–10 Gt CO_2. About 1 Mt of the gas is being separated annually by a standard amine absorption process and then injected underground under pressure; see Figure 3.6. The amount being sequestered in the Utsira Formation is equivalent to the output of a small (150 MW) coal-fired power plant. The incentive for this activity was a high carbon tax imposed by the Norwegian Government on industries that discharge carbon dioxide to the atmosphere. When the project began in October 1996, it marked the first instance of carbon dioxide being captured and stored because of concern over climate change. By early 2005, more than 7 Mt CO_2 had been injected. It has been claimed that the extra capital investment was paid back in less than 2 years solely on the basis of the savings in carbon tax. Elsewhere, in Alberta for instance, natural gas contains hydrogen sulfide in addition to carbon dioxide. Under such conditions, it is necessary to remove these two gases simultaneously for co-storage underground in order to avoid the added cost of separating out hydrogen sulfide and oxidizing it to sulfur.

Geological surveys of sedimentary basins are being carried out in various parts of the world (*e.g.*, Australia, Canada, Europe, Japan, USA) to establish the potential for storage of carbon dioxide. A comprehensive analysis of Australian saline aquifers has concluded that it would be possible to accommodate all of the carbon dioxide from the country's power stations for many hundreds of years at current emission rates. A demonstration project is planned

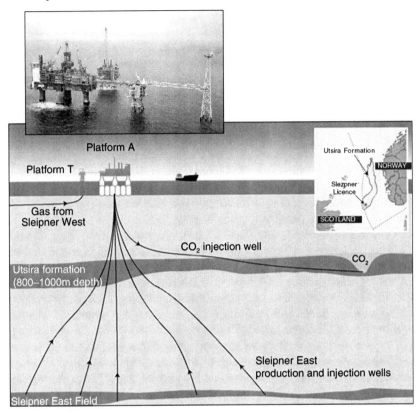

Figure 3.6 Schematic of storage of carbon dioxide at the Sleipner gas field.[4]
(Courtesy of Intergovernmental Panel on Climate Change).

at the Gorgon gas field, which is situated 130 km off the north-west coast of Western Australia, as part of a large liquid natural gas (LNG) project. Some of the Gorgon gas has carbon dioxide levels of more than 12 vol.% and it is proposed that up to 120 Mt of the latter will be buried in the Dupuy Formation below the existing oilfield on Barrow Island. At a distance of 72 km, this is the nearest landfall and lies directly between the Gorgon gas field and the mainland. Some opposition to this project has been expressed on environmental grounds as Barrow Island has a fragile eco-system.

In Canada, it would appear that the best areas for the geosequestration of carbon dioxide are to be found in the Western Canadian Basin, which covers much of Alberta and British Columbia. Preliminary estimates indicate a storage capability of tens to hundreds of gigatonnes. Western Canada is also rich in oil, gas, oil sands and coal – all of which may provide opportunities for storage sites. In Western Europe, further aquifers under the North Sea provide the most promising option; an initial survey identified a total capacity of ~800 Gt CO_2, which equates to about 800 years of power-plant emissions at present-day levels. Surveys in Japan have suggested the possibility of storing up to 90 Gt CO_2, more than half of which is off-shore. Various research groups in the USA are assessing prospects in California, Texas and the mid-west States. Most forecasts relate to the theoretical capacity of aquifers, based on surveys of their dimensions. This capacity will doubtless be much reduced when practical and economic considerations are taken into account.

3.3.1.4   Coal Seams

Many coal seams are considered to be unmineable because they are either too deep or too thin to be worked gainfully. On the other hand, there are often considerable quantities of coal-bed methane (CBM) associated with these formations. Coal can physically adsorb gases, which permeate through fractures ('cleats') in the beds and then diffuse into micropores within the solid itself. This adsorptive uptake of gas is different from that in other forms of geological storage where the pressurized gas is simply held in pores rather than adsorbed on the surface. By way of example, adsorption and desorption isotherms for nitrogen, methane and carbon dioxide on a sample of coal from the Powder River Basin in Wyoming, USA, are given in Figure 3.7.[6] The gases are confined by water that usually fills the fractures in the coal. To desorb a gas, its partial pressure must be reduced as this allows the gas and water to move through the coal-bed and up the wells. Methane is a by-product of the process by which plant material is converted to coal and up to $25 m^3$ (at 0.1 MPa and $0 °C$) of the gas may become trapped per tonne of coal at the prevailing pressures. Only recently has it been recognized that methane embedded in many unmineable coal seams may represent an enormous undeveloped energy

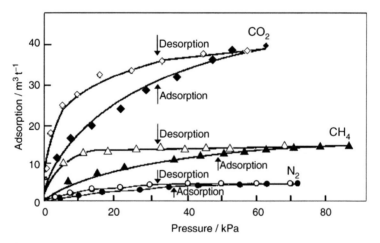

Figure 3.7   Adsorption–desorption isotherms for nitrogen, methane and carbon dioxide on a particular type of US coal.[6]
(Courtesy of Society of Petroleum Engineers, USA).

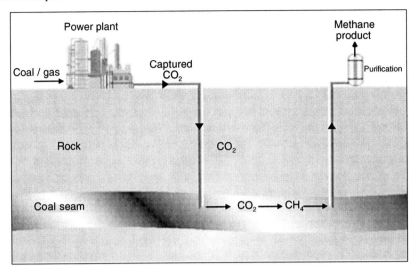

Figure 3.8 Schematic of storage of carbon dioxide in coal seams.[7] (Courtesy of the International Energy Agency).

resource. A substantial amount of CBM is already produced in the USA and in Australia the vast Queensland coalfields are being actively explored and mapped with a view to gas recovery.

Carbon dioxide has great affinity for coal. In general terms, coal can store at least twice the amount of carbon dioxide as methane on a molecular basis. This ratio will vary from coal to coal, and also with the pressure, temperature and physical conditions of storage. In volumetric terms, the ratio of the adsorbed amount of carbon dioxide to that of methane ranges from as low as unity for mature coals such as anthracite to 10 or more for younger, immature coals such as lignite. It has therefore been suggested that unworkable coal seams might be purged with pressurized carbon dioxide to provide not only a means of driving out and harvesting methane, but also an underground store for the waste gas. The carbon dioxide will flow through the cleat system of the coal, diffuse into its matrix, and then be adsorbed on the inner micropore surfaces to unlock gases with lower affinity for coal (*e.g.*, methane). Coal permeability varies widely and generally decreases with increasing depth of the deposit as a result of cleat closure due to the increasing effective stress of the overburden. Most CBM-producing wells in the world are found at less than 1000 m below ground level. An artist's impression of such a plant is shown in Figure 3.8.[7] Enhanced coal-bed methane (ECBM) recovery[§] is being undertaken in the San Juan Basin in north-western New Mexico, USA. More than 0.1 Mt CO_2 have been injected over a 3-year period. In the USA alone, it is claimed[8] that ECBM operations have the potential to provide 4.25 Tm^3 of technically recoverable methane and

---

[§]Enhanced with respect to the amount of methane recovered in the absence of carbon dioxide injection.

a further 0.8 Tm^3 of 'proven reserve', with the storage of correspondingly large quantities of carbon dioxide.

There have been relatively few ECBM trials in the world to date, but some estimates have been developed by different organizations. For example, based on a forecast of CBM production made by Canada's National Energy Board, it was concluded that 380 Mt CO_2 could be stored each year by 2025 – approximately half of Canada's current greenhouse gas emissions. A similar picture has emerged in Australia, China and elsewhere where coal permeability is sufficient to allow CO_2 ECBM operations. According to the IEA, the world potential for carbon dioxide storage via ECBM is 100–150 Gt, of which 40 Gt are considered to be viable at the present time.[9] To put this into perspective, it is interesting to compare these figures with the world energy-related emissions of carbon dioxide in 2004 that amounted to 26.6 Gt CO_2.[10]

All these projections must be treated with caution since the science involved in the uptake of gases by coal is imperfectly understood and complications may arise. As the concentration and prevailing pressure of carbon dioxide increase, *adsorption* on the coal surface is replaced by *absorption* (or dissolution) into the structure of the coal. The coal loses its brittle nature and becomes 'rubbery'; the latter, in turn, leads to plastic flow. The extent to which this change in state occurs is dependent upon the carbon dioxide pressure, the temperature and the nature of the coal. A probable consequence of plastic flow would be the sealing of capillary channels in the coal and thereby a reduction in its capacity to absorb more gas. Another pertinent factor is that coal swells as carbon dioxide is absorbed and this, too, would cause a major decrease of permeability and consequent gas uptake. For both of these reasons, present estimates of the capability of coal seams to absorb carbon dioxide should be treated as tentative.

There are several operational and safety issues to be resolved before this option can be adopted in a major way. Thermodynamically, carbon dioxide should react with carbon (coal) to yield carbon monoxide, which is a highly toxic gas sometimes found in coal mines. In practice, this reaction will not take place at an appreciable rate at ambient temperatures, although it could become significant at depths where the temperature is much higher. If the inadvertent ingress of air to a coal seam that was saturated with carbon dioxide led to a fire, then the potential might exist for vast quantities of toxic carbon monoxide to be formed via the reaction $CO_2 + C \rightarrow 2CO$, and subsequently liberated. It is known that the extraction of water from coal beds does allow air to enter and circulate more freely and that this sometimes results in underground fires. Such safety issues should be evaluated carefully. Other factors to be considered are:

- saturating coal seams with carbon dioxide may impart safety risks to neighbouring underground coal-mining operations;
- carbon dioxide injected into deep coal seams lying beneath shallow, workable coal beds may migrate upwards and render these resources unmineable;

- environmental concerns may arise, *e.g.*, many coal seams are also aquifers and while the water will have to be displaced to allow the methane to be released, the full consequences of its replacement by pressurized carbon dioxide are unknown;
- there is the associated problem of where to dispose of large amounts of saline water; also, its extraction may cause a fall in the level of the water table in the neighbourhood, with consequences for water supply;
- evidence exists from CBM workings in the USA that the upward seepage of methane through fissures in the rock can lead to contamination of drinking water, high concentrations of the gas in the basements of buildings, the killing of tree roots, and the poisoning of burrowing animals;
- coals with high sulfur contents may liberate toxic hydrogen sulfide.

The greater the amount of carbon dioxide injected, the more will return to the surface with the methane. This creates the need for a gas-separation plant of the type shown in Figure 3.2. The removal of carbon dioxide from oil is much simpler than from methane, which is why enhanced oil recovery is practised more widely than enhanced gas recovery. With an extensive coal seam, it will be necessary to drill injection and extraction wells every few hundred metres or so. These, together with the associated gas compressors, distribution pipes and pumps, access roads, *etc.*, will turn a rural landscape into an industrial site. When the operational/safety issues and the extent of industrialization become fully appreciated, not to mention the impact on the ecological habitat, there may well be local opposition to the recovery of coal-bed methane – with or without carbon dioxide injection and storage.

Underground *in situ* processing of unmineable coal seams, either by combustion or by gasification, is a possible future alternative to methane extraction and carbon dioxide storage. Combustion would produce hot gases that could be fed to a turbine generator, while gasification would yield syngas. Unfortunately, air is too dilute with nitrogen to sustain underground combustion, while syngas diluted with nitrogen is of limited value. Hence the use of oxygen is indicated, but this is a costly option. Nevertheless, this situation could change in a future Hydrogen Economy where hydrogen is manufactured on a large scale by electrolysis. There would then be substantial amounts of by-product oxygen for which there is no obvious market. One possibility would be to make use of this oxygen for underground combustion or gasification rather than to engage in conventional coal mining.

3.3.1.5 *Oil or Gas Shales*

Shales are fine grained, clay-like rock structures that cleave readily into thin layers. Many shales are coloured black on account of the bitumen that they contain. There are also gas shales in which considerable quantities of natural gas are occluded. Deposits of organic-rich oil or gas shale occur in many parts of the world. The trapping mechanism for carbon dioxide is similar to that for coal beds, namely adsorption on organic material. Enhanced production of shale gas via the injection of carbon dioxide has the potential (like ECBM) to offset the storage costs for the waste gas. The world-wide scope for such storage is currently unknown, but the large volumes of shale available suggest that its overall storage capacity may be substantial, even though these materials generally have low permeability to gases.

3.3.1.6   Basalts

Basalts are also widely distributed in nature. Commonly, they have low porosity, low permeability and low pore space continuity. These properties do not bode well for gas storage. Nonetheless, basalt may have some potential for the permanent mineral trapping of carbon dioxide, because the gas may react with silicates in the matrix to form carbonate minerals. This requires further investigation.

3.3.1.7   Supercritical Carbon Dioxide

For safe and long-term storage, carbon dioxide must be injected more than 800 m below the Earth's surface. At that depth, the gas becomes a supercritical fluid. Such fluids have the gas-like characteristic of low viscosity and the liquid-like characteristic of high density. Supercritical behaviour exists only when temperature and pressure both reach, or exceed, their respective values at the so-called 'critical point'. Every substance has its own critical temperature, above which the gas cannot be liquefied no matter how high the pressure. For carbon dioxide, the critical point lies at 31 °C and 7.4 MPa. Supercritical fluids have properties similar to those of liquid solvents and are employed commercially to extract soluble substances. For example, supercritical carbon dioxide is used to remove caffeine from coffee.

Given its greater density, supercritical carbon dioxide would require less storage volume underground than the gaseous form. It is also less mobile and has a higher solubility in other liquids, which would make sequestration more effective. Once injected, the supercritical carbon dioxide begins to displace the oil, brine or other fluids in underground porous rock formations. After a while, some of the carbon dioxide dissolves in these fluids to become trapped. One concern is that the injection may raise the pressure in an underground reservoir to a level that leads to faults or fractures in the overlying caprock. These events could result in gas leakage, which would compromise the storage effectiveness and possibly pose risks to both the environment and human health.

3.3.1.8   Salt Caverns

There are numerous natural underground cavities, such as caves in limestone formations, which might conceivably be used for the containment of carbon dioxide, although there is always the risk that pressurized gas will leak out along geological faults. The same is true of anthropogenic deep mineral mines of various types. This risk may be less for caverns formed artificially from salt domes that have been leached to recover the salt, since rock salt is essentially impermeable. Indeed, these caverns have been studied for the possible disposal of radioactive waste. Salt cavities have been used since 1971 in Kiel, Germany, for the seasonal storage of town gas (*i.e.*, gas derived from coal that contains around 50 vol.% hydrogen with varying amounts of methane, carbon monoxide and nitrogen). In the UK, pressurized industrial hydrogen (95% pure) has been held for many years in anthropogenic salt caverns on Teesside; see Section 5.2, Chapter 5.

Storage of carbon dioxide in salt domes differs from that of natural gas or compressed air. In the last two cases, the caverns are cyclically pressurized and depressurized on a daily-to-annual basis, whereas the containment of carbon dioxide must be effective on a centuries-to-millennia time-scale. A single cavern of 100 m diameter may hold as much as 0.5 Mt CO_2, while multiple arrays could be excavated in accordance with demand. Sealing of the cavities is important to prevent gas leakage and also possible roof collapse, which could liberate large quantities of gas. The advantages of this storage method include high capacity per unit volume and high injection flow rate. Disadvantages are the potential for gas release in the case of system failure and the environmental problems of disposing of the brine produced when forming the cavity. Salt caverns can also serve as temporary reservoirs for carbon dioxide between sources and more permanent sinks.

3.3.1.9 Overall Prospects

Geological studies in the mid-1990s indicated that, globally, there might be an opportunity to store underground between 1000 and 10 000 Gt CO_2. More recent analyses conducted by the IPCC[4] show, however, a greatly reduced capacity of only 675–900 Gt CO_2[¶]. Bearing in mind that the present world production rate is around 25 Gt per year, there is clearly scope for taking up a good fraction of emissions for many years to come. Notable storage projects that are in train or planned are summarized in Table 3.3.[4,11] Most experience has been gained with the first two of these projects (Sleipner and Weyburn). To date, both operations have proceeded successfully and monitoring has shown no signs of leakage from the respective reservoirs. To sequester all of the current annual emissions (world-wide) of carbon dioxide would require 25 000 operations, each the size of that at Sleipner. Evidently, numbers on this scale are impracticable in the short-term, but it does seem possible that sequestration technology could soon accommodate a few percent of the global production of carbon dioxide, as a contribution to arresting climate change. In the longer term, it is conceivable that many more repository sites will be developed and a significant fraction of the carbon dioxide liberated by combusting fossil fuels

---

[¶] This range does not agree well with the estimate (see above) of 800 Gt CO_2 for the North Sea alone. Such discrepancies indicate the present rudimentary state of knowledge.

Table 3.3   Some projects on sequestration.[4,11]

Project	Location	Carbon dioxide to be injected Daily/t	Planned total/Mt	Storage type	Status
Sleipner	North Sea	3000	20	Aquifer	Ongoing
Weyburn	Canada	3000–4000	20	EOR	Phase 1 complete
In Salah	Algeria	3000–4000	17	Gas field	Ongoing
Gorgon	Australia	10 000		Aquifer	In preparation
Frio	USA	177 for 9 days only		Aquifer	Pilot study
RECOPOL	Poland	1 for 10 days only		Coal seam	Pilot study
Salt Creek	USA	5000–6000	27	EOR[a]	In preparation
Minami-Nagoaka	Japan	40	0.01	Aquifer	Demonstration

[a] EOR = enhanced oil recovery.

could indeed be stored safely underground. The associated cost would ultimately be borne by the end-user of the processed fuel or electricity.

In summary, whichever geological storage option is adopted, there are a number of conditions that must be satisfied, namely:

- *kinetics*: the rate of injection of carbon dioxide into the reservoir should be sufficient to accommodate the rate of production at the source;
- *lifetime of the store*: it is essential that carbon dioxide does not leak back into the atmosphere at a rate that will exacerbate the atmospheric burden;
- *environmental impact*: no adverse effects should arise from carbon dioxide storage;
- *proximity to the sources*: the total cost of capture and transport from the sources to the storage locations must be acceptable.

In addition, it would be highly desirable if the chemical composition of the formation were alkaline so that, in the long term, the carbon dioxide would be tied up permanently as a basic carbonate.

3.3.2   Monitoring and Verification

Methods must be devised for assessing carbon dioxide losses during transfer operations and also the leakage rates from each type of store. Several techniques are available or under development but these vary in applicability, site specificity and detection limits. The formulation of protocols for monitoring the hazards associated with carbon dioxide capture and transfer does not appear to present fundamentally new challenges, as similar protocols are part of standard environmental health and safety practices for toxic gases.

The monitoring of underground stores is a somewhat different matter. Procedures and guidelines have still to be devised and this is a very important part of the overall risk management strategy for geological storage projects. Techniques are expected to evolve with time and will be subject to local regulations. Certain parameters such as injection rate and well pressure will be measured routinely. Periodic seismic surveys are known to be useful for tracking the underground migration of carbon dioxide. The sampling of groundwater and soil should be advantageous for directly detecting escape of gas. Newer techniques such as gravity and electrical measurements may prove equally useful. Carbon dioxide sensors with alarms can be located at the injection wells for ensuring worker safety. The intent is that leakage shall normally be almost zero, but in the event of an earthquake affecting the store there might be a sudden and explosive release of massive quantities of gas with potentially serious consequences. For instance, if the surrounding topography trapped the carbon dioxide, which is heavier than air, a dense cloud would form and suffocate humans and animals in its proximity. It is therefore inadvisable to locate gas reservoirs in regions of known seismic activity.

3.3.3 Mineral Carbonation

The weathering of alkaline rocks is essentially a natural form of carbon dioxide capture and storage, but is a very slow process. Carbon dioxide can form stable carbonates through reaction with minerals that contain magnesium or calcium. A typical example would be the transformation of serpentine, a common silicate mineral, to magnesium carbonate, silica and water, *i.e.*,

$$Mg_3Si_2O_5(OH)_4 + 3CO_2 \rightarrow 3MgCO_3 + 2SiO_2 + 2H_2O \qquad (3.1)$$

Estimates suggest that known deposits of magnesium and calcium silicates are sufficient to accommodate very large anthropogenic emissions of carbon dioxide. Although the process is thermodynamically favoured, the kinetics of carbonate formation with appropriate minerals are exceedingly slow at ambient temperatures and pressures. Research is being directed towards finding ways to speed up the reaction. A further priority is to identify a carbonation route that is industrially and environmentally viable and that will allow mineral sequestration to be implemented at an acceptable cost.

There are many engineering and technological issues to be resolved before external 'mineral carbonation' can become a fruitful pathway. These include: (i) mining and crushing of the ore to a fine powder without using too much energy in the process; (ii) bringing the crushed ore and the carbon dioxide together at a reaction site; (iii) determining the reaction conditions that are practical and economic; and (iv) finding suitable disposal sites for the carbonated waste products. In addition, the environmental impact of the whole operation must be carefully considered. Of course, if the mineralization takes place internally within the underground store, as suggested above, then the slow kinetics are no impediment and the above series of difficult steps would not be required.

Whereas the overall prospects for external mineral carbonation do not appear to be promising, the approach does offer a number of attractions. Foremost among these is the fact that the resulting carbonates are thermodynamically stable and the carbon will be locked up permanently with no possibility of escape. Raw materials for binding the carbon dioxide are present in vast quantities throughout the world. Readily accessible deposits exist in amounts that far exceed even the most optimistic estimate of coal reserves ($\sim 10\,000$ Gt) and therefore would be more than sufficient to cope with all the carbon dioxide expected to arise in the future. A single site operating on a scale similar to that of a typical mineral extraction process should, in principle, be able to handle the carbon dioxide output of several large power plants.

On a smaller scale of operations, there are some industrial wastes and mine tailings that are alkaline and potentially more reactive than virgin minerals towards carbon dioxide. These include pulverized fuel ash from coal-fired power stations. In this situation, development of a practical means to react the ash with some of the attendant carbon dioxide would be most beneficial. Indeed, this may prove to be the most expedient route to introducing the technology required for sequestration via mineral carbonation.

3.3.4   Ocean Storage

The world's inventory of carbon is divided among the landmass, the oceans and seas, and the atmosphere. Of these, the marine environment is by far the largest repository for carbon and the atmosphere is the smallest. It has been estimated that the oceans and seas contain over 90% of the total carbon and the atmosphere less than 2%.[12] The rate at which carbon has been accumulating in the atmosphere in recent times corresponds to 3.1–3.5 Gt per year, compared with anthropogenic releases of 6.0–6.5 Gt per year; see Table 1.2, Chapter 1. Thus, approximately half of the carbon dioxide produced by mankind is being sequestered naturally, mostly by the oceans. If the burning of fossil fuels were to cease, equilibrium between the atmosphere and the ocean would slowly be restored. Whereas the atmosphere contains ~ 750 Gt C, the oceans are thought to hold around 40 000 Gt C. Thus, addition to the latter of all of the anthropogenic carbon dioxide produced globally would ultimately be insignificant. It is because the atmosphere contains only a small proportion of the overall carbon inventory that the progressive build-up from human activities has become a major cause for concern.

The following question then arises: would it be sensible and acceptable to dispose of captured carbon dioxide in the oceans? Such action would merely be accelerating a natural process and may not therefore fall under the international convention on the dumping of industrial wastes at sea. This is a matter for marine biologists, lawyers and politicians to debate. In terms of the technology, there are two broad approaches that might be adopted:

  (i) to transport the carbon dioxide to suitable sites, by pipeline or ship, and then to inject it into the ocean – the 'direct approach';

(ii) to fertilize the oceans with additional nutrients so as to accelerate the uptake of carbon dioxide from the atmosphere – the 'indirect approach'.

Both of these methods have the potential to increase significantly the rate at which carbon is absorbed by the marine environment.

3.3.4.1 Direct Approach

Due to variations in the temperature and pressure of the ocean with depth, carbon dioxide can exist in both gaseous and liquid forms. At depths of less than about 500 m and depending on the temperature, the carbon dioxide tends to be gaseous, whereas at greater depths it is usually a liquid; see Figure 3.9(a). A crystalline CO_2-hydrate (a 'clathrate' compound[||]) is stable at depths from 200 to 400 m, as determined by the sea-water temperature.

The density of sea-water is almost independent of depth ($1.040-1.045 \, \text{g cm}^{-3}$) whereas that of liquid carbon dioxide increases with depth; see Figure 3.9(b). Down to about 2500 m, liquid carbon dioxide is less dense than sea-water and tends to float upwards. At 2500–3000 m, liquid carbon dioxide is neutrally buoyant (*i.e.*, a 'transition zone' in which it neither rises nor sinks), dependent on the sea temperature. Deeper than 3000 m, the liquid is denser than sea-water and sinks downwards to the ocean floor, where it accumulates as a lake, over which a solid layer of crystalline hydrates slowly forms as an ice-like combination of carbon dioxide and water. Within its stability range (low temperature, high pressure), solid CO_2-hydrate would inspire greater confidence as a permanent store than dissolved or liquid carbon dioxide, although there are few data to say how rapidly carbon dioxide would be leached out by sea-water.

Hence there are two major strategies for the direct injection of carbon dioxide into the ocean. One is to release the gas at mid-depth where a low concentration will dissolve and disperse as a rising plume to become part of the ocean cycle, but where a high concentration will liquefy and sink to the ocean floor. In this approach, storage would be made more effective by injection below the thermocline[**]. The second strategy is to send the carbon dioxide to a greater depth such that the gas settles out as a less mobile mixture of liquid and solid hydrates. The possibilities for ocean storage are shown schematically in Figure 3.10. Deeper discharge is preferable because it promises a more lasting form of storage, although on just what time-scale remains to be established. The re-dissolution rate will depend not only on the form of the carbon dioxide (liquid or crystalline hydrate), but also on the temperature, the depth of injection and the local water velocity. Concern has been expressed that the presence of underwater currents would lead to the carbon dioxide being dispersed and re-dissolving in the ocean, rather than being held captive permanently.

[||] A number of gases of small molecular size, *e.g.*, methane, chlorine, carbon dioxide and hydrogen sulfide, have the ability to be trapped in ice crystals to form 'clathrate' compounds.
[**] The thermocline is a narrow region where the temperature changes abruptly between surface warm water and deep cold water; it inhibits vertical mixing and therefore acts as a barrier to degassing back to the atmosphere.

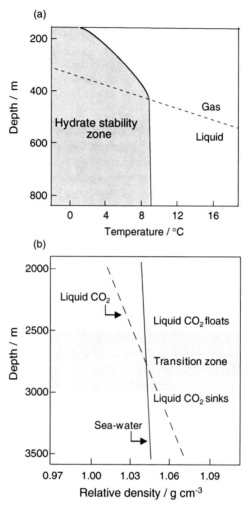

Figure 3.9 Behaviour of carbon dioxide in sea-water. (a) Zones of stability of carbon dioxide gas, liquid and hydrate as a function of temperature and depth. The liquid is stable at pressures (depths) below the dotted line and the hydrate is stable in the shaded area. (b) Density of liquid carbon dioxide and of sea-water at great depths.
(Courtesy of Intergovernmental Panel on Climate Change).

One of the major factors to take into consideration with respect to ocean disposal is that of environmental impact. Ocean water is slightly alkaline with a pH of ~ 8, whereas rainwater, containing dissolved carbon dioxide, is slightly acidic with a pH of ~ 6. When carbon dioxide from the atmosphere dissolves in the sea the pH is reduced marginally as the gas reacts with OH^- ions to form bicarbonate anions, *i.e.*,

$$CO_2 + OH^- \rightarrow HCO_3^- \tag{3.2}$$

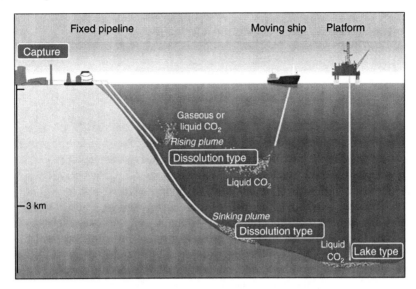

Figure 3.10   Schematic representation of ocean storage options. In 'dissolution type' ocean storage, the carbon dioxide rapidly dissolves in the ocean water, whereas in 'lake type' ocean storage, the carbon dioxide is initially a liquid on the sea floor, soon crystallizing as a hydrate. Given sufficient time, all forms of carbon dioxide – gas, liquid, hydrate – will dissolve in the water.
(Courtesy of Intergovernmental Panel on Climate Change).

Although bicarbonate is an acid anion, the sodium salt $NaHCO_3$ is slightly alkaline because it is formed from a strong base and a weak acid. The pH of its solution lies between 7.0 (neutral) and 8.0 (ocean water). A fall in pH by as little as 0.1 units can exert a profound effect on marine biology. The overall pH of the oceans is said to have fallen by this much already and could reduce to 7.5 by 2100 with further uptake of carbon dioxide. This pH value would then be lower than for hundreds of millions of years and the rate of change so fast that marine life may be unable to adapt. Given such a threat, there is absolutely no case for compounding the problem by using the ocean to dissolve sequestered carbon dioxide.

Even now, there is grave concern among marine biologists over the acidification of the oceans that has already taken place. The problem seems to arise because of the very slow equilibration between surface water and the ocean depths. Although globally the anthropogenic carbon dioxide added to the marine environment is insignificant, its impact locally in surface waters may be considerable. Acidification profoundly affects the viability of sensitive species such as corals and certain plankton. Shellfish also are vulnerable because acidification may lead to dissolution of their shells or inability to build them in the first place. These are important considerations when contemplating ocean storage of bulk quantities of carbon dioxide although little can be done about the gas that returns naturally from the atmosphere – other than moving to a low-carbon economy based on sustainable energy.

Over a period of time, dissolved carbon dioxide will also react with limestone, again to give the bicarbonate anion:

$$CO_2 + CaCO_3 + H_2O \rightarrow Ca^{2+} + 2HCO_3^- \qquad (3.3)$$

The reaction occurs naturally when acid rain-water percolates through limestone strata to form stalactites in underground caves; the acid rain dissolves the rock to form soluble bicarbonate, which then decomposes back to carbonate when it deposits on the stalactite. The same process takes place in the sea, both with limestone and with other alkaline sediments washed out by rivers. This serves ultimately to hold the pH of the ocean almost constant, but the equilibrium process between the river sediment and the ocean bulk is very slow. Reaction (3.3) does, however, provide a potential route for neutralizing the acidity of the injected carbon dioxide, via the deliberate addition of appropriate quantities of alkaline mineral. Such a procedure would be most suited to point sources of carbon dioxide (*e.g.*, power stations) that are located close to both the ocean and large deposits of the mineral.

The concern of ecologists and environmentalists over marine disposal of carbon dioxide stems from the current limited knowledge of deep-ocean ecosystems and the creatures that live there. At such depths, most organisms experience an environment of low and almost constant levels of carbon dioxide. They depend for their survival on a flux of organic materials and nutrients that are formed in the surface layers and sink to the bottom. Near volcanic vents, however, the concentration of carbon dioxide in the depths may be very much greater than elsewhere. The gases released from hydrothermal vents may be a mixture of carbon dioxide, methane and hydrogen sulfide, all of which form crystalline hydrates at low temperature and high pressure. Although the resultant pH of the ocean bed can be as low as 3.5–4.5, various species have evolved that can thrive under such conditions. Nevertheless and given the paucity of information on these ecosystems, the fact that this evolutionary process has occurred in localized regions is no justification for deliberately causing a sudden lowering of the pH elsewhere through the mass dumping of carbon dioxide on the ocean floor. Profound environmental consequences could well occur. More study is required to warrant such a course of action.

Recently, an entirely new procedure for ocean disposal has been proposed by scientists working at Harvard and Columbia Universities.[13] They suggest that the carbon dioxide should be pumped a few hundred metres into the porous sediment that covers the deep ocean floor. Under the prevailing conditions, the gas would be in dense liquid form and would soon convert to crystalline hydrate, which would then be locked into the structure of the porous sediment. It is surmised that the crystalline hydrate would dissolve only very slowly into the ocean over a period of hundreds of years and that no ecological damage would result. This proposition is seen to have material advantages over the alternative options. As discussed earlier, the drawback with geological containment in oil and gas wells or in aquifers is that the carbon dioxide will be in the form of high-pressure gas that has to be contained permanently under a caprock seal. Moreover, the number of suitable sites is finite and many of these will be distant from the source of the gas. It has similarly been noted that sea disposal presents the risk of ecological damage to marine life. The above scientists highlight the fact that deep ocean sites for containment in porous sediment are virtually unlimited and therefore would allow for the storage of all of the world's captured carbon dioxide indefinitely, with little or no ecological ramifications. Liquid carbon dioxide would be conveyed to the sequestration site by ship or pipeline and then injected into the sea floor by means of standard technology used in oil extraction. At present, this concept is little more than an interesting proposal; considerable work will be required to evaluate the prospects and determine costs. Meanwhile, it is expected that geological storage will continue to be developed.

When contemplating the injection of tens of thousands of tonnes of carbon dioxide each day from a single pipe or ship, it is clear that our understanding of the physical chemistry and hydrodynamics of the operation, and also of the ecological consequences, is inadequate. Recent plans to mount modest-scale demonstrations of ocean storage have been frustrated by regulatory authorities that have been concerned over localized environmental damage. Progress will probably be achieved through collaborative research programmes that involve chemists, biologists and hydrodynamicists, followed by small trials in coastal waters, with a gradual increase in scale as expertise and confidence build up. It may prove possible to undertake experiments in a deep fjord, for example.

3.3.4.2 Indirect Approach

The second, completely different, possibility for storing carbon dioxide in the marine environment seeks to enhance the rate of uptake from the atmosphere. This indirect approach depends on the fact that at least some of the gas dissolved in the upper layers of the ocean is photosynthesized to form phyto-plankton and algae, which are the basis of the marine food chain. For this process to occur, it is necessary to have not only carbon dioxide and sunlight, but also certain essential nutrient elements (nitrogen, phosphorus, iron) present in ionic form in the water. The rate of the photosynthesis reaction will be determined by the availability of these nutrients. Therefore, to accelerate this rate, 'seeding' of the ocean with nutrients, especially iron (which is the one likely to be in short supply), has been proposed. Such action would, in turn, reduce the partial pressure of carbon dioxide in the surface water and so facilitate its transfer from the atmosphere. The idea is still very much at a conceptual stage, but limited experiments have shown that seeding with iron leads to an increase in chlorophyll and plankton bloom. This phenomenon may not, however, persist. Plans for large-scale trials would probably run into even greater ecological objections than those for deep-sea disposal. Furthermore, the need to fertilize vast areas of ocean may prove to be impractical. Nevertheless,

this route does offer the advantage of abstracting carbon dioxide directly from
the atmosphere without having to separate it from other gases.

3.3.4.3   Atmosphere–Ocean Equilibrium

The natural transfer of carbon from surface water to the deep ocean is a slow
process in most parts of the world. This factor, rather than the equilibrium
between the atmosphere and the surface water, determines the rate at which
carbon dioxide is taken up. It is thought that the mixing of surface and deep
water can take decades, or even centuries. There are two mechanisms that
'pump' carbon dioxide from the atmosphere to deep waters, as follows.

(i) *The 'solubility pump'*. Carbon dioxide is highly soluble in the cold, dense
water that, in high latitudes, sinks to the bottom of the ocean. This
results in a 'thermohaline circulation' of sea-water, whereby the cold,
dense water of the North Atlantic ocean (rich in carbon dioxide) is
'conveyed' southwards nearly to Antarctica before eventually surfacing
in the Indian and Pacific oceans. There, dissolved carbon dioxide escapes
to the atmosphere again as the water warms. The time interval between
water sinking at high latitudes and resurfacing in the tropics is estimated
to be around 1000 years.

(ii) *The 'biological pump'*. Over 70% of the organic matter in surface water is
recycled through the food chain, but the balance is exported to the deep
ocean, mostly by sedimentation of decayed plankton and organisms. The
biological pump therefore transfers carbon from the surface and effectively sequesters it at great depth. Much of this organic material is
reconverted by bacteria back to carbon dioxide, which eventually returns
to the surface. Again, the time taken for the overall process can be 1000
years.

Mathematical models of the ocean carbon cycle indicate that, without either of
these two pumps operating, the concentration of carbon dioxide in the atmosphere would be twice as high as it is today.

   In summary, the oceans and seas have enormous potential to take up all of
the carbon dioxide that will result from the future global use of fossil fuels, but
the kinetics of the natural processes are slow. The practical problem is how to
transfer captured carbon dioxide (or gas in the atmosphere) to the marine
environment in a fashion that is not only affordable, but also acceptable to
marine biologists and environmentalists generally. Clearly, much more research
must be conducted in order to allay the fears of the ecological lobby.

3.3.5   Biological Storage

The estimated quantity of carbon and its distribution in the biosphere are listed
in Table 3.4.[14] Even if these data are only approximate, they are still of
considerable interest. As discussed above, the ocean is by far the greatest

Table 3.4 Estimated distribution of carbon in the biosphere.[14]

Location of carbon	Quantity/Gt
Intermediate and deep ocean	38 100
Surface ocean	1030
Bottom sediment	150
Terrestrial systems	1960
Recoverable fossil reserves	4000
Atmosphere	775
Total	46 015

repository for carbon, which is an argument in favour of those who advocate deep-sea disposal. By contrast, the amount of carbon sequestered in terrestrial vegetation is approximately half that potentially recoverable in fossil reserves. Therefore, if one wished to utilize half the remaining fossil fuels and store the resulting carbon dioxide biologically on land, it would still be necessary to have more than twice as much vegetation as that growing at present. Of course, this is a practical impossibility at a time when humanity is busy cutting down forests and replacing them with agricultural land. In any event, as the vegetation decayed there would be no control over how the carbon dioxide emissions would be distributed between the atmosphere, the oceans and the land, or over the kinetics of the redistribution processes. All that can be said is that the greater the amount of vegetation (especially trees), the greater will be the removal of carbon dioxide from the air. Biological storage is not permanent because all vegetation eventually dies and decays with the release of greenhouse gases – principally, carbon dioxide and methane.

Not all of the carbon designated as 'terrestrial carbon' in Table 3.4 is growing above ground. A considerable fraction resides in the roots of plants and trees and in the form of 'soil organic carbon'. The latter is also a potential storage medium for fossil-derived carbon dioxide, although at present it is by no means clear how it might best be utilized. Research on plant metabolic pathways is required to find new ways of accelerating carbon dioxide fixation rates and to develop novel technologies for managing vegetation and soils so as to enhance carbon uptake and retention by this route. It might be expected that the accumulation of greenhouse gases in the atmosphere over the past 100 years would have accelerated photosynthesis, but there is little evidence published to show that vegetation is growing faster than formerly.

3.3.6 Re-use of Carbon Dioxide

Carbon dioxide finds fairly extensive use in the food processing industry, as a coolant for perishable items (ice cream, meat products, vegetables, fruit, *etc.*), as a solvent (extraction of caffeine from coffee, as noted above) and for the carbonation of soft drinks. It is utilized also in the manufacture of certain chemicals, *e.g.*, urea, methanol, and both organic and inorganic carbonates.

Urea, for instance, finds application as a polymer material (urea–formaldehyde resin) and in agriculture as a slow-release nitrogen fertilizer. Carbon dioxide is employed to a lesser extent in the treatment of alkaline water, in fire extinguishing systems, and in welding where it acts as a shielding gas. The carbon dioxide stems mostly from fermentation processes or as a by-product of hydrogen manufacture. When used in the food industry, the elapsed time before the captured carbon dioxide is discharged back to the atmosphere is in the range of hours to days. In chemical applications, its retention is longer, but variable and rarely sufficient to be regarded as permanent storage. Moreover, the quantities required are far too small to make any measurable impact on the overall greenhouse gas problem. For these reasons, industrial uses of carbon dioxide have little to contribute to carbon immobilization.

3.4   Transport to Storage Site

Around 3000 km of pipeline for conveying carbon dioxide are said to exist throughout the world, of which 2400 km are in the USA. Collectively, these transport annually about 44 Mt CO_2 at high pressure. The gas is used predominantly for enhanced oil recovery.

The Sleipner (North Sea) gas project, described above in Section 3.3, has simply taken advantage of a conveniently located, deep aquifer for the storage of carbon dioxide. In most situations, however, the source of carbon dioxide and its intended storage site will not be in such close proximity and it will be necessary to supply the gas by pipeline over some distance. Whereas this may be acceptable in the case of enhanced oil recovery (given the credit gained through increased productivity) and also for the removal of carbon dioxide from natural gas, the economics may be different for simple disposal of waste carbon dioxide. A study has been made of a 100 km pipeline that would be capable of delivering 10 000 t of carbon dioxide per day from source to sink.[3] The estimated cost of this facility was found to be about one-third of that assumed for the total storage. Since unit costs vary approximately as the inverse square root of the flow rate, much smaller operations are unlikely to be financially viable under current conditions. This is a further justification, if one were needed, for large-scale steam-reformer and coal-gasification plants.

If transport to the storage site by pipeline proves to be impracticable or too expensive, there are alternatives to be evaluated. At atmospheric pressure, cryogenic carbon dioxide ('dry-ice') does not melt, but sublimes at $-78\,°C$. Thus, in principle, it can be carried as a solid in refrigerated trucks or ships. This would, however, require freezing at the production plant and then vaporization at the storage site before injection of the gas. Both of these activities are energy intensive. In addition, dry-ice is difficult to handle in quantity. A more feasible mode of transport depends on the fact that the gas can be liquefied under high pressure. At $22\,°C$, the vapour pressure over the liquid is around 6 MPa. It is therefore possible to convey carbon dioxide as a liquid in pressurized tankers, provided that the temperature is well controlled.

Not only does this method lead to reduced energy losses, given that the gas will usually be produced at high pressure anyway, but it also allows injection underground (or into the ocean) as a liquid. It is reported that a relatively small quantity of liquid carbon dioxide (about 0.3 Mt per year) is currently delivered by ship in Europe, and even lesser amounts are carried by road. The design of a carbon dioxide tanker is similar to that of existing LPG carriers, but the pressure is generally higher.

3.5 Institutional Issues

Although geological storage of carbon dioxide does appear to be technically feasible and indeed is being practised in a few places, its extension to cover the ultimate disposal of most of that produced from fossil fuels does raise numerous institutional and safety issues. In economic terms, storage space will have a financial value in an era of emission permits and/or carbon taxes. The worth of a particular facility will depend on many factors, *e.g.*, its location, size, ease of access and integrity. Owners of suitable underground sites, of whatever nature, will be able to charge customers for their services. Already this is the case where storage is in operation. A free market will probably develop, in which the disposer of the carbon dioxide will negotiate a price with the facility owner. This raises the question of who is the legal owner of each potential underground site. Most countries have defined mineral rights, but these do not generally extend to great depths where mining is impracticable. For oil and gas exploitation within territorial waters, governments normally invite tenders and sell franchises to the highest bidder. But who owns deep-lying saline aquifers that have no other obvious use? On-shore, are these the property of the landowner? Off-shore, will governments put such storage sites out to tender as with oil and gas franchises? Furthermore, who would approve deep ocean storage in international waters? Satisfactory answers have yet to be found.

Further legal issues relate to the status of captured carbon dioxide. Is it a waste that, under international treaty, cannot be disposed of at sea? Or is it a 'resource' when used, for example, to enhance oil recovery? And do the conventions relating to dumping at sea apply to storage under the sea-bed, particularly if the only access is through a pipe from the land? Hence there is much scope for international debate before agreement can be reached. In the meantime, some new power stations are being built 'capture ready' rather than incur the considerable capital costs of building capture units before the ground rules relating to storage have been agreed. It has even been suggested that stored carbon dioxide might be seen as a long-term resource to increase the greenhouse effect during the next ice age!

Other institutional questions revolve around the accounting rules to be employed.[9] It should be straightforward to measure and verify the quantity of carbon dioxide placed in a repository. This quantity is not, however, the same as that of the carbon dioxide produced in the first instance. Losses are incurred during capture, transport and injection into the storage site. Added complications are (i) the possible subsequent escape of some sequestered gas and (ii) the need to make allowance for the extra energy consumed (and therefore carbon dioxide produced) during capture, pressurization and pumping. These are matters that are being addressed by the International Energy Agency and on which there will again have to be global agreement.

Safety concerns relate to the integrity of the storage facility. This will vary enormously according to the local geology. The most stable sites will be exhausted gas fields where the natural gas has been held for millennia under an impervious dome of caprock. Even there, it will be necessary to ensure that former gas wells have been properly sealed. All other prospective storage locations will have to be investigated geologically for their suitability, with special attention to leakage pathways. What will be an acceptable leakage rate? This will have to be defined by statute before a site can be approved for storage – and how will it be validated in advance? Will there be an internationally agreed code of practice to formulate permissible leakage rates and other operational parameters? What body will lay down these rules and what court will adjudicate over disputes? In addition, there is the fear of sudden, possibly catastrophic, releases of carbon dioxide in bulk, for instance as a result of seismic activity. The gas is not a poison, but it is an asphyxiant. There is a well-documented occurrence of a sudden massive release of carbon dioxide in 1986 from the bed of Lake Nyos in Cameroon when some 1800 people died from asphyxiation. Finally, there is the question of who is accountable for the safety of the repository in the long term (centuries) when the initial owner has collected the storage fees and moved on. Presumably, the onus will fall on the relevant nation but it may not wish to accept such an indefinite liability. In international waters, the United Nations may have to establish an agency to take responsibility.

Some of the above issues have already been addressed in connection with the containment and disposal of highly toxic materials and radioactive waste. When the UK government attempted, in the 1980s and 1990s, to find deep geological repositories for the permanent storage of radioactive waste, it met such opposition from environmentalists and local authorities that it had to postpone the programme and continue to maintain the waste securely above ground. One of the concerns about deep disposal of any type is that it could be an irreversible step. By contrast, liquid or solid waste stored at or near the surface can be retrieved and re-packaged if necessary. The option of long-term surface containment is not so readily available for a gas, especially with respect to the amounts under discussion here. The problem with radioactive fissile material is that it will remain a hazard for thousands of years. The same is also true of carbon dioxide, even though the nature of the hazard is different. It is difficult to be sure of the reliability of geological sites over such long periods of time given the threat of earth tremors, ingress of underground water and other major disturbances.

How about deep-sea disposal? When Shell Oil proposed to scuttle a North Sea oil rig in the Atlantic Ocean, there was such a public outcry that it was forced to abandon the plan and dismantle the rig, piece by piece. This was an emotional response that lacked logic, since many steel ships of comparable size have been lost at sea. After this precedent, it is improbable that any other attempt will be made to use the ocean as a dumping ground for major objects. Liquid carbon dioxide in bulk quantities, surely, poses a greater danger to the environment than oil platforms and will probably run into the same fierce opposition, which may or may not be soundly based scientifically.

Overall, it would appear that geological storage is the most promising option for containing bulk quantities of carbon dioxide, although injection into sea-bed sediments remains to be evaluated. In economic terms, it has been estimated that capture, transport and storage underground of carbon dioxide will add 10–20% to the price of hydrogen derived from fossil fuels.[3] The larger part of the extra cost lies in the capture and compression of the gas. By comparison, transport and storage are thought to be relatively inexpensive. A great deal still remains to be done to achieve savings and further work is necessary on the institutional and safety fronts. Moreover, it is by no means clear that the public is yet ready to accept geological storage, particularly on land. There are no easy answers to the ultimate disposal of greenhouse gases.

3.6 The Way Ahead

With the huge global reserves of coal available as a primary source of energy, nations are now taking carbon capture and storage seriously. This is independent of how the coal is utilized, whether through direct combustion or through conversion to hydrogen. The problem with capture is that the bulk separation of carbon dioxide from other gases is costly. Consequently, more research and development has to be conducted on novel separation techniques, particularly on membrane technology as applied to the processing of hot, pressurized gases. Deep geological disposal poses a different set of problems. Experience of the technology is limited to just a dozen or so projects in which gas is injected into oil/gas reservoirs or saline aquifers. Many of the activities are still in their early stages and, although so far promising, it will be some time before the results can be fully evaluated. While these and future demonstration projects are on-going, there is much to be done in terms of geological mapping, matching large point sources of carbon dioxide to potential storage sites, and establishing the underlying institutional framework for the long-term monitoring and control of the repositories.

In terms of ocean disposal of carbon dioxide, further studies must be undertaken in the areas of physical chemistry, oceanography and marine biology to establish the fate of the carbon dioxide that is injected and its impact on the sea-bed ecology. This will necessitate both small-scale laboratory studies and pilot-scale demonstrations in carefully selected areas where the environmental consequences will be localized. Considerable effort will be required to demonstrate to a sceptical public that the practice would give rise to no adverse consequences. At the end of the day, should geological storage prove to be unsatisfactory or insecure, the ultimate choice may lie between release to the atmosphere or to the oceans, and the latter may be the lesser environmental evil. For this reason, if for no other, it is important to continue investigating marine disposal as a fall-back option.

Recently, the Asian Pacific Partnership on Clean Development and Climate, a consortium of six major countries around the Pacific Rim (Australia, Japan, China, India, Korea, USA), completed a major study on the influence of carbon capture and storage as a procedure for abating global carbon emissions.[15] The investigators allowed for expected growth in world population and for realistic improvements in the standard of living by 2050 to predict total energy usage up to that date. Next, they apportioned the energy requirement among the different fuels – fossil, nuclear, renewables – using their best predictions and so calculated the likely amount of carbon dioxide emissions. An assessment was then made of the extent to which these emissions might be reduced by possible advances in technology, adopted first by the participating countries and then globally. Finally, the impact of carbon sequestration in reducing emissions was considered. The results of the investigation give cause for concern. In the reference case, with no great advances in technology and no sequestration, the global greenhouse gas emissions grew by about 140% relative to 2001 levels (from 34.5 to 83.4 Gt CO_2). In the most favourable of the four scenarios examined, namely that in which advanced technology is diffused throughout the world and sequestration is applied from 2015 onwards to new coal- and gas-fired electricity plant in most countries, the global greenhouse gas emissions in 2050 were around 26% less than for the reference case. Cumulatively, over the period up to 2050, the emissions would be around 10% less than the reference case. Under no credible circumstances was it possible to hold emissions at current levels, still less to reduce them. The authors concluded that: (i) technology advances, coupled with extensive sequestration, will be insufficient to address the climate change problem; (ii) it will be necessary for governments to take far more active and aggressive steps to encourage economy in the use of energy, so as to decouple living standards from energy consumption and also to promote low-carbon energy sources; and (iii) given the challenges of reducing greenhouse gas emissions, particularly in large developing countries, there is a clear need to focus on strategies and technologies to adapt to the consequences of a warming climate.

Scenario studies, in which one peers into the future, are always open to criticism regarding the ground rules that are employed and the uncertainties associated with predictions. Nevertheless, the above global forecast suggests that there are severe limits to what can be achieved by, say, 2050 through the development of new energy-conversion technology and the implementation of carbon sequestration. The focus therefore has to be on energy conservation measures, the vigorous exploitation of renewable (sustainable) forms of energy, and adaptation to climate change.

References

1. R.M. Dell and D.A.J. Rand, *Clean Energy*, Royal Society of Chemistry, Cambridge, 2004.
2. J.J. Gale, *Overview of CO_2 Emission Sources, Potential, Transport and Geographic Distribution of Storage Possibilities*, Workshop on Carbon

Dioxide Capture and Storage, Intergovernmental Panel on Climate Change (IPCC), Regina, Canada, 18–21 November 2002.

3. *The Hydrogen Economy: Opportunities, Costs, Barriers and R & D Needs*, National Research Council and National Academy of Engineering of the National Academies, National Academies Press, Washington, DC, 2004.

4. B. Metz, O. Davidson, H. de Coninck, M. Loos and L. Meyer (eds.), *IPCC Special Report on Carbon Dioxide Capture and Storage*, Cambridge University Press, Cambridge, 2005.

5. *IPCC First Assessment Report*, 1990, Vols. 1–3. Vol. 1. *Scientific Assessment of Climate Change – Report of Working Group I*, J.T. Houghton, G.J. Jenkins and J.J. Ephraums (eds.), Cambridge University Press, 1990. Vol. 2. *Impacts Assessment of Climate Change – Report of Working Group II*, W.J.McG. Tegart, G.W. Sheldon and D.C. Griffiths (eds.), Australian Government Publishing Service, Canberra, Australia, 1990. Vol. 3. *The IPCC Response Strategies – Report of Working Group III*, Island Press, Washington DC, 1990.

6. F.M. Orr Jr., *Storage of Carbon Dioxide in Geologic Formations*, Distinguished Author Series, *Journal of Petroleum Technology*, Society of Petroleum Engineers, paper SPE 88842, September 2004.

7. *Putting Carbon Back into the Ground*, IEA Greenhouse Gas R&D Programme, International Energy Agency, Cheltenham, February 2001; see: www.ieagreen.org.uk.

8. V.A. Kuuskraa, in *Encyclopedia of Energy*, C.J. Cleveland (ed.), Elsevier, Amsterdam, 2004, **4**, 257–272.

9. S. Haefeli, M. Bosi and C Philibert, *Carbon Dioxide Capture and Storage Issues – Accounting and Baselines Under the United Nations Framework Convention on Climate Change (UNFCCC)*, IEA Information Paper, International Energy Agency, Paris, May 2004.

10. *Key World Energy Statistics*, 2006 Edition, International Energy Agency, Paris, 2006.

11. R. F. Service, *Science*, 2004, **305**, 962–963.

12. W.G. Ormerod, P. Freund and A. Smith (eds.), with revisions by J. Davison, *Ocean Storage of Carbon Dioxide*, 2nd Edition, March 2002, Report by the IEA Greenhouse Gas R&D Programme, Cheltenham, 2002.

13. Z. House, D.P. Schrag, C.F. Harvey and K.S. Lackner, *Proceedings of the National Academy of Sciences of the United States of America*, on-line publication, 7 August 2006.

14. *Carbon-free Production of Hydrogen from Fossil Fuels*, Global Climate and Energy Project, Stanford University, Stanford, CA, April 2004.

15. A. Matysek, M. Ford, G. Jakeman, A Gurney and B. Fisher, *Technology – Its Role in Economic Development and Climate Change*, Research Report 06.6, Australian Bureau of Agricultural and Resource Economics (ABARE), Canberra, Australia, July 2006.

CHAPTER 4

Hydrogen from Water

Water is a huge storeroom of hydrogen, but energy is required to split this source into its component elements. The water molecule is in fact very stable, with a heat of formation of $-285.83\,kJ\,mol^{-1}$ under standard conditions[†]; see Box 6.1, Chapter 6. Decomposition by direct heating only starts to become significant at very high temperatures, namely ~ 1 vol.% at 2000 °C. Moreover, it is difficult to separate the hydrogen and the oxygen at such high temperatures. Some research is in progress to produce hydrogen at lower temperatures by employing catalysts for the decomposition reaction together with ceramic membranes for efficient gas separation, but the work is very much in its early stages. Given these limitations, the splitting of water must be undertaken indirectly by processes that employ either chemical reactants or other energy forms of high thermodynamic potential (*e.g.*, electrons or photons). The chemical reduction of water to hydrogen by means of hydrocarbons, in the form of fossil fuels, is described in Chapter 2. Here, we discuss principally the use of renewable energy, via the medium of electricity or directly as solar radiation, to generate hydrogen from water.

In the short-to-medium term, the only economic route for the large-scale production of hydrogen is via the processing of fossil fuels. The alternative of water electrolysis can only be justified under rather special circumstances. In the longer term, however, the dependence on fossil fuels requires an ability to sequester the by-product carbon dioxide, which is hardly possible for widely dispersed, small-scale production of hydrogen to provide power for buildings or vehicles. The choice, then, lies between constructing a network of underground pipelines to distribute hydrogen produced centrally with sequestration or generating locally from non-polluting sources of energy that liberate no greenhouse gases. The latter would involve water decomposition effected by renewable energy in one form or another, or possibly by nuclear electricity. It is impossible to define a time-scale for a move from fossil fuels to renewable energy for the production of hydrogen. The practicality of this route will depend on many factors and the level of its uptake will vary from place to place according to the relative availability and the price of fossil fuels versus 'green'

[†]The negative sign indicates that heat is evolved in the reaction of hydrogen with oxygen.

electricity from renewables. In terms of energy efficiency and economics, it makes little sense to employ electricity – a premium form of energy – to obtain hydrogen for use in a fuel cell or, worse, to combust in an engine. The overall efficiency would be poor. Efficiency may not, however, be the predominant consideration. If, for reasons of petroleum shortages or pollution control, hydrogen is needed as a clean-burning fuel, for instance to power road vehicles, then conceivably such a requirement could take precedence over that of energy efficiency. The current activity in the USA aimed at developing a hydrogen infrastructure for fuel cell vehicles is largely motivated by a desire to reduce petroleum imports and, ultimately, to become nationally self-sufficient in transportation fuels. For the consumer, the prime consideration will almost certainly be the price of fuel. Under present economic conditions, there is no possibility that electrolytic hydrogen can compete with fossil fuels, other than in niche markets. Looking ahead by several decades, the situation could well change due to shortages and rising prices of fossil fuels, taxes or legislation on releasing greenhouse gases, developments in (and costs of) sequestering carbon dioxide, and improvements in the performance and cost of electrolyzers. Each of these different areas is currently receiving attention.

4.1 Electrolysis

Electrolysis is employed widely in the manufacture of both metals and chemicals. For instance, the method is used to extract the metals sodium, magnesium and aluminium from their fused salts. In the chemicals industry, one of the largest processes is the electrolysis of an aqueous solution of sodium chloride (brine) for the manufacture of chlorine and caustic soda, *i.e.*,

$$2NaCl + 2H_2O \rightarrow Cl_2 + 2NaOH + H_2 \qquad (4.1)$$

Chlorine gas is liberated at the positive electrode and caustic soda and hydrogen are formed at the negative electrode. The hydrogen, although pure, is usually seen as a by-product of low value that may be pumped to a nearby oil refinery or burnt to raise steam. (The terminology used for electrochemical cells is outlined in Box 4.1.)

The electrolysis of water to generate hydrogen or oxygen is practised in certain situations where the cost of electricity is not a prime consideration. Large-scale plants have been built in Brazil, Canada, Egypt and Norway, which are countries that have surplus hydroelectric capacity. Generally, alkaline electrolyzers are employed and the electrode reactions are as follows:

at the negative electrode:

$$4H_2O + 4e^- \rightarrow 2H_2 + 4OH^- \qquad (4.2)$$

at the positive electrode:

$$4OH^- \rightarrow O_2 + 2H_2O + 4e^- \qquad (4.3)$$

Box 4.1 Terminology Used in the Operation of Electrochemical Cells.

During the operation of an electrolysis cell, *i.e.*, a cell driven by the application of an external voltage, the positive electrode sustains an oxidation (or 'anodic') reaction with the liberation of electrons, while a reduction (or 'cathodic') reaction takes place at the negative electrode with the uptake of electrons [Figure 4.1(a)]. For this reason, the positive electrode is often known as the 'anode' and the negative electrode as the 'cathode'. The internal circuit between the two electrodes is provided by the electrolyte, in which negative ions ('anions') move towards the positive electrode and positive ions ('cations') move towards the negative electrode.

A fuel cell (see Chapter 6) operates in the reverse manner to an electrolysis cell, *i.e.*, it is a 'galvanic' cell that spontaneously produces a voltage. The anode of the electrolysis cell now becomes the cathode and the cathode becomes the anode [Figure 4.1(b)]. Nevertheless, the positive electrode remains a positive electrode and the negative electrode remains a negative electrode.

Therefore, to prevent confusion, it is better to avoid the use of the terms 'anode' and 'cathode' altogether and to adhere to 'positive electrode' and 'negative electrode' for the different types of electrochemical technology, which is the approach taken in this book.

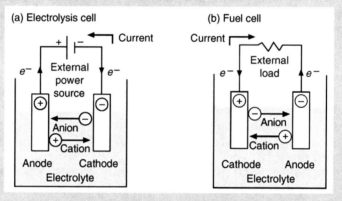

Figure 4.1   Terminology used in operation of electrolysis cells and fuel cells.

The hydrogen is principally used for the manufacture of ammonia-based fertilizers. With the advent of cheap natural gas as a source of hydrogen, however, the electrolytic route has largely become uncompetitive. Water electrolysis is also carried out to supply oxygen for life support in nuclear submarines that have the capability of remaining submerged for months at a time. This is a specialized application where there is no other option and therefore cost is immaterial.

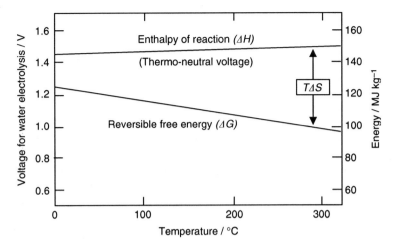

Figure 4.2 Voltage *vs.* temperature relationship for water electrolysis.
Note: under reversible conditions, the voltage and free energy are equivalent as related by Equation (4.5).

The thermodynamically 'reversible' voltage, $V°$, for water electrolysis under isothermal conditions and at standard temperature (298.15 K) and pressure (101.325 kPa) is 1.229 V; see Box 4.2[‡]. This value decreases almost linearly to 1.0 V at 573 K (300 °C); see Figure 4.2. The decrease in ΔG with increasing temperature is largely offset by an increase in the entropy term $T\Delta S$, so that the enthalpy of the reaction, ΔH, is almost independent of temperature. As large electrolyzers are essentially adiabatic[§] and little heat is absorbed from the surroundings, the energy corresponding to the entropy term is also supplied electrically. Under these circumstances, the entire enthalpy of reaction is supplied electrically and the reversible voltage is known as the 'thermo-neutral voltage'. At 25 °C, adiabatic operation increases the thermo-neutral voltage for water electrolysis to 1.47 V; see Figure 4.2. The electrical energy consumed in the reaction is then almost temperature independent at around 1.5 V.

The *practical voltage* required for water electrolysis, V_p, exceeds the reversible voltage, V_r, by an amount that is determined by the electrical losses in the cell, *viz.*, the resistive losses in the electrolyte (IR'_e), the overpotentials at the positive and negative electrodes (η_+ and η_-, respectively) and the total resistive ('ohmic') losses in the electrodes (IR'_t), *i.e.*,

$$V_p = V_r + IR'_e + \eta_+ + \eta_- + IR'_t \tag{4.6}$$

[‡] The reversible cell voltage at any temperature of operation is designated V_r, whereas $V°$ is specific to operation under standard conditions.
[§] An adiabatic system is one that is thermally isolated and experiences no loss or gain of heat.

Box 4.2 Thermodynamics of Water Electrolysis.

In thermodynamic terms, the total energy (or enthalpy, ΔH) required to split liquid water into hydrogen and oxygen is given by:

$$\Delta H = \Delta G + T\Delta S \qquad (4.4)$$

where ΔH is the enthalpy of the reaction [equal to the heat of formation of liquid water; under standard (STP) conditions, $\Delta H_f^\circ = -285.83\,\mathrm{kJ\,mol^{-1}}$], ΔG (*i.e.*, the change in Gibbs free energy) is the electricity requirement, and $T\Delta S$ is the heat absorbed from the environment at constant temperature T (ΔS is the change in disorder, or 'entropy', of the system in going from liquid water to hydrogen and oxygen). Equation (4.4) represents the Second Law of Thermodynamics, which may be expressed in words as: 'the change in Gibbs free energy (ΔG) of a system that is acting reversibly at constant temperature and pressure is equal to the change in heat content (ΔH) of the system less the heat ($T\Delta S$) that it absorbs during the reversible change'.

The change in Gibbs free energy may be expressed in terms of the reversible voltage (V_r) of an electrochemical cell (*i.e.*, at equilibrium when there is no current flow; also known as the 'open-circuit' voltage):

$$\Delta G = -nFV_r \qquad (4.5)$$

where n is the number of electrons involved in the reaction and F is the Faraday constant (the charge on one mole of electrons, *viz.*, 96 485 coulombs). Under standard conditions of temperature (298.15 K) and pressure (101.325 kPa), the reversible cell voltage for the decomposition of water (also known as the 'standard cell voltage', V°) is 1.229 V.

For an electrode reaction to proceed at an appreciable rate, the potential of the electrode must be changed in such a direction as will sustain the flow of current. The more the change in potential, the greater is the current flow and the greater is the 'irreversibility' of the electrode process. The degree of irreversibility is measured by the departure of the electrode potential from the reversible value under the same experimental conditions. An irreversible electrode is said to exhibit 'overpotential'.

For alkaline electrolyzers, the ohmic loss in the electrolyte (concentrated potassium hydroxide solution) shows a linear increase with current density and may be as much as 0.5 V at a current density of $400\,\mathrm{mA\,cm^{-2}}$. The other electrical losses are less dependent on current density, but are still of major significance and in total up to 0.4–0.8 V. When operating at 90 °C and atmosphere pressure, alkaline cells with base-metal electrodes (*e.g.*, mild-steel negatives and nickel-plated positives) require 2.1 V overall to yield a typical current density of $200\,\mathrm{mA\,cm^{-2}}$; see Figure 4.3. This equates to an energy consumption of 5 kWh per N-m^3 of hydrogen. By using negative electrodes activated with transition metal oxides, it is possible to lower the operating

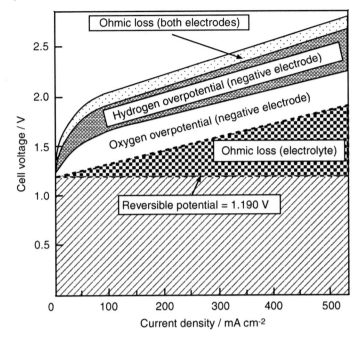

Figure 4.3 Performance (voltage *vs.* current density) of a basic (unactivated), unipolar electrolyzer running at 90 °C.

voltage to 1.7 V. Still further reductions are possible with precious metal electrocatalysts. Thus, in recent times, a great deal of research has been devoted to finding more active electrocatalysts, both in the context of reducing the operating cell voltage of electrolyzers and increasing that of fuel cells (see Chapter 6) while finding less expensive materials. Nevertheless, on comparing the relative costs of natural gas and electricity in most parts of the world, it will become clear that production of hydrogen with an electrolyzer is several times more expensive than with a steam reformer, although often the scale of production is quite different.

Some alkaline electrolyzers are operated at enhanced pressure in order to avoid the necessity of employing compressors subsequent to electrolysis – without compression, the gas storage volumes would be enormous. Raising the pressure from about 0.1 to 2.5 MPa leads to an increase in the reversible voltage of about 0.7 V, which corresponds to the free energy required to compress the product gases. As a consequence, the operating voltage of high-pressure cells is increased to 2.2–2.4 V, even when using the best electrocatalysts. This disadvantage is compensated to some degree by decreases in the overpotentials at the electrodes on account of the smaller size of the gas bubbles at high pressures. The effect of gas evolution and bubble blockage of the current flow is specific to the type of the electrolysis cell. Some electrolyzers are designed to have high circulation rates for the electrolyte so as to dislodge the gas bubbles from the electrodes. When this involves mechanical pumping, obviously the overall energy efficiency is reduced.

In order to lower the capital cost of the electrolyzer plant, it is necessary to operate at as high a current density as possible, *i.e.*, often more than $500 \, mA \, cm^{-2}$. Some commercial electrolyzers run at a current density as high as $3000 \, mA \, cm^{-2}$. The drawbacks are a greater cell resistance, the need for a higher applied voltage and, consequently, a lower electrical efficiency. Thus, a trade-off must be made between capital costs and operating (electricity) costs. This is an issue that may be overlooked when assessing the prospects for electrolysis.

The efficiency of an electrolyzer is defined as the ratio of the higher heating value (HHV) of the hydrogen produced to the electrical energy input[¶]. The ratio will vary considerably according to the generic type of electrolyzer employed, its size, its operating temperature and pressure, the current density and the choice of electrodes and electrocatalysts. With all these influential parameters, any general discussion of electrolyzer efficiency can only be semi-quantitative. At the ideal cell voltage of 1.229 V, the theoretical consumption of electricity is 2.94 kWh for each normal cubic metre of hydrogen produced. By measuring the electricity consumed and the hydrogen generated, it is then a simple matter to calculate the efficiency.

Conventional electrolyzers with alkaline electrolytes are 60–75% efficient, but the small-scale, best practice is now said to be closer to 80–85%. Larger units, or those operated at high current density, are less efficient. Thus, the figure of 80–85% may be optimistic since the US Department of Energy has in fact set a lower target for 2010, namely, a system efficiency of 78% for 5000 h of operation at a hydrogen delivery cost of US$2.85 per kg. The electrolyzers used in the European fuel cell bus (CUTE) programme (see Section 7.5, Chapter 7) were 76–77% efficient referred to the HHV of hydrogen.

Iceland, with its large reserves of hydropower and geothermal energy, is the first country to entertain seriously the electrolytic production of hydrogen in bulk quantity to serve as a fuel and energy vector; see Section 7.5, Chapter 7. The initiative has been motivated primarily by geopolitical considerations rather than by energy savings.

4.2 Electrolyzers

In order to minimize electricity consumption, it is important to choose an electrolyte of maximum conductivity. Usually, a concentrated aqueous solution of potassium hydroxide (30–40 wt.%) is employed. This must be prepared from very pure water, otherwise impurities will accumulate during electrolysis; the chloride ion, which is usually present in water, is particularly harmful in that it causes pitting of the protective films formed on metal surfaces in alkaline solutions.

[¶] For definition of higher heating value (HHV) and lower heating value (LHV), see Section 1.5, Chapter 1.

In principle, electrolyzers are well suited for use with electricity generated from renewable energy sources such as wind and solar. They may readily be matched to the size of the source and are able to operate over a wide power range. When the wind does not blow or the sun does not shine, the electrolyzer would theoretically go into stand-by mode, ready to start again when the electricity supply resumes. In practice, the situation would be more complicated than that because the current available may fluctuate from moment to moment. Conventional electrolyzers are designed to operate on grid-quality power, whereas the electricity from the aforementioned two renewable sources is intermittent and of variable quality. This disadvantage will necessitate the development of specialized power control and conditioning equipment. Experience to date in operating electrolyzers with a rapidly fluctuating power source is strictly limited.

Commercial electrolyzers are constructed in varying sizes, with power consumptions in the range of kW to multi-MW. A large industrial plant capable of producing 500 N-m^3 h^{-1} (45 kg h^{-1}) of hydrogen would demand about 2.3 MW of power. The capital cost of such large units is around US$600–700 per kW, with the prospect of mass production resulting in a future reduction to ~ US$300 per kW. On this basis, an electrolyzer consuming 2 MW of power will today involve an expenditure of US$1.2–1.4 million. Large though such a unit may be in terms of local generation of hydrogen, it is only 1/500th of the size required to couple to the output of a typical 1000 MW$_e$ power station. The idea of building 500 such facilities around a power station is daunting, quite apart from the capital outlay of some US$600–700 million and the high running costs for the electricity. In fact, under present conditions, the economics of operating large industrial electrolyzers is dominated by the price of electricity with the interest on the capital investment playing a lesser role.

By contrast, wind turbines or solar arrays of appropriate power output could readily be coupled to smaller electrolyzers, although this would be an even more expensive way of generating hydrogen. The capital cost per kW of these units is significantly greater than that of large industrial plant, while renewable electricity is more expensive than that derived from fossil fuels. Using such electricity, it is estimated that hydrogen can be obtained at US$7–10 per kg, *i.e.*, 3–5 times above that from fossil fuels.[1] Even with the mass production of small electrolyzers, should it occur, the overall economics would not change much until 'green' electricity becomes cheaper than the conventional alternative. Thus, if hydrogen from renewables is to make a significant impact in the overall energy scene, it will be necessary to make dramatic breakthroughs in the cost of both electrolyzers and the power to run them. Under existing conditions and with the exception of special circumstances and locations, a very high premium would have to be placed on hydrogen as a clean fuel and energy vector to justify its manufacture by this route.

A small alkaline electrolyzer (2 N-m^3 h^{-1}) that should be suitable for the local production of hydrogen when coupled to a renewable energy source is shown in Figure 4.4(a). At a rating of 13 kW, its efficiency is stated to be 45%.

(a)   (b)

Figure 4.4   Small-scale alkaline water electrolysis systems produced by (a) Hydrogen
Systems NV and (b) Hydrogenics.[2]
(Courtesy of the International Energy Agency Hydrogen Implementing
Agreement).

A rather larger system that is capable of producing $30 \, \text{N-m}^3 \, \text{h}^{-1}$ is illustrated in
Figure 4.4(b).

Electrolyzers, like batteries, may be constructed with either monopolar or
bipolar configurations. The monopolar (or 'tank-type') unit, shown diagram-
matically in Figure 4.5(a), consists of alternating positive and negative
electrodes that are held apart by microporous separators. The positives are
all coupled together in parallel, as are the negatives, and the whole assembly is
immersed in a single electrolyte bath ('tank') to form a unit cell. Alkaline cells
based on an aqueous potassium hydroxide electrolyte are often of this type. The
separator – employed to keep the electrodes apart, to prevent remixing of
hydrogen and oxygen products, and to absorb the electrolyte – was traditionally made of asbestos paper. This material is stable in concentrated potassium
hydroxide solution, but is a carcinogen and therefore undesirable. Recently,
new microporous polymer materials have been developed as more acceptable
alternatives.

A traditional, monopolar electrolyzer is built up by coupling tank units in
series electrically[‖]. By contrast, a bipolar unit uses a metal sheet (or 'bipole') to
join adjacent cells, as depicted in Figure 4.5(b). The electrocatalyst for the
negative electrode is coated on one face of the bipole and that for the positive
electrode of the adjacent cell is on the reverse face. A series-connected stack of
such cells forms a module that operates at a higher voltage and lower current

---

[‖] This is analogous to the construction of a typical 12 V automotive lead-acid battery, which is made
up of six cells connected in series.

(a) Monopolar

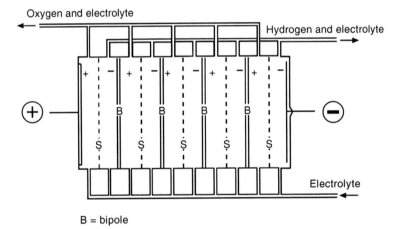

S = separator

(b) Bipolar

B = bipole

Figure 4.5 General arrangement of (a) monopolar and (b) bipolar electrolyzer modules. S denotes separator; B denotes bipole.
(Courtesy of Imperial College Press).

than the tank-type design; an example of such a unit is shown in Figure 4.6. To meet the requirements of a large plant, these modules are connected in parallel so as to increase the current. The principal advantages and disadvantages of bipolar electrolyzers are summarised in Table 4.1.

Alkaline water electrolyzers that operate at pressures up to 5 MPa are under development. At present, they are commercially available with a capacity range of $1-120 \, N\text{-}m^3 \, h^{-1}$. The goals are to optimize the units for energy efficiency and to make them responsive to fluctuating current supply that would occur with

Figure 4.6 Large-scale bipolar water electrolyzer.[2]
(Courtesy of the International Energy Agency Hydrogen Implementing
Agreement).

Table 4.1 Features of bipolar electrode stack electrolyzers.

Advantages	*Disadvantages*
Compact design	Parasitic electrical bypass currents
No busbars required	Consequent reduced electrical efficiency
Low-cost electrical equipment	Inhomogeneous current distribution
Higher operating voltage	Increased corrosion rate of electrodes
Lower operating current	

wind or solar energy input. Naturally, operation at elevated pressures introduces technical challenges with respect to cell design and materials, and also
raises the capital cost substantially. Such units will have to be subject to
evaluation and certification of the pressure vessel for safe operation.

A concentrated solution of potassium hydroxide at temperatures approaching
100 °C is not an easy material to handle, especially in high-pressure systems. It
poses some severe materials science problems that arise from stress-corrosion
cracking of steels and degradation of gaskets and sealants. In order to avoid such
difficulties, many years ago in the USA General Electric introduced a new type of
electrolyte in the form of a solid polymer. This material, which is manufactured
by the DuPont Corporation and sold under the tradename of Nafion®, is a
perfluorosulfonic acid polymer that is highly conductive for hydrated protons. It
is also available as thin sheets to form a membrane that serves as a separator. The
reactions for water electrolysis in an acidic medium [corresponding to reactions
(4.2) and (4.3) for an alkaline medium] are as follows:

at the negative electrode:

$$4H^+ + 4e^- \rightarrow 2H_2 \tag{4.7}$$

at the positive electrode:

$$2H_2O \rightarrow O_2 + 4H^+ + 4e^- \tag{4.8}$$

Thus, hydrated protons are conducted through the polymer membrane electrolyte from the positive to the negative electrode.

Appropriate electrocatalysts of platinum or its alloys are coated on each side of the membrane to form a so-called 'membrane electrode assembly' (MEA). A metal mesh contacts each electrocatalyst in order to complete the electrical circuit. Electrolyzers based on this technology offer major advantages over their alkaline counterparts in terms of (i) higher turn-down ratios, *i.e.*, ratio of maximum to minimum gas flow, which is a measure of the operating range to meet variations in demand; (ii) greater safety through the absence of potassium hydroxide electrolyte; (iii) a more compact design due to higher current densities; and (iv) higher operating pressures. Present research is directed towards the development of improved cell stacks and materials in order to increase the efficiency and capacity of the system. The limited lifetime and high cost of membranes are other important considerations, as also is the use of expensive precious-metal electrocatalysts.

A schematic of a design for a bipolar proton-exchange membrane (PEM) stack is shown in Figure 4.7. This type of unit is also known as a solid polymer electrolyte (SPE) electrolyzer. The conducting bipole is made of carbon (graphite) and is corrugated or ribbed, to form channels through which the gases produced during electrolysis escape from the cell stack. A thin sheet of titanium protects the carbon positive electrode from oxidation. The MEAs are similar to those developed for PEM fuel cells (see Section 6.4, Chapter 6) and draw heavily on the latter for their design technology. New techniques for applying the electrocatalysts to the membranes have also been formulated. These advances have resulted in enhanced performance at reduced loadings, which is especially important when using expensive electrocatalysts. Efforts are also being made to find electrocatalysts that are less prone to poisoning and are able to operate at lower temperatures. In general, PEM electrolyzers are best suited to small plants, particularly those powered by renewable energy (*e.g.*, wind, solar) where the electricity supply is variable, while alkaline counterparts are more practical in larger systems that are connected to the grid.

Drawing also on fuel-cell research, there is the possibility of developing electrolyzers that are incorporated in furnaces and operate at high temperatures (700–1000 °C), so-called 'steam electrolyzers'. Such units employ a ceramic electrolyte that is conductive towards oxide ions (O^{2-}) and is of the type that has been developed for high-temperature fuel cells; see Section 6.5, Chapter 6. As shown in Figure 4.2, high-temperature operation favours electrolysis (*i.e.*, improves efficiency), since some of the energy required to split water is now supplied by heat. In addition, the overpotentials at the electrodes are reduced. Thus, steam electrolysis is advantageous from both a thermodynamic and a kinetic standpoint. These twin benefits substantially reduce the electricity demand (at least 25% of the electricity can be saved), but material

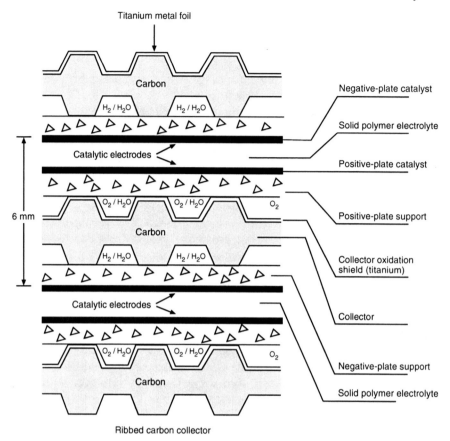

Figure 4.7   Cell stack configuration of a PEM electrolyzer.

specifications are more severe. Little is yet known about scale-up issues, the durability of the solid electrolyte and electrodes, and the capital costs. More-over, meeting the technical challenges of designing plant to operate at 900 °C *and* at high pressure is a daunting prospect. Of course, corresponding amounts of high-temperature heat must be supplied. This condition makes steam elec-trolysis a potential new application for a dedicated combined heat and power system, a solar–thermal plant or even – on a huge scale – a high-temperature nuclear reactor. When using an oxygen ion-conducting electrolyte, the corres-ponding water-splitting reactions are as follows:

at the negative electrode:

$$2H_2O + 4e^- \rightarrow 2H_2 + 2O^{2-} \tag{4.9}$$

at the positive electrode :

$$2O^{2-} \rightarrow O_2 + 4e^- \tag{4.10}$$

In the electrolysis of water (or steam), a significant fraction of the voltage that must be applied is associated with the liberation of oxygen at the positive electrode. This is a fundamental consequence of the free energy of formation of water from hydrogen and oxygen. A novel idea has been advanced to reduce the operating voltage of ceramic steam electrolyzers by flowing natural gas past the positive electrode. This serves to reduce chemically the oxygen to carbon dioxide. Thus, the electrode reaction becomes:

$$4O^{2-} + CH_4 \rightarrow CO_2 + 2H_2O + 8e^- \quad (4.11)$$

Thermodynamically, this corresponds to a lower free energy and, therefore, a lower voltage requirement. The approach essentially replaces one unit of electricity with one equivalent-energy unit of natural gas. Experiments on a small cell have shown a voltage reduction by as much as 1 V when compared with straightforward steam electrolyzers. For example, the application of only 0.5 V to a cell at 700 °C resulted in a high current density of $1\,A\,cm^{-2}$. Depending on the current density, the system efficiency has been estimated to be in the range 50–80%. Since natural gas is so much cheaper than electricity, this is a promising concept. Nevertheless, much remains to be done to design a practical system. It is likely that technical experience can be gained from the extensive work that is being undertaken on high-temperature fuel cells; see Section 6.5, Chapter 6.

There is some interest in using nuclear-generated electricity, with or without nuclear heat, to power an electrolyzer. Insofar as nuclear energy does not use fossil fuels, it may be regarded as 'renewable' or at least 'low carbon'. Countries such as France (which derives much of its electricity from nuclear reactors) and possibly Japan might find this option attractive. In the case of conventional pressurized water reactors, which operate at around 300 °C, the temperature would be too high for an aqueous electrolyzer and below that required by a steam electrolyzer. A high-temperature gas-cooled reactor (HTGCR) operating at 900 °C would be ideal to couple to a steam unit, but such reactors are not yet commercially available, despite prototypes having been built and operated in the 1960s. Ongoing development work in the nuclear industry holds some promise for a renaissance of HTGCR technology, but the employment of this system to power a steam electrolyzer is beyond the foreseeable time horizon. Ultimately, fusion power may provide a source of cheap electricity that could be used to support a Hydrogen Economy in a sustainable fashion, but again this is too speculative and far off for serious consideration at present.

The electrolysis of water provides one molecule of oxygen for every two molecules of hydrogen. In discussions of hydrogen energy, the fate of this oxygen has largely been ignored. By contrast, in one of the few current applications for water electrolysis, namely, for life support in nuclear submarines, it is the oxygen that is the desired product and the hydrogen that is an unwelcome by-product. Much the same is true of chlor-alkali electrolysis for the manufacture of chlorine and caustic soda, where hydrogen is a by-product of limited value, see reaction (4.1).

What use can be made of electrolytic oxygen? There are many small-scale requirements for cylinder oxygen, *e.g.*, in hospitals, in welding and in the treatment of sewage. These markets are well served by established gas companies that provide oxygen prepared from air by cryogenic distillation. It is doubtful whether distributed electrolyzer operations, such as those based on renewable energy sources, could break into this business. The steel industry is a major consumer of bulk quantities of oxygen but, again, there is no shortage of supply. If the day ever arrives when large numbers of MW-scale electrolyzers are installed for hydrogen manufacture and tonnage amounts of oxygen become available from this source, then the gas might be fed to a plant for the partial oxidation of fossil fuels (see Section 2.3, Chapter 2) and thereby create a substantial demand. Another potential outlet for bulk quantities would be for the underground gasification of coal. The utilization of by-product oxygen in such applications, rather than allowing it to go to waste, would enhance the overall economics of electrolyzer operation and thereby reduce the cost of hydrogen.

4.3   Water Splitting with Solar Energy

As noted at the beginning of this chapter, solar energy may be employed to generate hydrogen from water. There are several ways in which this may be accomplished, as summarized in Figure 4.8[**].

4.3.1   Photovoltaic Cells

The most obvious route for the conversion of solar energy to electricity is via photovoltaic (PV) cells, which are commonly known as 'solar cells'. Technically, this is straightforward but, because of the high cost of solar cells, it is likely to be expensive on anything other than a small scale for some time to come. Although the efficiency-to-cost ratio of the cells has improved dramatically over the past 20 years, a further leap forward is required before photovoltaic electricity will become competitive with conventional forms of generation; at present, it is 5–10 times more expensive.[3] Recent sharp increases in the price of oil and natural gas, together with progressive reductions in the price of solar cells, have combined to reduce the gap. An encouraging feature is that PV modules based on monocrystalline silicon have fallen from $\sim$ US\$70 per peak watt (W_p) in 1976 to around US\$4 today. This has been due to steady improvements in their performance and the growth in cumulative production from less than 1 MW_p to almost 3000 MW_p.[3] In the USA, the ultimate goal is to reduce the commercial cost of solar cells per W_p to around US\$0.40, with intermediate targets of US\$3.0 and US\$1.50 by 2010 and 2020, respectively. To achieve these objectives, it will almost certainly be necessary to adopt a new technology for cell manufacture. Monocrystalline ('single-crystal') silicon – currently the preferred PV material – is an intrinsically expensive material that

---

[**] The production of hydrogen from biomass – an indirect exploitation of solar energy – is discussed in Section 2.8, Chapter 2.

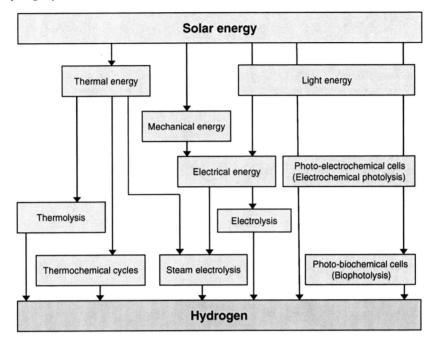

Figure 4.8 Solar-to-hydrogen conversion options.

will have to be replaced. Multicrystalline ('polycrystalline') silicon and amorphous silicon are cheaper materials to produce, but their light-conversion efficiencies are lower. Other semiconductors that have been employed in solar cells are gallium arsenide, GaAs, cadmium telluride, CdTe, and copper indium gallium diselenide, $Cu(In,Ga)Se_2$ The principles governing the operation of PV cells are outlined in Box 4.3.

Developmental work is being undertaken on organic (polymer)-based photovoltaics (OPVs), which are especially attractive for low-cost, scalable, power generation. Their attributes include:

- the ability to be fabricated, patterned and packaged as much larger sheets of solar cells (via standard printing, coating and laminating techniques) than is possible with silicon technology; moreover, such sheets can be made flexible so that the cells can be moulded to fit on to curved surfaces and thus allow easy integration into existing buildings and devices;
- high optical absorption coefficients, which means that only a very thin (100–200 nm) film is required to absorb most of the incident light and therefore less material is involved in the production of cells;
- transparent or coloured options; many of the polymers are brightly coloured and there is the capability to tune the colour through tailoring of the chemical structure, which is an important feature in terms of design and aesthetics;
- ecological and environmental benefits.

Box 4.3 Physical Principles of Photovoltaic Cells.

Semiconductors are materials with conductivity between that of metals and insulators. In solids, the electron energy levels form bands of allowed energy that are separated by forbidden bands [Figure 4.9(a)]. The separation between the outermost filled band (the 'valence band') and next highest band (the 'conduction band') is known as the 'energy gap' or 'band gap energy', E_g, of the material. In essence, E_g is the amount of energy required to liberate electrons from their covalent bonds in the crystal lattice of the semiconductor. The energy gap is fairly small (~ 1 eV) for a semiconductor, but is large for an insulator (~ 10 eV) and zero for a metal (the valence and conduction bands overlap). The energy level at which there is a 0.5 probability of finding an electron is known as the 'Fermi level', E_F. In a metal, this level is very near the top of the valence band. In a semiconductor, however, the Fermi level is located at about the middle of the energy gap and since this gap is small, appreciable numbers of electrons can be excited from the valence band to the conduction band, to produce a moderate current.

There are both positive and negative charge carriers in a semiconductor. When an electron moves from the valence band into the conduction band (*e.g.*, by absorption of energy from a photon of sunlight), it leaves behind a vacancy or so-called 'hole', in the otherwise filled valence band [Figure 4.9(b)]. This hole (electron-deficient site) appears as a positive charge and acts as a charge carrier in the sense that a nearby valence electron can transfer into the hole, thereby filling it and leaving a hole behind in the electron's original place. Thus, the hole migrates through the material. In a pure crystal that contains only one element or compound, *e.g.*, silicon, there are equal numbers of conduction electrons and holes. Such combinations of charges are called 'electron–hole pairs' and a pure semiconductor that contains such pairs is an 'intrinsic semiconductor'.

The electrons and holes in an intrinsic semiconductor recombine after a short time and their energy is wasted as heat. To obtain useful electricity, therefore, the band structure and conductivity of semiconductors must be modified. This is achieved by the addition ('doping') of controlled micro-quantities of impurities, as shown schematically in Figure 4.9(c) and (d). In the case of silicon semiconductors, trivalent impurities (*e.g.*, boron, aluminium) that have an E_F just above the valence band, accept electrons from the silicon and so leave holes in the valence band [Figure 4.9(c)]. Such impurities are referred to as 'acceptors' and the resulting 'extrinsic' semiconductor is known as a '*p*-type' because the charge carriers are positively charged holes. Conversely, pentavalent dopants (*e.g.*, arsenic, phosphorus) that have an E_F just below the conduction band are 'donors' and provide electrons to the conduction band of silicon [Figure 4.9(d)]. This produces an '*n*-type' extrinsic semiconductor, so-called because the majority of charge carriers are now electrons, whose charge is negative.

Box 4.3 Continued.

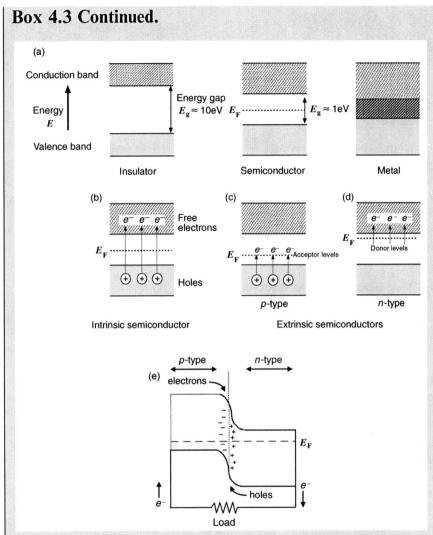

Figure 4.9 (a) Schematics of the band structure and conductivity of insulators, semiconductors and metals. (b)–(e) Operating principles of semiconductor devices.

When p-type and n-type semiconductors are brought into contact, to form a 'p–n homojunction', the band structure at the interface is distorted so the Fermi level is the same on each side of the interface [Figure 4.9(e)]. A region that is depleted of mobile charge carriers is formed at the interface (the 'depletion region') and an electric field of the order of 10^4–10^6 V cm^{-1} is built up across the interface – the 'photovoltaic effect'. This internal electric field creates a potential barrier that prevents further electron–hole recombination. Thus, when photons are absorbed by the semiconductor, the electric field attracts electrons released on the p-side of the junction to the

> ## Box 4.3 Continued.
>
> *n*-side, while holes are swept in the reverse direction. This charge separation induces a voltage across the device and when the two sides of the junction are connected together via an external circuit current is able to flow and, thereby, produce electrical energy [Figure 4.9(e)]. The resulting device is known as a 'photovoltaic cell' or more commonly a 'solar cell'.

The principal disadvantages of OPVs are (i) poor solar conversion efficiency (at present around 5%) compared with inorganic cells and (ii) the sensitivity of the polymers to oxygen and/or water vapour that leads to instability and degradation of performance. Research is being directed towards overcoming these two limitations. By using organic materials of improved purity and with better encapsulation techniques, working lives of more than 10 000 h have already been achieved.

4.3.2   Solar–Thermal Process

Thermolysis is the term employed to describe a chemical process by which a substance is decomposed through the application of heat. When the required heat is obtained directly from the Sun, the reaction is often referred as a 'solar–thermal process'. The high temperature required for breaking down water directly into hydrogen and oxygen can be achieved by focusing the Sun's rays from a large number (up to thousands) of individual mirrors on to a thermal receiver mounted on top of a central, tall 'solar tower' of similar design to that constructed in Australia by the Commonwealth Scientific and Industrial Research Organisation (CSIRO) for the solar–thermal reforming of natural gas; see Figure 2.4, Chapter 2. The key scientific challenges are (i) to find catalysts that will reduce the dissociation temperature of water and hence improve the efficiency of the process and (ii) to provide an improved means for separating the two gases so as to prevent their recombination.

Much of solar radiation lies in the infrared part of the spectrum and is of too low energy to be utilized in conventional PV cells (or in photo-electrochemical reactions; see below) so that it is wasted. To harness this thermal energy and thereby improve the efficiency of solar energy conversion, Licht[4] has proposed the use of dielectric filters to separate the radiation received by the solar tower into an infrared component to heat pressurized water to at least 300 °C and an optical/ultraviolet component to generate electricity through PV (or photo-electrochemical) cells. The electricity would then be used to split water by high-temperature steam electrolysis (see above) with the entropy term $T\Delta S$ of Figure 4.2 now being supplied thermally. This concept is illustrated schematically in Figure 4.10. Thermodynamic calculations suggest that it should be possible to reach overall efficiencies approaching 20% for the conversion of solar energy to hydrogen. Nevertheless, some formidable technological problems have to be overcome for a practical system to be realized.

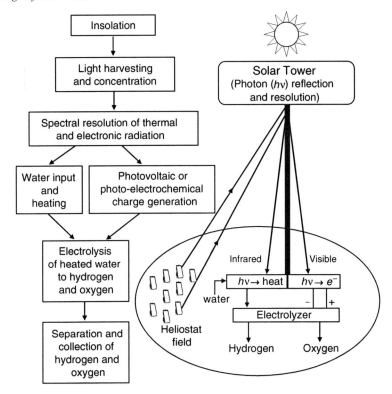

Figure 4.10 Schematic of thermally enhanced solar water-splitting.[4]
(Courtesy of Elsevier Ltd).

An alternative approach is to employ the thermal energy from the solar concentrator to generate electricity directly by means of a PV cell based on a semiconductor with a low band gap (*e.g.*, 0.6 eV) that is sensitive to infrared radiation. Such 'thermo-photovoltaic' cells are already in small-scale specialized use by the military, although with the heat provided by a propane burner, rather than from the Sun. This is an alternative to the thermoelectric effect for converting heat to electricity.

4.3.3 Photo-electrochemical Cells

Photovoltaic (PV) cells are physical devices that operate on the principles of solid-state physics. Another class of device – one that is capable of splitting water – is based on 'photo-electrochemical' reactions, which take place at electrodes that are light-sensitive. Photo-electrochemistry may serve to generate d.c. electricity (via dye-sensitized solar cells) and this can then be used to electrolyze water (as with PV cells). Alternatively, light illuminating an electrode may reduce water directly to hydrogen – a process known as 'photolysis'. These two processes are described next.

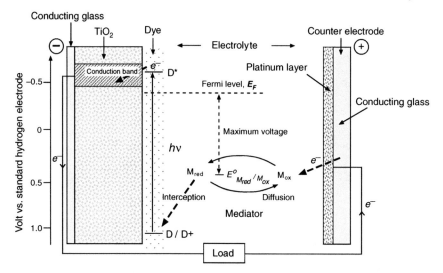

Figure 4.11   Operating mechanism of a dye-sensitized titania solar cell.

4.3.3.1   *Dye-sensitized Solar Cells*

Titania (TiO_2) absorbs light energy in the ultraviolet rather than in the visible part of the spectrum and therefore is of limited value as a PV cell. Accordingly, considerable research effort has been directed towards improving its efficiency for solar energy conversion by finding a means to shift the spectral response into the visible region. This is achieved in a dye-sensitized solar cell (DSSC) (also known as a 'Grätzel cell' after the pioneer of the technology) by means of a subterfuge and involves separating the optical absorption and charge-generating functions. A dye, usually ruthenium-based and which is capable of being photo-excited, is adsorbed on the surface of highly porous titania, where it acts as an electron-transfer sensitizer. The coated titania then serves as the negative electrode in a photo-electrochemical cell.

The operating mechanism of a DSSC is illustrated in Figure 4.11. The dye (D), after having been excited (D*) by a photon of light (*hv*), transfers an electron to the conduction band of the titania ('injection process') and itself becomes oxidized (D^+) in the process. The cell electrolyte contains a 'redox mediator' (symbol M), *i.e.*, a substance that can be oxidized and reduced electrochemically[††]. Positive charge is transferred from the dye to the mediator (M_{red}) and the former is thereby returned to the reduced state ('interception process'). The oxidized mediator (M_{ox}) diffuses to the positive counter-electrode, where it is reduced by the electrons travelling around the external circuit. The theoretical maximum voltage that such a cell can deliver is the difference between the redox potential of the mediator and the Fermi level (see Box 4.3) of the titania semiconductor.

[††] Redox-mediated reactions are very common in biochemistry. More specifically, photo-induced redox-mediated reactions form the basis of photosynthesis in nature.

In practice, the titania electrode is present as a thin film (~ 10 μm) of tiny crystals that have diameters of just a few nanometres and that are sintered together. This film is deposited on a transparent, conducting glass substrate (generally, coated with tin oxide) which acts as a current collector. The titania is ultra-porous and has a very high surface area. This assists the uptake of dye from solution. The dye-coated electrode is then assembled into a cell with a counter electrode, also made from conducting glass, and the intervening space is filled with electrolyte and the mediator (typically, the iodide–triiodide couple, $I^- - I_3^-$, dissolved in acetonitrile or some other organic solvent). A small amount of platinum is deposited on the counter electrode to catalyze reduction of the mediator.

The titania layer has a very high surface-to-volume ratio, which results in an effective surface area for light absorption that is 1000 times greater than for a dye monolayer on a non-porous, solid surface. As a result, the light-harvesting ability of a dye when adsorbed on a mesoporous film is greatly increased and thus results in an improvement in cell efficiency. The use of sintered meso-porous titania as the photo-electrode was the breakthrough that established DSSC technology as a credible alternative to solid-state photovoltaics and raised the DSSC efficiency from 1% (for cells having a non-porous titania surface) to $\sim 7\%$ in the laboratory. This has subsequently been extended to 11%, while 5% has been obtained in the field.

Compared with silicon cells, the dye-sensitized titania cell has lower efficiency under conditions of strong solar radiation, but provides somewhat better per-formance at low illumination levels as, for example, on cloudy days. This reveals that the DSSC is less sensitive to light intensity variations than conventional photovoltaic devices, which is an important advantage for application in con-sumer electronic devices. The principal attractions of DSSCs lie in their ease of fabrication into large sheets, their mechanical flexibility and the low cost of polycrystalline titania compared with monocrystalline silicon. Their performance is also relatively insensitive to temperature, *e.g.*, no significant change is observed between 20 and 60 °C, whereas conventional solar cells experience a decrease of approximately 20% in efficiency over the same temperature range. Moreover, the output of DSSCs has been shown to be stable over a long lifetime; the dye can sustain $10^7 - 10^8$ cycles without significant decomposition and lifetimes of 10–20 years have been extrapolated from accelerated tests. The technology is still under active development. There are, in fact, many different possibilities for photo-electrochemical cells and this appears to be a promising area of research for cost-effective renewable energy in the future. The main challenge is to improve the efficiency of DSSCs to the level of traditional silicon cells. World-wide, several companies are engaged in this task. Efforts are also being made to reduce manufacturing costs by replacing the conducting glass support, which is the most expensive component of the cell, with cheaper polymer films.

4.3.3.2 Direct Hydrogen Production

As mentioned earlier, photo-electrochemical (PEC) reactions may be employed not only to generate electricity, but also to decompose water to hydrogen; a brief description of the basic principles is given in Box 4.4. Effectively, this photolytic approach combines a photovoltaic system and an electrolyzer into a single monolithic device. Prospectively, this could result in a higher overall system efficiency and a lower capital cost compared with a separated system. The configuration offers the following advantages: (i) the junction required for efficient charge separation of photo-generated electrons and holes is very easily formed by simply immersing the semiconductor in an appropriate electrolyte solution; (ii) the liquid electrolyte offers the capability of a readily conformable and strain-free junction; and (iii) the conversion of light energy directly into hydrogen eliminates the need for external wires and a separate electrolyzer. Although PEC cells have achieved encouraging efficiencies for hydrogen generation, there are still technical and cost issues to be addressed before such devices could become commercially viable.

The known materials that are active in water splitting are not responsive to a wide portion of the solar radiation spectrum; they work best in the ultraviolet region and thus yield relatively low efficiencies. Titania is a good example of a semiconductor subject to this limitation. Thus, one of the primary aims of research in this field is to find new photo-electrodes that can operate, either individually or in combination, across the solar spectrum and thereby allow more efficient conversion of sunlight to chemical fuels. To this end, extensive effort is being directed towards the perfection of multi-junction PEC cells. These consist of thin layers of several semiconductors that each have a different band gap and are mounted on top of one another so as to take advantage of the entire solar spectrum. In a typical three-junction design, the ultraviolet, visible and long-wavelength components of the incoming radiation are successively absorbed by semiconductors that have band gaps of about 1.8, 1.6 and 1.4 eV, respectively. Another option is to explore a wide range of candidate semiconductors in a combinatorial fashion (as practised in drug screening) so as to seek out possible new compounds or solid solutions that are both satisfactory in performance and cheap to fabricate.

4.3.3.3 Tandem Cells

A strategy for improving the efficiency of DSSCs is to develop a tandem cell. There are several possible configurations. One is the multi-junction design described above. Another is to connect electrically a hydrogen PEC cell in series with a DSSC. A schematic of such an arrangement is shown in Figure 4.14. The photoactive electrode (*e.g.*, TiO_2, WO_3 or Fe_2O_3) in the front cell absorbs the high-energy ultraviolet and blue light in sunlight to liberate oxygen, while radiation of longer wavelength in the green-to-red region of the spectrum passes to, and is absorbed by, the DSSC. This boosts the flow of electrons that are then fed back to a counter electrode in the PEC cell to produce hydrogen. With such an arrangement, photon-to-hydrogen efficiencies of up to 12% have been reported. The development of tandem cells is a fairly recent activity and one with expectation of leading to commercial PEC cells and DSSCs.

Box 4.4 Photo-electrochemical Generation of Hydrogen.

The operating mechanism of a conventional electrolysis cell is illustrated in Figure 4.12(a) and that of a photo-electrochemical (PEC) cell for light-driven electrolysis of water in Figure 4.12(b). In the examples shown, each cell has a sulfuric acid (H_2SO_4) electrolyte. The devices are similar in concept. The main difference is that the external d.c. source in the former is replaced by an internal, photon-induced, current generator in the latter. Thus, on irradiation with sunlight, a PEC cell operates as a form of solar cell that can split water into hydrogen and oxygen. In other words, the energy of the photon is converted into chemical energy rather than into electrical energy as in a photovoltaic cell. A semiconductor serves as the photoactive component and is prepared as a thin film deposited on a conducting substrate. The latter also is a thin transparent film that is composed of fluorine-doped tin oxide or tin-doped indium oxide on a glass substrate to allow the passage of light. Platinum or platinized graphite is employed as the counter electrode. If, as shown in Figure 4.12(b), electrons are forced to leave the illuminated electrode through its back contact, then an *n*-type semiconductor is being used and is the counterpart of the positive electrode in the electrolysis cell. (Note: with a *p*-type semiconductor, the flow of charge is reversed.)

On the absorption of a photon by a *n*-type semiconductor, one electron is excited from the valence band and promoted to the conduction band [see Figure 4.9(b)] to leave a net positive charge – a hole (h^+) – in the valence band. The electron in the conduction band is forced to the back contact and is transferred to the counter electrode. At the electrode|electrolyte interface of this latter electrode, the electrons react with protons to produce hydrogen, while the simultaneously created holes in the valence band of the semiconductor (the minority carriers) generate oxygen by oxidizing water. Thus, the electrode half-reactions in the PEC cell are as follows:

at the negative electrode:

$$4H^+ + 4e^- \rightarrow 2H_2 \qquad (4.12)$$

at the positive electrode:

$$2H_2O + 4h^+ \rightarrow O_2 + 4H^+ \qquad (4.13)$$

The corresponding reactions for an electrolyzer operating with an acid electrolyte are as follows:

at the negative electrode:

$$4H^+ + 4e^- \rightarrow 2H_2 \qquad (4.14)$$

at the positive electrode:

$$2H_2O \rightarrow O_2 + 4H^+ + 4e^- \qquad (4.15)$$

Box 4.4 Continued.

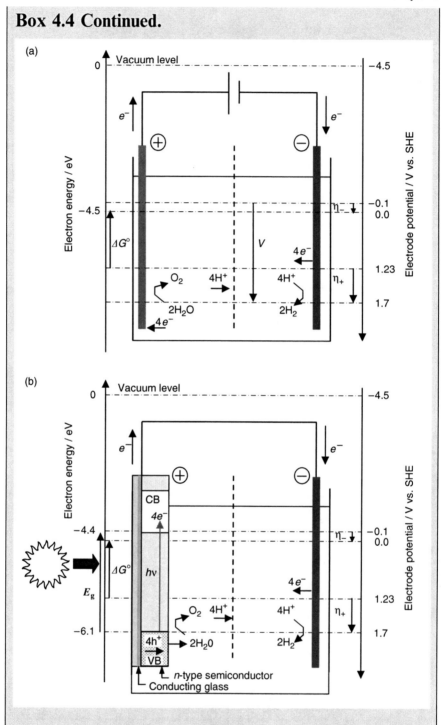

Figure 4.12   Schematics of (a) an electrolysis cell and (b) a photo-electrochemical cell.

Box 4.4 Continued.

The net cell reaction is the same for both types of device:

$$2H_2O \rightarrow O_2 + 2H_2 \qquad (4.16)$$

This reaction requires four absorbed photons, each with an energy that is equal to or more than the band gap energy, E_g, of the semiconductor. In practice, a greater number of charge carriers (both electrons and holes) are necessary as some are lost by recombination, both within the bulk and on the surface of the semiconductor. As discussed in Box 4.3, the band gap energy defines the minimum energy that one of these photons must have to promote an electron from the valence band to the conduction band, *i.e.*,

$$E_g = E_{CB,b} - E_{VB,t} = h\nu_{min} = hc/\lambda_{max} \qquad (4.17)$$

where $E_{CB,b}$ is the energy at the bottom of the conduction band, $E_{VB,t}$ is the energy at the top of the valence band, h is Planck's constant, ν_{min} and λ_{max} are the minimum frequency and the maximum wavelength of the absorbed light, respectively, and c is the speed of light in vacuum.

Although the rules set by Nature are inviolate, different branches of science have formulated different concepts for, in principle, the same phenomenon. Thus, immersing a semiconductor into an electrolyte forces the vocabulary of more recent solid-state physics to be joined with that of traditional (19th century) electrochemistry and thermodynamics. The energy E in electronvolts (eV) with its zero point at the vacuum level is given on the left-hand side of Figure 4.12(a) and (b). By convention, the zero point chosen for the electrode potential is that of a standard hydrogen electrode (SHE; in practice: $H_2SO_4 \approx 0.5\,M$; H_2 pressure=101.325 kPa; temperature=298.15 K). Thus, electrode potentials, E,[‡‡] are given in terms of volts (V) *vs.* SHE. The relationship between the solid-state scale on the left-hand side of Figure 4.12(a) and (b) and the electrochemical scale to the right, is expressed by:

$$\boldsymbol{E} = -eE - K \qquad (4.18)$$

where the constant K is equal to 4.5 eV; for simplification purposes, the negative sign in the symbol for the electron has been removed both here and in the following Equations (4.19) and (4.20).

It is also necessary to understand the effects of irreversibility imposed on the system by the fact that hydrogen has to be produced at a practical rate. Specifically, the influence of the reaction kinetics at the electrode|electrolyte interfaces, and also that of the transport of the reactants through the electrolyte, on the performance of the PEC cell must be taken into account. These phenomena can be visualized as ohmic losses ('IR' drop') in the cell, as

---

[‡‡] In the scientific literature, it is common practice to use the same symbol, E, for both 'energy' and 'electrode potential'. This obviously leads to confusion in discussions, such as here, where both parameters are under consideration. To avoid this, bold font $\boldsymbol{E}$ has been adopted as the symbol for energy.

Box 4.4 Continued.

expressed by Equation (4.6) in Section 4.1, Thus, E_g must be sufficiently large to drive reactions (4.12) and (4.13) according to the relationship:

$$E_g = \Delta G + e\,IR'_e + e\eta_+ + e\eta_- + e\,IR'_t = h\nu_{min} = hc/\lambda_{max} \qquad (4.19)$$

or, in terms of the practical cell voltage, V_p:

$$V_p = E_g/e = V_r + IR'_e + \eta_+ + \eta_- + IR'_t = h\nu_{min}/e = hc/e\lambda_{max} \qquad (4.20)$$

As mentioned in Section 4.1, the reversible voltage, V_r, for water electrolysis under standard conditions is 1.299 V, which corresponds to a standard free energy of $\Delta G° = 1.299$ eV.

The energetics of a PEC cell that is performing close to ideal operation for water electrolysis are shown in Figure 4.12(b). For simplicity, the ohmic losses in Equation (4.19) have been set to zero; the inclusion of these would make E_g even larger. With increasing E_g, the absorption edge of the semiconductor will be shifted towards shorter wavelengths [see Equation (4.17)], and thus fewer photons will be absorbed from the sunlight. The band gap of the semiconductor must be at least 1.6–1.7 V for water splitting to be successful, but not over 2.2 V for the efficiency to be acceptable. This requirement rules out most metallic oxides and sulfides. Titania (TiO_2), which was the first semiconductor electrode found to be capable of driving water electrolysis, has a band gap of 3.2 eV (anatase polymorph) and therefore absorbs only ultraviolet light below 385 nm. Although a PEC cell with this electrode can have a fairly high quantum yield in this spectral range, the registered solar energy efficiency is low – at best, around 1%.

It should be emphasized that the use of a PEC cell for water electrolysis imposes a very specific condition on the semiconductor electrode. As shown in Figure 4.12(b), the positions of $E_{VB,t}$ and $E_{CB,b}$ must be selected so that there is sufficient overvoltage to drive the evolution of oxygen and hydrogen. The data presented in Figure 4.13 illustrate how this stipulation excludes a number of semiconductors for effective use in a PEC cell. An external voltage bias can be applied to support the PEC cell, but this will result in a decrease in efficiency. The semiconductor must be stable in aqueous solution over long periods and not be subject to corrosion in the electrolyte. A practical way to collect the generated hydrogen must also be found. Again, these are criteria that are hard to meet.

For a PEC cell to operate successfully, it is necessary for the minority charge carriers (*i.e.*, holes for an *n*-type and electrons for a *p*-type semiconductor) to be sufficiently long-lived to be transferred to the semiconductor|electrolyte interface without recombining with the majority carriers or being trapped at a defect. This requirement for good efficiency implies either that the ideal semiconductor electrode is an ultra-pure single crystal (as in conventional semiconductor technology) or that it is composed of small particles (nm dimensions), *i.e.*, a so-called 'nanostructured electrode'. In the former case, an electrical field is generated in the semiconductor by

Box 4.4 Continued.

redistribution of charge carriers at the electrode|electrolyte interface, *i.e.*, in much the same way as in the *p–n* junction shown in Figure 4.9, and the electrons and holes are separated in the field created. In the alternative form of semiconductor, the nanostructure is porous so that the electrolyte penetrates through the electrode to the back contact, as occurs in a dye-sensitized solar cell. For instance, in a nanostructured *n*-type electrode for the generation of oxygen, the holes have short distances to travel to the electrolyte and are therefore rapidly transferred to the nanoparticle|electrolyte interface to react in accordance with reaction (4.13). Moreover, there is no need for an electrical field to achieve charge separation. Rather, this is determined by the relative lifetime of the electrons and holes. Thus, if the hole is rapidly consumed by reaction (4.13) and the lifetime of the electron in the conduction band is long, then the dissociation of water will be favoured.

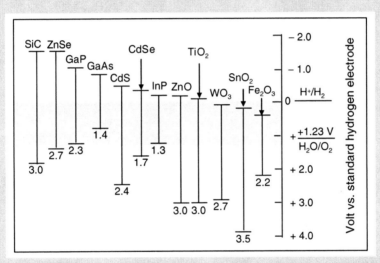

Figure 4.13 Position of valence and conduction bands for several semiconductor materials in contact with aqueous electrolyte at pH 1.[2] (Courtesy of the International Energy Agency Hydrogen Implementing Agreement).

Given that a suitable semiconducting electrode material can be found, what advantages can a PEC system offer compared with an electrolyzer driven by a module of photovoltaic (PV) solar cells? With a PEC system at an estimated theoretical efficiency of 25% (corresponding to $E_g \approx 1.7\,\mathrm{eV}$), the current density will be around $20\,\mathrm{mA\,cm^{-2}}$. As seen from Figure 4.3, the overpotential $(\eta_+ + \eta_-)$ falls off rapidly below $30\,\mathrm{mA\,cm^{-2}}$. Assuming that the overpotential for water electrolysis is the same in both a PEC cell and an electrolyzer, it follows that the required band gap of the chosen photoactive semiconductor in the PEC cell can be kept as low as 1.6–1.8 eV. Compared with an electrolyzer driven by a solar PV module, each PEC cell

Box 4.4 Continued.

in an assembly works independently and thus will allow resistance-free passage of current between it electrodes; see Figure 4.12(b). Moreover, an eventual failure in one cell will not affect the operation of others. These properties derive from the fact that all photons absorbed by the photoactive electrode are consumed within each cell unit for water electrolysis. No external electrical work is done and thus no external connections are needed. (The extreme version of the device is found in nanoparticle suspensions where each semiconducting nanoparticle has a nano-sized platinum electrode on its surface, but difficulties arise in separating the co-generated hydrogen and oxygen.) An applied voltage of about 2.2 V is necessary to operate a PV-driven monopolar electrolyzer (clearly, higher voltages would be necessary with bipolar designs). Since the voltage delivered by a single PV cell is typically in the range 0.6–0.8 V, a minimum of three PV cells must be connected in series to drive the electrolyzer. As a consequence, the efficiency of solar electrolysis per unit panel area is reduced by a factor of at least three. Furthermore, sets of series-connected PV cells must be coupled in parallel to deliver a practical current to the (external) electrolyzer. All the associated wiring would be avoided when using PEC cells. For grid-connected installations, PEC systems offer a further advantage in that PV modules would incur additional electrical losses due to the transformation of the generated d.c. to a.c. and then back to d.c. to run the electrolyzer.

The fundamental principles of the photo-electrochemical cell for light-driven electrolysis of water are well known. Its future as a device for the production of hydrogen from water rests solely on the ability to find the correct semiconducting material for the photoactive electrode.

4.3.4 Photo-biochemical Cells

The operation of tandem cells bears a close similarity to the processes that take place in photosynthesis where there are also two photosystems connected in series. In the first, light is absorbed by chlorophyll and this acts as a mediator to oxidize water to oxygen [reaction (4.21)], while in the second, the organic compound nicotinamide adenine dinucleotide phosphate (NADP) is reduced by electrons to a state generally designated NADPH.

$$2H_2O + 2h\nu \rightarrow 4H^+ + 4e^- + O_2 \qquad (4.21)$$

Along with adenosine 5′-triphosphate (ATP), NADPH is an important intermediary in the photosynthetic fixation of carbon dioxide. The protons and electrons react with carbon dioxide via these two mediators to produce sugars. The overall photo-biochemical process taking place in green plants is represented by:

$$nCO_2 + 2nH_2O + ATP + NADPH \rightarrow n(CH_2O) + nH_2O + nO_2 \qquad (4.22)$$

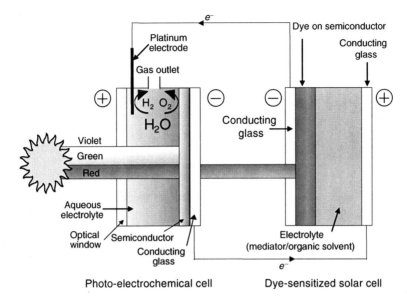

Figure 4.14 Operating principles of a tandem cell for enhanced hydrogen production.

where n is defined according to the structure of the resulting carbohydrate. These carbohydrate products are then variously used to form other organic compounds such as the building material, cellulose.

A few natural bacteria, *e.g.*, blue–green algae (cyanobacteria), have the propensity to decompose water under the action of sunlight and release hydrogen freely into the air. This is a photo-biochemical process, which is often referred to as 'biophotolysis'. In theory, such algae can produce hydrogen with an efficiency of up to 25%, but in practice the performance is very much lower, *i.e.*, often less than 1%. The problem is that, like photosynthesis, this is a tandem PEC process in which oxygen is also released and this inhibits the hydrogen-producing enzyme (hydrogenase), so that only small amounts of hydrogen are actually generated. The role of hydrogenase is to manage the dark-to-light transition that the microorganism faces daily; it catalyzes the reduction of protons by electrons to form hydrogen. Since the enzyme is very sensitive to oxygen and is only synthesized after several hours of dark pre-incubation under anaerobic conditions, the system is biologically not designed for continuous operation. To overcome this limitation, two-stage 'indirect biophotolysis' processes are being investigated, namely photosynthetic fixation of carbon dioxide with concomitant generation of oxygen, followed by dark anaerobic fermentation with evolution of hydrogen. The system proposed by the Hawaii National Energy Institute is shown schematically in Figure 4.15. The main components are the following: (i) an open pond for the production of algal biomass; (ii) an algal settling (harvesting) tank, which becomes anaerobic by endogenous respiration and thereby activates/induces the hydrogenase

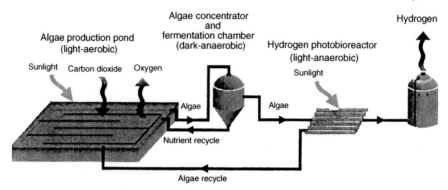

Figure 4.15   Schematic of developmental system for the biophotolytic production of hydrogen.

enzyme; (iii) a fermentation tank for dark production of about one-third of the hydrogen [note: the tanks in (ii) and (iii) are combined in Figure 4.15]; (iv) a photo-bioreactor for completion of the hydrogen production from the remaining stored carbohydrates and extra-cellular fermentation products via a light-driven anaerobic reaction; (v) a gas-separation unit (carbon dioxide from hydrogen); and (vi) treatment, storage and other support systems (not shown in Figure 4.15).

Research is also in progress to moderate the release of oxygen and to evaluate both the general feasibility and cost-effectiveness of such photo-biochemical processes. One avenue of study is the genetic engineering of bacteria to give a species that uses less of the hydrogen for its own metabolism and reproduction, and thus releases more as free gas. Other work is seeking to gain an understanding of how the hydrogenase enzyme functions and then endeavouring to prepare simpler (and more effective) chemical analogues – so-called 'biomimetic' catalysts. This line of investigation is still in its infancy, but holds hope for the future. Finally, the technological development of efficient and inexpensive photo-bioreactor designs will have to be undertaken before such light harvesting can become a practical proposition.

4.4   Thermochemical Hydrogen Production

Because of the stability of the water molecule and the very high temperatures required to split it thermally, as discussed above, attempts have been made to accomplish the process at a more moderate upper temperature (*i.e.*, less than 1000 °C) by means of an indirect route. The general idea is to decompose water by reacting it with one or more chemicals that are regenerated via a series of cyclic thermochemical reactions. In this way, the hydrogen and oxygen evolution steps are separated. Clearly, on practical and efficiency grounds, the fewer reactions involved the better. The system may be viewed as a 'chemical black box' in which the net effect is the conversion of heat and water into hydrogen and oxygen, with the chemicals themselves being regenerated and unchanged at the end of the cycle. This concept is sound thermodynamically (apart from possible side-reactions), but does introduce severe problems in terms of plant engineering and materials compatibility. It has the attraction, however, that the required heat could be provided by a low-carbon source such as a high-temperature nuclear reactor or solar heat. Most of the cycles require heat at greater than 850 °C for at least one step and this rules out conventional water-cooled nuclear reactors; only high-temperature gas-cooled reactors would be suitable and these have yet to be commercialized. Solar heat can, in principle, be obtained at temperatures of 1500 °C or more and so should be suitable for this purpose.

Scientists working in the early 1970s at the European Commission's nuclear laboratory (Ispra, Italy) were among the first to suggest and study the production of hydrogen via thermochemical cycles. They investigated several different schemes with a view to coupling them to a dedicated high-temperature nuclear reactor. One of the preferred candidates was the 'iron–chlorine' cycle, which involves three distinct reaction steps, as follows:

$$6FeCl_2 + 8H_2O \rightarrow 2Fe_3O_4 + 12HCl + 2H_2 \qquad (4.23)$$

Endothermic : 850 °C

$$2Fe_3O_4 + 3Cl_2 + 12HCl \rightarrow 6FeCl_3 + 6H_2O + O_2 \qquad (4.24)$$

Exothermic : 200 °C

$$6FeCl_3 \rightarrow 6FeCl_2 + 3CL_2 \qquad (4.25)$$

Endothermic : 420 °C

Since then, laboratories around the world have taken up the research and have collaborated under sponsorship from both the European Commission and the International Energy Agency. Many different cycles have been proposed and investigated; here we mention just a few as illustrative of this area of endeavour.

A typical early process (known as the 'UT-3 cycle') was based on the use of calcium bromide. The cycle consists of the following four reactions:

$$CaBr_2 + H_2O \rightarrow CaO + 2HBr \qquad (4.26)$$

Endothermic : 725 °C

$$CaO + Br_2 \rightarrow CaBr_2 + 1/2 O_2 \qquad (4.27)$$

Heat neutral : 550 °C

$$Fe_3O_4 + 8HBr \rightarrow 3FeBr_2 + 4H_2O + Br_2 \qquad (4.28)$$

Exothermic : 250 °C

$$3FeBr_2 + 4H_2O \rightarrow Fe_3O_4 + 6HBr + H_2 \qquad (4.29)$$

Endothermic : 575°C

All the individual steps in this cycle have been demonstrated at the laboratory bench level and, to some degree, have been linked together. Many difficulties remain, however, and it is now acknowledged that a four-step process is just too complex when conducting chemical reactions in bulk at temperatures as high as 725 °C. There are problems with respect to reactor design, the handling of acid gases, the filtration and separation of solids, and the corrosion of plant materials. Heat transfer, mass transfer and energy conservation are also difficult when the consecutive reactions have to be carried out at diverse temperatures.

Recent screening programmes have reviewed the numerous proposed cycles to identify those with the greatest potential for large-scale production of hydrogen. Arising from this exercise, the sulfur family of reactions has been selected for priority study. These are all based on the cycling of sulfur between the S(IV) and S(VI) valence states, *i.e.*, between sulfur dioxide/sulfurous acid (SO_2/H_2SO_3) and sulfur trioxide/sulfuric acid (SO_3/H_2SO_4), respectively. The reduction step, common to all schemes, is the high-temperature dissociation of sulfuric acid to water, sulfur dioxide and oxygen, as represented by reaction (4.31) below. This reaction takes place only at 1150 °C when conducted non-catalytically, but can be reduced to 850–900 °C with a suitable catalyst. Various pathways have been proposed to oxidize the sulfur dioxide back to sulfur trioxide and, in the process, liberate hydrogen.

4.4.1   Sulfur–Iodine Cycle

The sulfur–iodine cycle has been studied extensively by the nuclear industry since the mid-1970s. It employs iodine (I_2) to promote the oxidation of S(IV) to S(VI), as follows:

$$I_2 + SO_2 + 2H_2O \rightarrow 2HI + H_2SO_4 \qquad (4.30)$$

Exothermic : 120°C

$$H_2SO_4 \rightarrow SO_2 + H_2O + 1/2O_2 \qquad (4.31)$$

Endothermic : 850–900°C

$$2HI \rightarrow I_2 + H_2 \qquad (4.32)$$

Endothermic : 300–450°C

Two immiscible phases are first formed. The upper phase contains almost all of the sulfuric acid and the lower, dense phase holds most of the hydrogen iodide (HI) and iodine. These are separated and the upper phase is then decomposed via reaction (4.31), while the hydrogen iodide in the lower phase is converted to hydrogen and iodine according to reaction (4.32). The products of these two reactions (SO_2 and I_2, respectively) are then recycled to reaction (4.30).

Pilot-plant operation in Japan identified the following difficulties:

- handling of corrosive acids;
- loss of chemicals in side-reactions and unintentional release of sulfur dioxide from the cycle;
- clogging of pipes due to solidification of iodine;
- shortage of physicochemical data;
- maintenance of a consistent production rate.

Clearly, further studies have to be conducted on the chemistry and chemical engineering aspects of each of the process steps. In particular, a method must be found to reduce the excesses of water and iodine that are presently required over and above the reaction stoichiometry and thereby minimize material flows and attendant energy losses.

4.4.2 Westinghouse Cycle

The Westinghouse cycle is a variant of sulfur–iodine and is a hybrid concept that requires both thermal and electrical energy. When the product of reaction (4.31) is cooled, sulfurous acid is formed and this is electrolyzed at around 80 °C to produce hydrogen and sulfuric acid for recycling, *i.e.*,

$$H_2SO_3 + H_2O \rightarrow H_2 + H_2SO_4 \tag{4.33}$$

A key feature of this process is that the electrolysis of sulfurous acid can theoretically be conducted at a very much lower reversible cell voltage (0.17 V) than that for water (1.23 V) and therefore the theoretical power requirement is greatly reduced. Experimental results from Westinghouse show, however, that electrode and cell losses can significantly increase the energy input so that, for example, operating cell voltages of around 0.55–0.6 V are necessary for electrolyzing a 50 wt.% solution of sulfurous acid. This appears to limit overall cycle efficiency to around 30%, in contrast to the 51% that was first predicted for this method of hydrogen production. It should be stressed that efficiency is a difficult metric to evaluate between various published papers and reports in this field because the underlying assumptions are not always clearly stated.

The decomposition of sulfuric acid at 850–900 °C [reaction (4.31)], a component of all the sulfur cycles, places severe specifications on the materials of plant construction due to the extremely corrosive nature of the species at high temperatures. This is one of the major issues that is being addressed by the European HYdrogen THErmochemical Cycles (HYTHEC) programme. The initiative has been established to evaluate the sulfur−iodine and Westinghouse cycles for the tonnage production of hydrogen.

4.4.3 Sulfur−Ammonia Cycle

The sulfur−ammonia cycle appears to be a previously unexplored cycle that has been put forward by the Florida Energy Centre in the USA. The novel step in the process is the photocatalytic oxidation of ammonium sulfite, $(NH_4)_2SO_3$, to ammonium sulfate, $(NH_4)_2SO_4$. The latter is then used to split water.

$$(NH_4)_2SO_3 + H_2O \rightarrow H_2 + (NH_4)_2SO_4 \quad \text{(photocatalytic reaction)} \quad (4.34)$$

followed by:

$$(NH_4)_2SO_4 \rightarrow 2NH_3 + H_2SO_4 \quad \text{(thermal decomposition)} \quad (4.35)$$

$$H_2SO_4 \rightarrow SO_2 + H_2O + 1/2\ O_2 \quad \text{(thermal decomposition)} \quad (4.36)$$

$$2NH_3 + SO_2 + H_2O \rightarrow (NH_4)_2SO_3 \quad \text{(condensation)} \quad (4.37)$$

While the challenging dissociation of sulfuric acid at high temperature remains, there appears to be some significant simplifications in terms of the chemistry of the remainder of the cycle:

- the photocatalytic oxidation step uses only part of the solar spectrum, so that the balance is available for providing the thermal input to the cycle;
- the thermal decomposition of ammonium sulfate is reported to be achievable at the moderate temperature of 277 °C, *i.e.*, about 50 °C below the boiling point of sulfuric acid; this may be useful with respect to cycle integration and the utilization of waste heat from other parts of the process;
- the dissolution of sulfur dioxide and ammonia in water [reaction (4.36)] is easy to perform due to the high solubilities of these species and their affinity for each other through the formation of a solution of ammonium sulfite.

On the other hand, the test data available to date suggest that a number of matters remain to be resolved before this cycle can realistically be considered as a serious competitor to the sulfur−iodine and Westinghouse alternatives. Further experimental and engineering development work is required to address the following issues:

- validation of the fundamental chemistry underlying the individual unit reactions;
- assessment of whether this scheme is more, or less, susceptible to side-reactions than the other cycles;
- development of effective and stable materials/catalysts for the photocatalytic oxidation of ammonium sulfite that have a broad absorption

band, promote rapid reaction rates and yield efficient hydrogen evolution;
- design of reactors for the photocatalytic step that are capable of being integrated with sources of concentrated solar energy.

In Australia, CSIRO intends to demonstrate the technology through the design of a pilot-scale sulfur–ammonia unit that can be integrated with its established solar tower; see Figure 2.4, Chapter 2.

4.4.4 Metal Oxide Cycles

In an attempt to reduce the number of steps in the thermochemical cycle to just two and at the same time avoid the use of highly corrosive reagents, attention has focused recently on simpler cycles in which water is reduced to hydrogen by metals or by metal oxides in a lower valency state:

$$M/MO_{red} + H_2O \rightarrow MO_{ox} + H_2 \tag{4.37}$$

In the second step of the cycle, the product oxide in its higher valency state is thermally decomposed to liberate oxygen and revert to its reduced form:

$$MO_{ox} \rightarrow M/MO_{red} + 1/2O_2 \tag{4.38}$$

These are examples of so-called 'redox reactions'. The redox couples that have been evaluated consist either of oxide pairs of multivalent metals (Fe_3O_4–FeO, Mn_3O_4–MnO) or systems of metal oxide and metal (*e.g.*, ZnO–Zn). The thermodynamically decisive step is the reduction of the oxidized metal oxide to generate the reduced form, thereby releasing oxygen. Unfortunately, most of the simple cycles require much higher temperatures than the earlier alternatives discussed above and are therefore suitable only for solar furnaces and not for nuclear reactors.

An example of this class of thermochemical cycle is that based on iron oxides:

$$Fe_3O_4(l) \rightarrow 3FeO(l) + 1/2O_2 \tag{4.39}$$

Endothermic : $> 1600\,°C$

$$3FeO + H_2O \rightarrow Fe_3O_4 + H_2 \tag{4.40}$$

Exothermic

The fact that the oxides are liquid at 1600 °C introduces serious processing problems that have to be solved.

The temperature of reaction (4.39) can be lowered by several hundred degrees by replacing magnetite (Fe_3O_4) by nickel manganese ferrite ($(Ni_{0.5}Mn_{0.5})Fe_2O_4$), which is partially reduced to an oxygen-deficient state. In the second step of the cycle, this reduced oxide reacts with water at $\sim 800\,°C$ to regenerate the fully oxidized ferrite and liberate hydrogen. The process was demonstrated within the European Union project HYDROSOL at the solar furnace of the German Aerospace Center, Cologne. The reactor works at a level of $15\,kW_{th}$ in a quasi-continuous way. There are two separate reaction chambers where the two reaction steps are carried out in parallel. This is ongoing and highly challenging research.

Solar–thermal processes and nuclear–thermal processes are likely to be conducted at dramatically different levels of operation. Solar energy is, by its nature, diffuse ($\sim 1\,kW\,m^{-2}$) and can be collected over only a limited area. The output of any individual solar tower will probably be restricted to $\sim 1\,MW$ of heat, at most. By contrast, nuclear reactors tend to have an output of up to $1\,GW$ of electrical energy and rather more of heat. Hence there is expected to be a difference of at least three orders of magnitude in the scale of operation between the two heat sources. This creates difficulties for each of the two cases. For a solar-based operation, because of its modest size, the chemical reaction scheme will need to be exceedingly simple to yield cost-competitive hydrogen and this introduces major limitations of choice. When taking the nuclear route, the requirement to scale up such complex processes by several orders of magnitude, so as to couple to a nuclear reactor, will involve a formidable task in chemical engineering. We must await developments, but at the present time the prospects for developing a commercially realistic set of thermochemical cycles to split water do not appear promising.

4.5 Concluding Remarks

Whereas Chapter 2 focused on the production of hydrogen from fossil fuels, this chapter has discussed the splitting of the water molecule to generate hydrogen. The size of the operation may range from that of a present-day natural gas steam reformer producing $\sim 150\,000\,N\text{-}m^3$ of hydrogen per hour (see Figure 2.1, Chapter 2) to a small electrolyzer providing just a few $N\text{-}m^3$ per hour [Figure 4.4(a)], *i.e.*, five decades difference in output. The source of energy is similarly varied in both type and magnitude – from coal delivered by the trainload or natural gas supplied by a 60 cm diameter pipe at high pressure, to a single wind generator on a farm or solar arrays in a rural village. With such a wide variety of energy sources, levels of operation and diverse possibilities for hydrogen as an energy vector, it is expected that at least some commercial applications will arise.

Apart from photovoltaic electricity, all of the solar-based approaches to water splitting are somewhat futuristic. At this stage of their development, when new materials and concepts are still being investigated, it is difficult to make a valid assessment of their respective prospects. Given that in many situations photovoltaic electricity is not yet affordable, and nor is the use of hydrogen as an energy carrier or fuel, there is a long road ahead before electrolytic hydrogen from a solar source becomes competitive with conventional fuels. Nevertheless, there are many avenues of materials science and catalysis to be explored and

Table 4.2 Summary of hydrogen production routes.

Energy source	Technology	Positive features	Negative features
Natural gas	Steam reforming (benchmark technology)	• Proven technology • Large scale • Cheap feedstock	• Air emissions • Limited reserves of natural gas
Oil derivatives	Partial oxidation	• Proven technology • Wide range of feedstocks	• Air emissions • More costly than natural gas
Coals	Gasification	• Proven technology • Reserves plentiful • Cheap feedstock	• Air emissions • Need for sequestration of carbon dioxide • High sulfur levels in coal
Renewables (wind, solar PV, hydro, tidal, geothermal)	Water electrolysis Photo-electrochemistry	• Small/medium scale • Negligible emissions • Distributed generation • Versatile energy options • Produces high-purity hydrogen	• High costs • Little scope for large-scale production with wind and solar
Nuclear electricity	Water electrolysis	• No emissions • Large scale	• Doubts about scaling-up electrolyzers and the future of nuclear power
High-temperature heat (solar or nuclear)	Thermochemical cycles	• Potential route to hydrogen with no emissions	• Highly speculative prospects – both technically and commercially
Biomass	Gasification	• Reserves plentiful • Proven technology • Cheap feedstock • Small net emissions of carbon dioxide	• Moderate scale only • Doubts about sustainability of energy crops
Algae	Biophotolysis	• Widespread resource	• Immature – not proven • Unknown costs

time in which to conduct this fundamental research while sustainable energy sources are being introduced on an ever-growing scale.

At present, tonnage quantities of hydrogen are used in petroleum refining and in chemical and fertilizer manufacture. This chemical market for hydrogen will expand significantly as refineries are progressively forced to move to lower grade and 'dirtier' fuels, and as the demand increases for highly refined, sulfur-free products. As an energy vector for storing renewable energy, hydrogen will be derived from a multitude of comparatively small units. It is anticipated that it would then be supplied to fuel cells, either to provide power (and possibly heat) to individual buildings or to propel electric vehicles. Although each individual unit may be small, the collective market would become substantial if the technology were to be commercialized. A summary of the various routes for producing hydrogen together with their positive and negative features is given in Table 4.2. Inevitably, this is a cursory survey.

Each potential application of hydrogen needs to be evaluated with respect to its practicality, reliability, sustainability and cost. Some of the detailed criteria to be employed in such an assessment are as follows:

- required scale of hydrogen production;
- the purity demanded by the application;
- ease and flexibility of operation;
- compatibility with existing infrastructure and systems;
- stability and durability of components;
- energy efficiency (reflects on both operational and carbon dioxide disposal costs);
- environmental impact;
- safety and hazard management;
- possible markets for by-product oxygen (produced by electrolysis);
- affordability (capital and running costs) in relation to competitive technology.

Many of the suggested openings for hydrogen that appear to be impractical or too costly today many prove otherwise in 50 or 100 years time. A solution that is non-viable in one region of the world, or is impractical for a particular end-use, may well be acceptable in other circumstances. Proponents of hydrogen energy should aim first to seek out opportunities where there exists a favourable juxtaposition of need, available technology and economics. In this way, it may prove possible to inaugurate the concept of hydrogen as both an energy vector and a sustainable fuel.

References

1. *The Hydrogen Economy: Opportunities, Costs, Barriers and R&D Needs*, National Academies Press, Washington, DC, 2004.

2. A. Luzzi, L. Bonadio and M. McCann, *In Pursuit of the Future: 25 Years of IEA Research Towards the Realization of Hydrogen Energy Systems*, International Energy Agency Hydrogen Implementing Agreement, International Energy Agency, Paris, June 2004.
3. *Basic Research Needs for Solar Energy Utilization*, Report of the Basic Energy Sciences Workshop on Solar Energy Utilization, 18–21 April 2005, Office of Science, US Department of Energy, Washington, DC, 2005.
4. S. Licht, *Int. J. Hydrogen Energy*, 2005, **30**, 459–470.

CHAPTER 5

Hydrogen Distribution and Storage

To be useful as a future fuel, hydrogen – manufactured by whatever means and from whatever source – has to be conveyed to the point of use and stored there until required. The distribution and storage of hydrogen are intricately bound together and depend on both the scale of operations and the intended application. In general, the storage of hydrogen for stationary applications (*e.g.*, for heating and air conditioning of buildings, dispersed electricity generation and industrial processes) is simpler than that needed for the propulsion of various forms of road transportation, where there are more severe constraints in terms of acceptable mass and volume, speed of charge–discharge and, for some storage systems, heat dissipation and supply. Finding a satisfactory solution for the on-board containment of hydrogen is one of the major challenges facing the development of fuel cell vehicles as an integral component of a projected Hydrogen Economy.

5.1 Strategic Considerations

One of the key issues for the widespread introduction of hydrogen is whether the gas will be manufactured in a limited number of large central plants or in a multitude of strategically located, small units. In reality, each approach has its merits as determined by the primary energy source employed (fossil or renewable), the use envisaged for the hydrogen and the amounts required. Centralized production is most appropriate for large fossil-fuel steam reforming or coal gasification operations. In principle, it allows the possibility of sequestrating the carbon dioxide that is formed as a by-product. Tonnage quantities of hydrogen will then have to be supplied to customers, either as a cryogenic liquid (liquefied hydrogen, LH_2) by tanker delivery or as a gas through a newly installed national grid of hydrogen pipelines – both of which are expensive procedures. A simpler transitional option, as mentioned in Section 2.1, Chapter 2, may be to blend hydrogen produced centrally (with sequestration) into natural gas in existing pipelines, thereby reducing the overall carbon-to-hydrogen ratio of the transported fuel. The resulting gas mixture is known as 'Hythane®'. The viability of this method of conveyance would be determined by (i) technical parameters concerning the embrittlement of steel pipes, leakage from gaskets, flanges, valves, *etc.*, and the development of practical dispensers and (ii) the calorific value and burning characteristics of Hythane® in existing gas appliances. It should be pointed out, however, that domestic gas was originally derived from coal (so-called 'coal' or 'town' gas). Composed of around 50 vol.% hydrogen and 35 vol.% methane, this was delivered both safely and satisfactorily through steel pipes. When natural gas was introduced, all domestic burners had to be modified to accommodate the higher calorific value of this new fuel. Insofar as Hythane® resembles coal gas, it might then be necessary to re-convert burners to accept lower calorific gas, which would be a retrograde step.

Localized, small-scale production of hydrogen is best suited to using renewable energy to generate electricity that is then employed to electrolyze water. Initially, this would be a limited option, but in the longer term, as renewables assume a greater share of the market, it may become more important. These forms of energy are themselves often well scattered (wind turbines, solar panels, *etc.*) and give rise to little or no carbon dioxide. The disadvantage of this route is that the hydrogen is likely to cost significantly more than that produced in bulk from fossil fuels. It is possible to construct small natural gas reformer units, appropriate for the distributed supply of hydrogen, but again the economies of scale would be lost. The question also arises over what to do with the carbon dioxide produced in such local production units. There appears to be no single answer to the dilemma of making hydrogen readily available across a nation at low, competitive cost with accompanying sequestration of carbon dioxide. The best approach will vary both with the location and with the application for the hydrogen. Moreover, as the availability and price of fossil fuels changes progressively with the passage of time, so too will the most economic technology. Another important factor will be the evolution of public opinion towards carbon dioxide emissions and the regulations or financial penalties that may be imposed for carbon release. A world-wide renaissance of nuclear power would be another relevant consideration. Altogether, this is an imponderable situation and all energy options must be kept open for the present.

Having regard to these imponderables, the USA has advanced a matrix scenario based on three levels of hydrogen production, namely, in large central facilities (1080 t per day), mid-sized plants (21.6 t per day) and distributed units (0.43 t per day).[1] This represents a factor of 50 in size between each of the three categories. The matrix is completed by assigning various feedstocks or primary energy sources and the associated production technologies to the different scales of operation; an outline is given in Table 5.1. We might also add solar–thermal to the mid-sized or distributed options, although both this and the large-scale nuclear route depend upon future advances in thermal water splitting; see Sections 4.3 and 4.4, Chapter 4. Other speculative solar options for dispersed hydrogen generation are photo-electrochemical cells and photo-biochemical systems; see Section 4.3, Chapter 4; these were presumably discounted by the US academies that compiled the above report.

Table 5.1 Assignment of feedstock or primary energy source, together with production technologies, to various scales of hydrogen production.[1]

Feedstock or primary energy source	Central plant (1080 t H₂ daily)	Mid-sized plant (21.6 t H₂ daily)	Distributed plant (0.43 t H₂ daily)
Natural gas	Steam reforming	Steam reforming	Steam reforming
Coal	Gasifier		
Nuclear energy	Thermal splitting of water		
Biomass		Gasifier	
Solar photovoltaics			Electrolysis
Wind			Electrolysis
Grid-based electricity			Electrolysis

The US committee charged with developing a framework for evaluating a Hydrogen Economy[1] then proceeded to calculate the costs involved in manufacturing the hydrogen and in handling the accompanying carbon emissions when using existing and projected technologies, with or without carbon sequestration. In brief, the 'current technology' costs for central production facilities (natural gas and coal) were estimated to be US$2.0–2.2 per kg of hydrogen delivered. Around half of that sum was for production and the remainder for distribution and dispensing. By comparison, corresponding costs for the various technologies in mid-sized and small-scale operations were approximately three times as high, with the exception of (i) natural gas reforming at the local level, which was estimated to be only about 80% higher than that of central reforming, and (ii) distributed solar-photovoltaic/electrolysis backed up by grid-based electricity, which was over four times as expensive as the base case, *i.e.*, US$9–10 per kg H₂. (Note: as stated in Section 2.1, Chapter 4, the US government target for electrolytic hydrogen in 2010 is US$2.85 per kg, so there is little prospect of photovoltaic electricity meeting this goal.) Bearing in mind that the present-day cost of hydrogen obtained from large natural gas steam reformers is significantly greater than that of fossil fuels themselves, it is clear that much development work has still to be undertaken if hydrogen is ever to become competitive as a fuel.

An interesting forecast arising from the US study[1] was that carbon sequestration would constitute only a minor part of the overall cost and would probably be almost balanced by that of purchasing the carbon emission permits which would otherwise be needed. Looking into the future, it was concluded that projected advances in production technology would bring about some significant reductions in the cost of hydrogen per kilogram delivered, but in no case would medium- to small-scale generation compete with large operations. This conclusion was especially true for electrolytic hydrogen based on renewable electricity. Of course, these predictions relate to US conditions in 2004 and may well vary from country to country and from time to time, as dictated by the availability and price of fossil fuels. Nevertheless, the findings probably apply fairly generally and do provide some indication of the task ahead if hydrogen, particularly when obtained through the use of renewable energy, is to become competitive in an era when fossil fuels are still relatively plentiful.

A further factor has to be considered, namely, the purity of the hydrogen. As a fuel for internal combustion engines, purity is not a prime consideration and hydrogen from almost any source will be suitable, provided that sulfur is removed. With low-temperature types of fuel cell, however, purity is a critical parameter since the electrocatalysts are subject to poisoning by many contaminants, several of which are found in fossil fuels; see Section 6.3, Chapter 6. In this regard, hydrogen produced by the electrolysis of water is much purer and may prove to be the preferred source for this application, despite its higher cost.

5.2 Distribution and Bulk Storage of Gaseous Hydrogen

When hydrogen is produced at a central facility, it can either be stored in bulk before dispatch to customers or first distributed and then stored locally on-site until needed. Whatever form of delivery is employed, whether as gas in cylinders, by pipeline or as liquid hydrogen, it is first necessary to compress the gas. To do this, work has to be done and energy expended. Hydrogen may be compressed adiabatically (*i.e.*, with no heat exchange with the surroundings) or isothermally or in an intermediate multi-stage hybrid process. The energy required as a percentage of the higher heating value (HHV) of hydrogen is given in Figure 5.1 for various pressures up to 80 MPa. The latter is the approximate pressure at which the gas occupies the same volume as liquid hydrogen. Broadly, the energy of compression lies between 5 and 15% of the

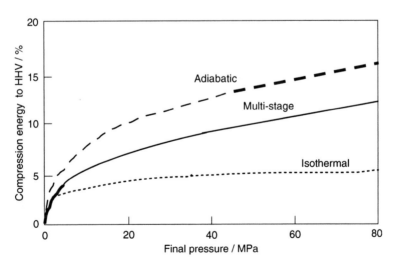

Figure 5.1 Energy required to compress hydrogen as a percentage of the energy content of the gas.[2]
(Courtesy of the European Fuel Cell Forum).

HHV, as determined by the final pressure and whether the process is carried out adiabatically or isothermally. In practice, a multi-stage process will probably be adopted, with an input energy equal to about 10% of the HHV. To this must be added electrical losses in the electricity supply system (say, 5%) and mechanical losses in the compressors (another 5%). Thus, in round numbers, the total energy wasted will be ~20%. Since this loss is in the form of electricity, the overall inefficiency in terms of primary energy will be far greater; its actual value will depend on the type of power plant that is producing the electricity. For hydrogen liquefaction, the energy deficit is even more severe because of the heat evolved in the ortho−para-hydrogen conversion (see Section 1.5, Chapter 5) and through evaporation due to heat gain from the surroundings.

5.2.1 Gas Cylinders

Large industrial users of hydrogen store the gas in substantial cylindrical tanks at comparatively low pressures (5−7 MPa); an example at a steam reforming plant is given in Figure 5.2(a). At these pressures, the compression energy equates to 6−7% of the energy stored in the hydrogen. Because of the low volumetric energy density of hydrogen (12.8 MJ N-m^{-3})[†], the comparatively low pressures employed and the consequent size and mass of the tanks, such storage facilities are static installations.

For transport and distribution purposes, hydrogen is stored in smaller cylinders at higher pressures. Tube trailers are used to supply industrial users of merchant hydrogen, where the quantities involved are considerable; an example is shown Figure 5.2(b). The cylinders are a permanent fixture on the vehicle and are discharged *in situ*, not off-loaded. Smaller users, such as laboratories, employ cylinders with storage capacities of just a few cubic metres (pressurized to 20 MPa) that can be manhandled with simple trolleys.

Industrial consumers of bought-in hydrogen usually store the gas in banks of vertical or horizontal cylinders at pressures in the range from 20 MPa up to a maximum of 80 MPa. Such high pressures necessitate thick-walled and heavy steel containers. In the past, failures have been experienced through hydrogen embrittlement, which induces stress fractures in the steel. Cylinder manufacture is now therefore subject to strict standards and codes of practice. This form of storage is modular with little economy of scale.

For portable and mobile applications (*e.g.*, employing hydrogen as a fuel for road vehicles), where both weight and volume are vital considerations, it is desirable to have storage densities as high as possible. Conventional steel cylinders are not satisfactory with respect to either of these two criteria; they are limited in terms of both their gravimetric and volumetric energy densities. Accordingly, a new family of cylinders has evolved in which modern lightweight materials take the place of steel. The basic principle is to use an internal liner, often aluminium, that is strengthened by an outer winding of

---

[†] This relates to the HHV for hydrogen and to gas volumes at normal temperature and pressure (NTP), *i.e.*, 273.15 K and 101.325 kPa.

(a)

(b)

Figure 5.2   (a) Low-pressure, large-volume, static hydrogen tanks.
(Courtesy of Linde AG).
(b) Tube trailer for distribution of hydrogen gas in bulk.
(Courtesy of Air Products and Chemicals, Inc.).

fibre-reinforced composite, *e.g.*, fibres of various grades of glass, aramid or carbon embedded (either singly or in combination) in different resin matrices. For the highest pressures, carbon fibres are generally employed. The cylinder is protected by an impact-resistant external shell, also made from a fibre–resin composite. There are two principal designs. The most common form – due to its ease of fabrication and, therefore, lower cost – is the 'hoop-wound' arrangement, in which the composite wrapping is applied only along the parallel section (vertical walls) of the cylinder. A photograph of a tractor trailer loaded with vertical, hoop-wound composite cylinders for the delivery of merchant hydrogen is shown in Figure 5.3(a). These cylinders have a storage capacity of $0.25 \, \text{N-m}^3 \, \text{kg}^{-1}$ ($3.17 \, \text{MJ} \, \text{kg}^{-1}$), which is more than a 50% improvement on conventional steel counterparts. In the alternative 'fully-wound' configuration, which is designed to withstand the highest pressures, the liner is made from a polymeric material of high molecular weight and is strengthened by wrapping the cylinder ends, and also the sidewalls, with a continuous filament of composite. The latter technology has been adopted by Quantum Technologies for its TriShield™ all-composite cylinder. A schematic of the design, together with its attributes, is presented in Figure 5.3(b) and a unit fitted to a car is shown in Figure 5.3(c). These containers can sustain an internal pressure of at least $55 \, \text{MPa}$, which offers a theoretical volumetric energy density of $6.9 \, \text{MJ} \, \text{N-dm}^{-3}$. When such cylinders are close-packed together, the effective energy density of the assembly (taking account of the space between the units) reduces to a practical value of 3–$4 \, \text{MJ} \, \text{N-dm}^{-3}$.

Despite the above advances, the intrinsic problems with present-day composite cylinders are their high cost, the energy penalty of compressing gas to very high pressures (together with the associated problem of cooling the compressed gas) and safety issues such as rapid loss of hydrogen in the event of an accident. In addition, improvements in technology are required to meet the US targets for storage tanks in fuel cell cars of 5.4 and $9.7 \, \text{MJ} \, \text{N-dm}^{-3}$ by 2010 and 2015, respectively; see Section 5.6. The corresponding US gravimetric specifications of 7.2 and $10.8 \, \text{MJ} \, \text{kg}^{-1}$ will be even more difficult to achieve. Development work on composite cylinders for hydrogen storage and distribution is continuing with the focus on new synthesis techniques for producing high-strength, low-cost fibres and matrices (binders) that are stronger and totally impermeable to hydrogen. The US industry is standardizing on a $10\,000$ psi ($\sim 70 \, \text{MPa}$) cylinder with a target weight of 110 kg. When this can be achieved, it will provide a storage capacity of $0.70 \, \text{N-m}^3 \, \text{kg}^{-1}$, which is equivalent to $8.8 \, \text{MJ} \, \text{kg}^{-1}$.

Conventional vessels made from high-tensile steel are not suitable for storing hydrogen on road vehicles. Even at the maximum pressure of $80 \, \text{MPa}$, which necessitates special cylinder manufacture, the theoretical value for the volumetric energy storage ($10 \, \text{MJ} \, \text{dm}^{-3}$) is approximately that of liquid hydrogen, but does not start to compare with that of methane at the same pressure ($32 \, \text{MJ} \, \text{dm}^{-3}$), petroleum ($\sim 34 \, \text{MJ} \, \text{dm}^{-3}$) or liquid propane ($25 \, \text{MJ} \, \text{dm}^{-3}$).[2] Moreover, these ultra-high-pressure steel cylinders are far too heavy. Undoubtedly, compressed hydrogen gas poses a major storage problem for

(a)

(b)

Impact-resistant dome
• lightweight
• energy absorbing
• cost-competitive

Manual valve or electrical valve
or in-tank regulator

Compressed H₂

Polymer liner
• lightweight
• corrosion resistant
  (hydrogen embrittlement)
• permeation barrier
• flexible in size
• cost-competitive

Carbon-fibre reinforced shell
• corrosion resistant (acids, bases)
• fatigue/creep/relaxation resistant
• lightweight

Reinforced external protective shell
• gunfire safety
• impact safety
• cut/abrasion resistant

(c)

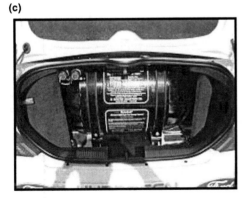

Figure 5.3   Composite cylinder storage of compressed hydrogen.
(a) Hoop-wound cylinders on a trailer.[3]
(Courtesy of the Materials Research Society).
(b) Schematic of fully-wound cylinder; (c) fully-wound cylinder installed in
a car (TriShield™ technology).
(Courtesy of Quantum Technologies Worldwide).

Table 5.2   Hydrogen storage benchmarks for road vehicles[a].

	Gravimetric target		Volumetric target	
Organization	*wt.%*	*MJ kg^{-1}*	*kg dm^{-3}*	*MJ dm^{-3}*
US Department of Energy				
2010 target	6.0	7.2	0.045	5.4
2015 target	9.0	10.8	0.081	9.7
International Energy Agency	5.0	6.0	0.050	6.0
Japan	5.5	6.6	–	–

[a] It is assumed that these target figures are for mid-sized cars and light-duty commercial vehicles to provide a driving range of 300 miles (480 km).

transportation applications in terms of both weight and volume. Even if the above-mentioned lightweight designs made from carbon-fibre composites are employed, the cylinder volume required for a given vehicle range would still be many times greater than that of a conventional petrol tank holding a corresponding amount of energy. And, of course, the weight would also be much greater. Several official bodies have set targets for hydrogen storage vessels to be employed on road vehicles, especially cars, so as to provide a driving range of 300 miles (480 km) between refuelling stops. Some of these are summarized in Table 5.2.

Some calculations have clearly demonstrated the poor efficiency of distributing energy as compressed hydrogen, at 20 MPa in standard steel cylinders, compared with petroleum in tankers. For example, taking in each case a truck with a gross weight of 40 t and assuming some technical development in steel gas cylinders, the hydrogen that can be delivered would contain 57 GJ of energy, whereas an equivalent petroleum tanker would carry 1252 GJ, *i.e.*, 22 times as much. In other words, for every petrol tanker on the road that is transporting fuel there would need to be 22 hydrogen trucks. For typical delivery distances of 100–400 km, the diesel fuel consumed by the truck would equate to 20–80% of the energy contained in the hydrogen. Apart from the energy efficiency and cost considerations, the larger number of heavy vehicles would add to congestion and emissions on the highway. These figures would be reduced by a factor of up to 3.5 if high-pressure (70 MPa), lightweight, composite cylinders were used routinely. Such designs have the potential to meet gravimetric targets, but not volumetric ones; in short, the cylinders are still too bulky.

A novel method to improve the amount of hydrogen that can be stored in composite cylinders involves 'cryo-compression' of the gas. This depends on the fact that gases are denser at cryogenic temperatures than at ambient temperature. Also, they adsorb more readily on to materials with high surface areas. A medium-pressure composite cylinder (20–40 MPa) is filled with activated carbon as an adsorbent and then enclosed in an insulated jacket of liquid nitrogen (77 K). Compressed hydrogen is introduced into the cylinder where it cools, densifies, and is adsorbed on the surface of the carbon. The resultant storage capacity is several times that of the same cylinder at ambient temperature.

Of course, the cylinder must be kept at 77 K at all times, which necessitates regular topping up of the system with liquid nitrogen. Failure to do so would result in activation of a pressure release valve or, worse, explosion of the cylinder through progressive heating.

In the context of hydrogen-fuelled vehicles, research is also being undertaken on the use of hollow glass microspheres, of diameter 5–200 μm and wall thickness 0.5–20 μm, to store hydrogen at high pressure. The technique involves three steps: charging, transferring and discharging. First, the glass spheres are placed in a pressure vessel and charged with hydrogen (at 35–70 MPa pressure and at elevated temperature, ~300 °C) by permeation through their walls. Next, the microspheres are cooled to room temperature and transferred to the low-pressure vehicle tank. Finally, the microspheres are heated to 200–300 °C for controlled release of hydrogen to run the vehicle. At present, this method of storage is little more than a concept. The main problems anticipated would be the difficulty of producing uniform microspheres, the slow leakage of hydrogen at ambient temperatures and the need for a suitable on-board heat supply to release the hydrogen. In addition, the glass shows a tendency to disintegrate during charge–discharge cycling. Nevertheless, there is scope for further research.

5.2.2 Pipelines

Natural gas pipelines cannot be used safely to distribute pure hydrogen without first being modified to ensure against progressive degradation of the materials employed in the construction of these networks. Hydrogen is not compatible with high-carbon steel, or in fact with high-density polyethylene, both of which have been utilized in natural gas pipelines. The resilience of other materials incorporated in these installations should also be considered. Examples are brass for valves, natural and synthetic rubber for mechanical joint seals and meter diaphragms, lead and jute for sealing compounds, and cast aluminium for meter housings and regulator parts. When high-carbon steel pipelines are subjected to pressure cycling in pure hydrogen, embrittlement and decarburization of the steel and welds can occur that, in turn, lead to an accelerated propagation of cracks. This phenomenon can be prevented by internal coating of the pipe or by the addition of small quantities of other gases (carbon monoxide, sulfur dioxide or oxygen). The latter approach is, however, self-defeating if the object is to convey pure hydrogen. The alternative is to manufacture special pipelines from low-carbon steel, which is not subject to attack by hydrogen. Since welds are particularly susceptible, technology that eliminates entirely the need for joining sections together (*e.g.*, 'spoolable' composite pipes) may also help solve the embrittlement problem.

Several factors make hydrogen pipelines more expensive to install than those for the transport of natural gas. The hydrogen molecule is smaller than the methane molecule and therefore diffuses more readily through materials, which necessitates the use of special gaskets and flanges. Since the volumetric energy content of hydrogen at NTP ($12.8\,MJ\,N\text{-}m^{-3}$) is much lower than that of methane ($39.7\,MJ\,N\text{-}m^{-3}$), a pipe of larger diameter is necessary to convey the same amount of energy. Alternatively, the same gauge of pipe could be selected, but with three times the gas velocity, which implies a requirement for larger and more powerful compressors. Over a long distance, there will also be a need to re-pressurize regularly, perhaps every 100 km or so; this will represent a further loss of energy and hence will add to the operating overheads. For chemical uses of hydrogen, the energy consumed in pumping the gas is of no great significance since its cost is absorbed in that associated with manufacture of the industrial product. By contrast, efficient and affordable delivery is crucial for the widespread deployment of hydrogen as a fuel.

Natural gas is transmitted across country at high pressure (3.5–10 MPa) and then the pressure is dropped to near-ambient (0.1 MPa) for local distribution in pipes of lesser specification. Many of the potential applications for hydrogen will demand gas at higher pressure, *e.g.*, for re-fuelling vehicles. Such re-pressurization would naturally have a negative impact on the overall economics. If this is not done, however, the existing local gas network would be unsuitable for delivering hydrogen at 2–3 MPa pressure, as it was not designed for this purpose and would therefore have to be replaced by a new infrastructure made from hydrogen-compatible materials. It has been estimated that the capital investment required for a new hydrogen network would be 40–50% higher than that of an equivalent system for natural gas. The operating costs per unit of energy delivered would also be around 50% higher, although still relatively low compared with the hydrogen production costs. Given the above issues, it is concluded that existing natural gas grids are best kept for their intended purpose, although they might conceivably be employed for distributing Hythane® or syngas, but only at low pressure.

Some local or regional hydrogen delivery systems already exist in various parts of the world to convey hydrogen from one works to another for use in chemical operations; see Table 5.3.[4] Altogether, there are approaching 1500 km of hydrogen pipelines in Europe and at least 1100 km in North America, in addition to smaller networks in China, Singapore, South Korea and Thailand. Clearly, there is now considerable experience of transporting hydrogen at high pressure safely through custom-built pipelines and the requisite technology is

Table 5.3   Some existing hydrogen pipelines.[4]

Company	Location	Hydrogen flow/Mm³ daily	Pipe Length/ km	Pipe Diameter/ cm	Pipe Pressure/ MPa	Hydrogen purity/%	Years of operation
Air Liquide	France Belgium Netherlands	0.48	1100	10	10	99.995	Since mid-1980s
Chemische Werke Hüls	Ruhr, Germany	2.8	220	10–30	2.4	95	Since 1938
ICI	Teesside, UK	0.57	16		5.0	95	Since 1970s
Praxair	Texas	2.8	480	20			Since 1970s
Air Products	Texas Louisiana	1.1	415	10–30	0.3–5.3	99.5	Since 1970s

well established. Nevertheless, codes and standards for hydrogen pipelines must be further developed to ensure adequate safety and to simplify the process of obtaining permits. Improved leak detection or sensor technology will be essential to ensure safe operation and conformance to standards.

5.2.3 Large-scale Storage

The large-scale storage of hydrogen draws on experience with natural gas, which is often stored seasonally in depleted oil and gas fields, as well as in aquifers. This is a high-volume, low-cost option. The ability of a field to act as a reservoir is dependent on the nature of the rock strata. Porous, permeable rock is essential, together with a fine-pore 'cap' filled with water to retain the gas. The conditions are much the same as those for holding carbon dioxide (see Section 3, Chapter 3), although the storage periods for natural gas are obviously much shorter, namely, days or months rather than millennia. The object in storing natural gas underground is to meet fluctuations in demand while maintaining primary production at a steady rate. Consumption of gas in winter may be at least twice that in summer. Generally, the storage site is chosen to be close to the end-users so that it provides a buffer that can be drawn down at short notice. The facility then smoothes the demand – both on the production wells, which may be at a considerable distance from the consumers, and on the pipeline that conveys the gas from field to market.

Short-term storage in depleted gas fields is considerably cheaper than in aquifers, for the following reasons. The geology has already been established and the field surveyed. There is confidence that the gas will not leak from the store without the requirement for a detailed appraisal. Since the reservoir will already be full of gas at low pressure, there will be no need to provide an initial charge (so-called 'cushion gas') so that most of the gas that is injected can be recovered. Finally, much of the necessary equipment and the transport pipeline are likely still to be in place. By contrast, aquifers provide none of these advantages and much work will be involved in their evaluation and inauguration as gas stores.

Hydrogen is rather more difficult to contain than natural gas on account of the smaller size of its molecule and its higher diffusion coefficient, both of which are factors that would tend to facilitate escape from the store. Fortunately, the pores in the caprock are sufficiently fine that the water they retain is not readily displaced, provided that the gas pressure is not excessive. The high diffusion coefficient of hydrogen should even assist the filling and emptying of the reservoir.

Another form of underground store is a cavern in rock-salt strata, either formed naturally or by solution mining. The latter process involves sinking two pipes to the appropriate level. Water is sent down one pipe to dissolve out the salt and the resulting brine is returned via the other pipe to the surface, where it is evaporated in lagoons. Rock salt strata offer the advantage of being almost impermeable to gases. For many years, Imperial Chemical Industries stored

95% pure hydrogen at 50 MPa pressure in old salt mines at Teesside, UK. These were formed by the leaching of 20–30 m thick layers of rock salt at a depth of 350–600 m. Salt caverns can be filled and emptied much faster than stores that rely on permeation through porous rock, a distinct operational advantage. On the other hand, they suffer from the same limitation as aquifers, namely, the initial charge of gas at atmospheric pressure (cushion gas) is irrecoverable.

The concept of using a 'lined rock cavern' for the storage of natural gas in areas where salt deposits are not available has been under development in Sweden for more than 10 years. This is an engineered structure that is hewn out of the rock and takes the form of a vertical cylindrical cavity that is 20–50 m in diameter, 50–115 m in height, and lined with reinforced concrete of ∼1 m thickness to provide strength. Mild-steel plates (thickness 12–15 mm) are attached to the inside of the concrete and welded together so as to provide an impermeable gas barrier. The cavern serves as a pressure vessel to hold natural gas at 15–25 MPa. A demonstration facility of 40×10^3 m^3 capacity has been established at Skallen and has been in commercial operation as part of the Swedish gas grid since early 2004. Two far larger structures are being built in the USA to contain 74×10^6 and 148×10^6 m^3 of natural gas, respectively. The principal advantage of rock caverns is that they can be constructed in regions where depleted gas/oil fields, aquifers and salt domes are not found, so permitting flexibility of storage nearer to the market. The possible use of such caverns as repositories for hydrogen has yet to be explored.

In former days, coal gas was stored in 'gasometers', which acted as reservoirs between the gasworks and the cities that they served. These devices are still in use to some extent for natural gas. A gasometer is a large-volume, inflatable container that floats on a water seal. As the gas is held at atmospheric pressure only, the storage capacity is strictly limited. Given the success of this storage method for coal gas that contains ∼50 vol.% hydrogen, similar low-cost units should prove suitable for the local storage of the latter gas alone. On the other hand, many people would see these installations as 19th century eyesores and it may therefore prove difficult to obtain planning permission to erect them close to residential centres. Large-volume cylinders storing gas at high pressure (*e.g.*, of the type shown earlier in Figure 5.2) would be less visually intrusive, especially if mounted horizontally.

5.3   Liquid Hydrogen

Because of its low density (70.8 g dm^{-3}), liquid hydrogen has a volumetric energy density (10.1 MJ dm^{-3}) at its boiling point (20 K) that is poor compared with its gravimetric energy density (142 MJ kg^{-1})[‡]. The problem of storage is therefore one of bulk rather than mass.

Large-scale liquefaction of hydrogen was developed in the USA in the 1970s to provide fuel for space rockets. The low specific mass of hydrogen is vital in

---

[‡]These data refer to the higher heating value (HHV) of hydrogen.

Figure 5.4 LH$_2$ production facility; road tanker for conveying LH$_2$ to user. (Courtesy of Linde AG).

this application. The National Aeronautics and Space Administration (NASA) required a continuous supply of liquid hydrogen (LH$_2$) as fuel for the Apollo Moon landings and subsequent missions. A spherical, vacuum-jacketed container, 20 m in diameter, was built to hold over 3×10^6 litres of the cryogenic liquid, which was shipped across the country from the production site to Cape Canaveral in cryogenic railcars or road trucks. Nowadays, LH$_2$ is transported primarily by road in tankers with capacities of 40 000–60 000 litres (Figure 5.4). Modern trucks can carry up to 10 times more hydrogen than the tube trailers used for conveying compressed gas. The cryogenic liquid may also be delivered via inland waterways by barge or sent overseas in tank ships. The latter are similar to LNG tankers, apart from the fact that even better insulation is needed to minimize losses through evaporation.

The liquefaction of hydrogen is a multi-step process. The feed supply of hydrogen, which should have a pressure of at least 2 MPa, is first cooled to liquid nitrogen temperature (77 K). At this temperature, most impurities are condensed out; alternatively, they may be removed in a preliminary treatment by means of pressure swing adsorption; see Section 2.3, Chapter 2. The hydrogen gas is then cooled further, by multiple expansion through nozzles (Joule–Thomson cooling)[§], to 20 K where LH_2 condenses out. Often, five or six expansion stages are required to reach this temperature. By means of magneto-caloric processes[¶] conducted at these low temperatures, ortho-hydrogen (parallel nuclear spins) is transformed to para-hydrogen (anti-parallel nuclear spins), which is a lower energy state; see Section 1.5, Chapter 1.

The cryostats that hold the LH_2 are double-walled and are insulated by vacuum enclosures that contain multiple silvered reflectors, which are separated from each other by thermal insulators. The latest designs are capable of holding LH_2 for up to 2 weeks with very little loss, as determined by their size, design and number of insulating layers. Inevitably, liquefaction plants and storage tanks of such sophistication are expensive to purchase. The construction cost of facilities with outputs of 4–5 t per day is said to be around US$25–30 million. Two of the major manufacturers are Air Liquide and Linde. The large operations that service NASA are capable of producing 15 000–35 000 litres of LH_2 per hour (25–60 t per day). Other plants of varying capacity have been built in Europe, Japan and India.

The energy required to liquefy hydrogen is considerable. For large-scale production, the input is equivalent to around 30% of the energy value of the hydrogen itself and rises to an even greater percentage for smaller operations. This, together with the high capital cost of the cryogenic plant, is a major disincentive to conveying and storing hydrogen as a liquid. To be balanced against these limitations is the advantage that a far greater quantity of hydrogen can be carried compared with compressed gas and fewer delivery trucks are therefore needed. For any particular situation, a cost–benefit analysis of the alternatives has to be made, although transport as LH_2 is normally employed only for sizeable quantities over long distances where conveying compressed gas would be uneconomic. At present, neither route appears to be commercially viable for distributing hydrogen to filling stations for use in vehicles. With the prospect of ever-rising petroleum prices, however, this state of affairs could change in the future.

From an automotive engineering standpoint, there are practical difficulties in terms of designing, at acceptable cost, LH_2 cryostats for on-board vehicular use that are both appropriate and safe. Nevertheless, some progress has been made towards this goal in Germany, notably by the automotive company BMW and its hydrogen supplier Linde; see Figure 5.5. There are also aspects of vehicle refuelling to be addressed, especially the development of transfer lines for LH_2 that can be operated safely by the general public and without too much loss

---

[§]The cooling of a gas by allowing it to expand without gaining any external heat.
[¶]The reversible change in temperature that accompanies the change in magnetization of a ferromagnetic or paramagnetic material; also known as the 'thermo-magnetic effect'.

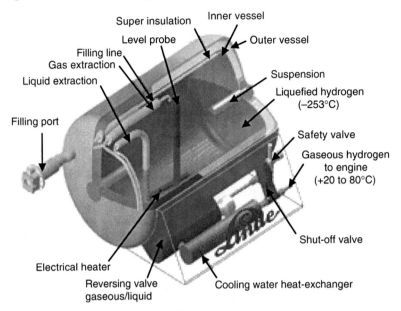

Figure 5.5 Automotive design of LH$_2$ storage tank.
(Courtesy of Linde AG).

through evaporation. Boil-off on standing would also be a problem with the relatively small tanks that would typically be used in cars. Compared with gas cylinders, LH$_2$ tanks have the attractive features of lower mass and smaller volume, which translate into higher energy-storage capacity and therefore longer vehicle range between refuelling. This mode of hydrogen containment does, however, incur a cost penalty that may prove to be unacceptable.

5.4 Metal Hydrides

Compressed gas and liquid hydrogen are not the only options available for the storage of relatively limited quantities of hydrogen. An alternative is to employ a hydrogen carrier material such as a metal hydride. Ideal materials would undergo simple, inexpensive, treatment processes, either at a fuelling station or on-board a vehicle, to release hydrogen for use in fuel cells. Certain metals and alloys have the ability to absorb and release gaseous hydrogen, reversibly, via the formation of hydrides, *i.e.*,

$$M + x/2H_2 \rightleftharpoons MH_x + \text{heat} \tag{5.1}$$

The hydrogen molecule first dissociates into its two atoms, which are chemisorbed on the surface of the metal/alloy and then diffuse into the bulk lattice. The dissolved atoms can take the form of a random solid solution or react to produce a hydride of fixed stoichiometric composition. The quantity of

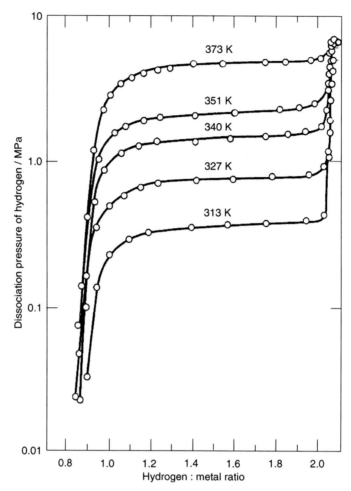

Figure 5.6 Dissociation pressure curves for vanadium hydride VH_2.[5] (Courtesy of the Materials Research Society).

hydrogen absorbed is expressed in terms of the hydride composition, either on a molar basis (MH_x) or on a weight-% basis. The rates of absorption and desorption can be controlled by adjusting the temperature or pressure. (Note: heat is liberated during formation of the hydride and therefore must be added in order to effect its subsequent decomposition with discharge of the hydrogen.) To be useful as a hydrogen store, the metal or alloy should react with the gas at, or near, ambient temperature and at not too great a pressure. The enthalpy of absorption should not be too high otherwise heat transfer becomes a problem, especially in large hydride beds. Finally, the system should be capable of sustaining a practical number of absorption–desorption cycles without deterioration.

As an example of a simple metal hydride that forms reversibly at near-ambient temperatures and modest hydrogen pressures, Figure 5.6

shows equilibrium isotherms for vanadium hydride, VH_2. The horizontal portion of each isotherm corresponds to a two-phase region where VH_2 co-exists with VH. The dissociation pressure in this region ranges from around 0.35 MPa at 313 K (40 °C) to around 5.0 MPa at 373 K (100 °C). Thus, VH_2 has a favourable operating regime as regards both pressure and temperature. Moreover, the enthalpy of decomposition is moderate at 40 kJ per mole H_2. On the other hand, since only one of the two hydrogen atoms is readily reversible, the effective hydrogen mass content is reduced from 3.8 wt.% to around 1.9 wt.%, a value that is far too low to be useful for storing hydrogen on vehicles.

The relationship between the equilibrium dissociation pressure (P) of a hydride and the absolute temperature (T) is expressed by the van't Hoff equation, *i.e.*,

$$\ln P = \Delta H / RT - \Delta S / R \qquad (5.2)$$

where ΔH is the enthalpy of reaction (the heat that has to be supplied to decompose the hydride), R is the gas constant and ΔS is the entropy of reaction; see Box 4.2, Chapter 4. A logarithmic plot of the dissociation pressure against the reciprocal of the absolute temperature should therefore be linear with a slope that is a measure of the heat that has to be supplied to decompose the hydride. A series of such plots for various hydrides over the temperature range −20 to 400 °C is presented in Figure 5.7. The hydrides can be broadly classified into three categories, namely high-temperature (HT), medium-temperature

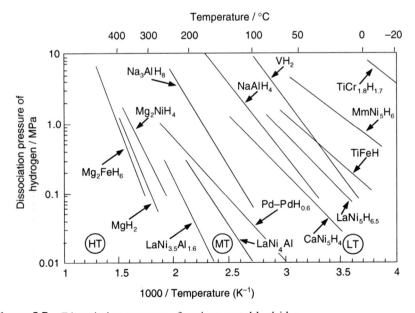

Figure 5.7 Dissociation pressure of various metal hydrides.

(MT) and low-temperature (LT) hydrides. The LT category is the most useful in that such hydrides offer hydrogen storage closest to ambient temperature and are therefore convenient for supplying fuel cells. In this case, the required enthalpy of hydrogen desorption would be provided by the waste heat from the fuel cell, provided this were at a sufficiently high temperature. Proton-exchange membrane fuel cells (see Section 6.4, Chapter 6) are generally employed for vehicles and their waste heat is at less than 100 °C, which places a restriction on the hydride store that could be used.

One of the attractions of metal hydrides for hydrogen storage is that although their gravimetric hydrogen content is only modest, on a volumetric basis the content may be as high as, or even higher than, that provided by LH_2. This rather surprising fact is a direct consequence of the high packing density of the hydrides and the low density of LH_2. The main obstacles to the use of hydride beds to store hydrogen, especially in vehicles, lie in their excessive mass and high cost and in the slow kinetics of hydrogen uptake and release. For stationary applications, however, the mass is not an important factor, whereas the high volumetric density is a positive attraction.

Compared with vanadium hydride, discussed above, magnesium hydride (MgH_2) has a significantly greater hydrogen content (7.6 wt.%) and, moreover, the absorption of both hydrogen atoms is reversible. Magnesium is also a relatively inexpensive metal. The problem with this material is that it only decomposes above about 300 °C, which is not too practicable for many applications. In addition, the enthalpy of dissociation is high (75 kJ per mole H_2), which necessitates the supply and removal of considerable quantities of heat as the hydride is decomposed and reformed, respectively. Titanium–iron alloy, TiFe, forms a hydride that is usable at ambient temperature and has a low enthalpy of decomposition (28 kJ per mole H_2) but, like VH_2, has a small reversible hydrogen content of only 1.5 wt.%.

During the course of hydriding, most metals and alloys undergo a lattice expansion, which causes them to fracture. Starting with a bulk ingot, they rapidly convert to a fine powder as hydride–dehydride cycling continues and this enhances the subsequent reaction kinetics. Allowance must be made for the considerable increase in volume as breakdown takes place. Subsequently, the powder is likely to consolidate in the bed and 'fines' can become entrained in the gas and cause blockages in the filter system. Some metals, including magnesium, have to be activated before the first hydriding takes place. Activation involves the application of an over-pressure of hydrogen (*i.e.*, many times that of the equilibrium dissociation pressure) during the initial cycle. This limitation can be circumvented, at the expense of reduced gravimetric hydrogen content, by employing the alloy Mg_2Ni to form the hydride Mg_2NiH_4 (3.6 wt.% H_2), which is readily activated, has good reaction kinetics and cycles reversibly at a lower temperature than MgH_2 itself; see Figure 5.7. Some recent research indicates that the kinetics of hydrogen release from MgH_2 can be much improved by the addition of a small quantity of $LiBH_4$ that functions as an activating agent, although the required temperature of around 300 °C is still high. Investigations are continuing to elucidate the mechanism of this effect and to explore the action of other additives.

Small hydride beds are commercially available and are generally based on rare earth/nickel alloys, as used in rechargeable nickel–metal-hydride batteries.[6] Numerous potential applications have been identified, *e.g.*, chemical heat pumps and refrigerators, purification of hydrogen, isotope separation, gettering of vacuum systems, thermal switches, chemical compressors for hydrogen, and sorption cryo-coolers. These uses have not so far assumed the same level of commercial importance as that of metal hydrides in batteries.

The present shortcomings of metal hydrides as media for the storage of hydrogen become more severe as the size of the systems is increased. To be at all energy efficient, the heat liberated during hydriding would have to be stored and recycled for undertaking the dehydriding reaction. With a multi-kilogram hydride bed, the kinetics of hydride decomposition and reforming (*i.e.*, the rate of fuel availability and recharging, respectively) are determined by the accompanying heat flow through the powder. In an early (1970s) trial with 400 kg of iron–titanium alloy, which absorbed 5.5 kg of hydrogen (approximately equivalent in energy content to 21 litres of petrol), it took 5 h to recharge the bed because of poor thermal conduction. Since that time, other experiments and demonstrations have been conducted with vehicles that were powered by hydrogen-burning internal combustion engines and were equipped with metal hydride storage. Although the feasibility of this approach has been established, operational difficulties associated with the kinetics of heat and hydrogen transfer have invariably been encountered and have yet to be resolved. These problems compound the drawbacks of the large mass and volume of the storage units and their high capital cost. To bring the issue of heat transfer up to date, it has been calculated that if a typical hydride bed capable of holding 5 kg of hydrogen (to provide a driving range of ~ 500 km) were to be charged in 3 min (an acceptable time for refuelling a vehicle), it would then be necessary to remove heat at a rate of ~ 500 kW, as dictated by the enthalpy of absorption. Clearly, the development of an effective means of heat transfer presents a formidable engineering challenge.

It has been claimed that the requirement to supply heat to desorb the hydrogen from the alloy is a positive safety feature, compared with storage as pressurized gas or as liquid hydrogen, since desorption cannot happen spontaneously. Although true, this is a dubious argument because in the event of an accident, any ingress of air may lead to combustion of the finely-divided hydride.

Several criteria have been established for the realization of a practical hydrogen storage bed:

- the alloy should be readily prepared, easily activated to form the hydride and inexpensive;
- the enthalpy of hydride formation should be as low as possible;
- the hydride should have a high hydrogen content per unit mass;
- the dissociation pressure of the hydride should lie in the range 0.1–1.0 MPa at near-ambient temperature;

- the system should exhibit favourable and reproducible reaction kinetics;
- the reactant bed should have a high thermal conductivity;
- the system should be capable of cycling many thousands of times without degradation;
- the alloy should not be poisoned by gaseous impurities;
- the system should be safe on exposure to air and should not ignite.

This is a demanding set of conditions and, although many hundreds of alloys have been studied, none have yet been found to be fully satisfactory.

The conclusion from this discussion is that while small hydride beds may well find diverse technological applications, large (multi-kilogram) beds are still beset with many problems. Whereas these may be solvable for stationary installations, the prospects of employing metal hydrides to store hydrogen fuel aboard road vehicles are not considered to be good. In particular, the performance and cost targets set by the US Department of Energy for hydride stores on fuel cell cars are particularly challenging.

5.5   Chemical and Related Storage

Hydrogen forms stable covalent compounds with carbon (hydrocarbons, alcohols) and with nitrogen (ammonia). These materials are moderately convenient to carry and store, but decompose only at elevated temperatures and, in most cases, are not easily reconstituted. There are also various lightweight ionic compounds that contain hydrogen as part of a stable anion; examples are $LiBH_4$, $LiAlH_4$ and $NaAlH_4$. These two classes of chemical, covalent and ionic respectively, each have a relatively high mass-% of hydrogen. The former are generally referred to as 'simple hydrogen-bearing chemicals', and the latter as 'chemical hydrides' to differentiate them from metal hydrides.

Hydrogen storage materials may, in turn, be categorized either as reversible ('two-way' or 'round-trip') carriers or as irreversible ('one-way' or 'one-shot') sources of hydrogen. In a round-trip system, the chemical is loaded on to a vehicle, where it is thermally treated to yield hydrogen. It is then returned to a central facility for recharging with hydrogen. A one-shot carrier is a hydrogen-rich material that is decomposed or reformed on board the vehicle to produce hydrogen, together with a by-product that can be discarded (*e.g.*, nitrogen from ammonia or carbon dioxide from methanol). These carriers therefore offer the advantage of not having to be returned to a central facility for reprocessing. Ideally, the by-product must pose no environmental issues, possess virtually no value and be readily disposable[||]. In the following discussion, we consider first some 'simple hydrogen-bearing compounds' as possible storage media and then 'complex (anionic) chemical hydrides'.

[||] It is arguable that carbon dioxide, as a greenhouse gas, is not environmentally benign and disposable. Indeed, in this context, there is little advantage in decomposing methanol to yield hydrogen; it might just as well be used directly as a fuel for an internal combustion engine.

5.5.1 Simple Hydrogen-bearing Chemicals

Certain organic chemicals contain significant atomic proportions of hydrogen that can be recovered and therefore they may be considered as prospective hydrogen carriers. One of the best known of these is cyclohexane (C_6H_{12}), which can be decomposed catalytically to yield benzene (C_6H_6) and hydrogen:

$$C_6H_{12} \rightarrow C_6H_6 + 3H_2 \qquad (5.3)$$

In theory, this reaction may be carried out cleanly at a moderate temperature of several hundred degrees, depending on the catalyst, although at around 500 °C there may be 'cracking' of the molecule to give a variety of by-products, *e.g.*, methane, ethane, ethene, propane, propene, *etc.*

Cyclohexane is an example of a reversible (round-trip) carrier since the chemical is manufactured by the hydrogen reduction of benzene over a nickel catalyst at 150–200 °C. It is a volatile liquid that boils at 81 °C and could, in fact, be used directly as a fuel. Alternatively, it could be employed to transport and store hydrogen and then decomposed when the gas is required. The hydrogen content is 7.1 wt.%. The cyclohexane contained in a vehicle's fuel tank would be vaporized and passed over a heated catalyst to generate hydrogen. This could then feed either an internal combustion engine or a fuel cell. The benzene by-product would be condensed, collected at the service station and returned to a factory for re-conversion to cyclohexane. In practice, there would be difficulties in matching the hydrogen supply to the varying demand of the engine or fuel cell. Hence it might be necessary to have either a small, supplementary buffer reservoir of hydrogen or a secondary power source for the vehicle. If on-board catalytic decomposition of cyclohexane proved to be impracticable, the chemical could serve as an intermediary to permit local hydrogen generation at a service station or perhaps even in the home. This would then introduce the problem of how to pressurize the hydrogen and store it on the vehicle. The use of methylcyclohexane (C_7H_{14}) has also been proposed. This compound dehydrogenates to toluene (C_7H_8), with a yield of 6.1 wt.% H_2, but poses just the same operational problems.

Methanol (CH_3OH) may be seen as either a fuel in its own right (*e.g.*, to burn in an engine or a direct methanol fuel cell; see Section 6.4, Chapter 6) or as a carrier for hydrogen. Again, this chemical may be decomposed catalytically, but at around 250 °C:

$$CH_3OH \rightarrow 2H_2 + CO \qquad (5.4)$$

As the carbon monoxide produced is highly toxic, it could not be discharged to the atmosphere, but would first have to be 'shifted' to carbon dioxide [see reaction (2.5), Chapter 2] and this presents major operational difficulties. Even more problematic is the steam reforming of methanol to hydrogen and carbon dioxide directly. For vehicular use, the incorporation of a steam reformer seems impractical.

The thermal decomposition of methanol [reaction (5.4)] is not such a straightforward option as that of cyclohexane, since it may not be feasible to separate the two gases, certainly not on-board a vehicle, although it may be possible for stationary applications, as discussed in Chapter 2. With an internal combustion engine, it would be far simpler to burn the methanol directly. Methanol is an inferior fuel to petroleum (its gravimetric and volumetric energy densities are both 50% smaller; see Table 1.4, Chapter 1) and it has the undesirable features of being both water-miscible and toxic. On the other hand, it does have the merit of releasing less carbon dioxide per unit of energy produced. Furthermore, methanol can be obtained from syngas (see Section 2.3, Chapter 2) by reaction over a catalyst of copper and zinc oxides held at high temperature and pressure. In the commercial process for methanol manufacture, the syngas derives from the steam reforming of natural gas, although it can be made from biomass if so desired.

Hydrogen produced via the decomposition of methanol would be suitable for use in fuel cells, except for the fact that carbon monoxide impurities in the hydrogen would poison the electrocatalysts. The addition of a purification unit would of course add further complexity to the overall fuel-cell system, to such an extent that the adoption of methanol as a hydrogen carrier is no longer considered to be viable for vehicular applications. Serious attempts are being made to develop fuel cells that run on methanol directly; see Section 6.4, Chapter 6. A similar poisoning problem has to be tackled which, in this case, arises from deactivation of the electrocatalyst by intermediate species formed during methanol electro-oxidation.

In addition to the above organic chemicals, some simple inorganic compounds have also been suggested as carriers and stores for hydrogen. Two examples are liquid ammonia (NH_3) and hydrazine hydrate ($N_2H_4 \cdot H_2O$). Ammonia contains 17.7 wt.% of accessible hydrogen and is a liquid at room temperature under a few atmospheres pressure. This makes transport in cylinders a realistic route, just as propane and butane are widely distributed as fuels today. Ammonia is decomposed catalytically at elevated temperatures to yield hydrogen and nitrogen. Hydrazine hydrate, which has a recoverable hydrogen content of 8.0 wt.%, is an endothermic compound and is therefore easier to decompose than ammonia. Indeed, it has been known to decompose explosively. Although not easy to manufacture or handle in bulk, hydrazine has been used as a rocket fuel and many years ago was employed successfully in an experimental fuel cell that powered a small car; see Section 7.5, Chapter 7.

Ammonia offers attractions as a hydrogen carrier in that it is already a bulk chemical of commerce, it is inexpensive and large-scale distribution by pipeline or tanker truck is widely practised. Nevertheless, difficulties lie in the fact that liquid ammonia has to be held under pressure, the gas is highly toxic and decomposition can only be effected catalytically at high temperatures. Researchers in Denmark have suggested a novel way of circumventing these problems, namely the conveyance of ammonia as an inert solid rather than as a volatile liquid. The chemical is formed into a metal ammine, such as $Mg(NH_3)_6Cl_2$, which holds 51 wt.% NH_3 and thus 9.1 wt.% H_2. This compound is safe to handle and can be compacted into dense tablets that have a low vapour pressure of ammonia at ambient temperature and, on a volumetric basis, contain 60–70% more hydrogen than liquid hydrogen. When the tablets are heated, ammonia is desorbed between 80 and 350 °C and may then be passed over a suitable catalyst at elevated temperature for breakdown to hydrogen and nitrogen. The desorption process is said to be smooth and the residual $MgCl_2$ can be recycled to form more of the ammine complex. This concept for storing and transporting hydrogen merits further consideration, although care would be needed to ensure that the hydrogen was completely free from ammonia, which is poisonous both to fuel cells and to people. In addition, the temperatures required for catalytic decomposition are too high to use the waste heat from a proton-exchange membrane fuel cell and the hydrogen produced would have to be cooled to 80–100 °C before it could be fed to the fuel cell. These objections would not arise when using a high-temperature fuel cell. Other metal ammine complexes that might be even more suitable for storing ammonia are being investigated.

The feasibility of ammonia borane (NH_3BH_3) as a hydrogen carrier is also being examined. This compound is analogous to (and isoelectronic with) ethane (CH_3CH_3), but is a solid rather than a gas. It releases up to 12 wt.% H_2 at 100–200 °C to leave a polymer $(NHBH)_n$ that can, in principle, be reconverted to ammonia borane. In practice, however, the latter process presents a major chemical challenge. Attempts are being made to bring down the decomposition temperature of ammonia borane to 100 °C so as to exploit the waste heat from a proton-exchange membrane fuel cell to effect the reaction.

5.5.2 Complex Chemical Hydrides

There is a class of inorganic metal hydrides that are ionic rather than metallic in nature. For instance, the elements boron and aluminium form the hydride anions $[BH_4]^-$ and $[AlH_4]^-$, respectively. When combined with alkali metal cations, soluble ionic salts are formed, *e.g.*, $LiBH_4$, $NaBH_4$, $NaAlH_4$. These compounds are generally known as 'complex chemical hydrides'. Lithium and sodium borohydrides are used in organic chemistry as reducing agents. For hydrogen storage, the aluminohydrides (the so-called 'alanates') are preferred to the borohydrides.

Thermal decomposition of the alanate $NaAlH_4$ takes place in two steps, as follows:

$$3NaAlH_4 \rightarrow Na_3AlH_6 + 2Al + 3H_2 \tag{5.5}$$

$$Na_3AlH_6 \rightarrow 3NaH + Al + 3/2H_2 \tag{5.6}$$

The reactions are reversible, but only at temperatures above the melting point of $NaAlH_4$ (183 °C) and at hydrogen pressures of 10–40 MPa, which are impracticable. Fortunately, research has shown that the temperatures for discharge and recharge of hydrogen may be reduced significantly in the presence of a titanium catalyst. Titanium-catalyzed $NaAlH_4$ has thermo-dynamic properties that are comparable with those of classical low-temperature hydrides, *e.g.*, $LaNi_5H_6$ and TiFeH (Figure 5.7). The first step in the decomposition, at 50–100 °C, corresponds to the release of 3.7 wt.% H_2 and the second step, at 130–180 °C, to a further 1.9 wt.%. Moreover, even if only the first reaction step can be utilized, the gravimetric hydrogen storage of $NaAlH_4$ is still greater than that offered by most metal hydrides. By contrast, Na_3AlH_6 requires higher temperatures for hydrogen liberation and might be useful for applications other than in fuel cells, such as heat pumping and heat storage. The study of catalysed $NaAlH_4$ is still at a preliminary stage and there is considerable scope for the development of alternative catalysts and optimization of their performance. Two of the many problems to be addressed are the pyrophoricity of the alanates (*i.e.*, their propensity to ignite spontaneously) and their production cost.

The alkali metal borohydrides contain more hydrogen than the alanates, *e.g.*, 18.5 and 10.6 wt.% for the respective lithium ($LiBH_4$) and sodium ($NaBH_4$) analogues, but are more stable and therefore less useful as accessible hydrogen stores. For example, $LiBH_4$ starts to decompose only above 300 °C, whereas $NaBH_4$ does not decompose until 350–400 °C. For borohydrides, the main issues to be addressed are as follows.

- Can the experience gained with catalyzed $NaAlH_4$ be extended to the borohydrides?
- Can borohydrides be destabilized to achieve hydrogen release below 100 °C?
- Can the cost of the borohydrides be substantially reduced?

Improved dehydrogenation kinetics and reduced thermodynamic stability are being sought through fundamental investigations of the crystalline structure, chemical binding status, particle size and morphology of the borohydrides, and also the distribution of the catalyst.

Although $NaBH_4$ is at present too stable for providing hydrogen practically through a thermal activation process, it does release the gas on reaction with water, together with formation of sodium borate ($NaBO_2$):

$$NaBH_4 + 2H_2O \rightarrow NaBO_2 + 4H_2 \qquad (5.7)$$

This is an irreversible reaction, but has the advantage that 50% of the hydrogen comes from the water. In effect, $NaBH_4$ is a 'water-splitting' agent. Based on the mass of $NaBH_4$, the hydrogen available is 21 wt.% – a remarkably high figure. Of course, allowance has also to be made for the mass of the water, which reduces the theoretical hydrogen content to 10.9 wt.%. For a solution of 30 wt.% $NaBH_4$ in water, with 3 wt.% of sodium hydroxide present, the effective hydrogen content will be 6.6 wt.%. This is still an attractive yield; it is better than that obtained from any of the practical metal hydrides. Several of these so-called 'irreversible chemical hydrides', *e.g.*, CaH_2, LiH, $LiAlH_4$, $LiBH_4$, KH, MgH_2 and NaH, are being evaluated for their reactivity with water. One approach to preparing the storage medium is to mix dry $NaBH_4$ in particulate form with light mineral oil and a dispersant to produce an 'organic slurry'. The oil coats the particles and protects them from inadvertent contact with water during handling and also moderates the reaction rate of the hydride when water is introduced. More development work is required to control the reaction kinetics so as to yield hydrogen at the desired rate.

For road transportation applications, instead of refuelling with hydrogen gas at the service station, the vehicle would have its tank emptied and refilled with fresh hydride slurry[**]. The spent solution would then be returned to a processing plant for regeneration of the hydride. The expense of this operation is a major hurdle to overcome. It has been suggested that the cost of $NaBH_4$ itself needs to fall by a factor of ~ 50 for this hydride to be acceptable for vehicular use. From the standpoint of both mass and volume, however, the system appears superficially to be attractive as a hydrogen-storage scheme for fuel cell vehicles. Not only is the chemical reaction a prolific source of hydrogen, but also the operation of discharging the spent sodium borate solution and replenishing with fresh $NaBH_4$ is likely to be rapid. DaimlerChrysler has demonstrated that an $NaBH_4$ system, designed by Millennium Cell in the USA, can provide a minivan (the *Natrium*) with a range of 480 km. Much will depend on the logistics and cost of the transport and reprocessing operations to regenerate the $NaBH_4$ using fresh hydrogen.

Metal nitrohydrides, known as amides $[M(NH_2)_n]$ or imides $[M_{2/n}(NH)]$, are a third class of complex chemical hydrides that, recently have been attracting interest as promising materials for hydrogen storage. For example, lithium amide ($LiNH_2$) contains 8.7 wt.% H_2. At $\sim 300\,^{\circ}C$, this compound reacts with lithium hydride (LiH) to yield lithium imide (Li_2NH) with the release of 6.5 wt.% H_2. In a second step at $\sim 430\,^{\circ}C$, the Li_2NH reacts with more LiH to liberate a further 5 wt.% H_2 and leave a residue of lithium nitride (Li_3N). Research is in progress to examine the possibility of reducing the desorption temperature of the first step by subjecting the amide to mechanical activation (*i.e.*, ball-milling) and also by introducing additives such as magnesium hydride (MgH_2). Similar investigations are being conducted on the amides of calcium, magnesium and sodium.

In summary, storing and transporting hydrogen via simple (discussed above) or complex chemicals would avoid many of the problems associated with distributing pure molecular hydrogen. Furthermore, if such chemicals could be conveyed via existing or new low-cost infrastructures, they could significantly reduce hydrogen delivery costs. To achieve this, a carrier should:

- provide high hydrogen capacity with respect to both mass and volume;
- offer a transformation process for discharging hydrogen that is both highly efficient and inexpensive;
- support a simple and low-energy process for recharging with hydrogen, either *in situ* or at a regional site;
- be safe and environmentally benign.

This is a challenging, but not impossible, set of specifications.

[**]This is similar to the procedure proposed for the zinc–air traction battery, which has been investigated as a candidate power source for electric vehicles.[6]

5.5.3 Nanostructured Materials

In recent years, there has been burgeoning interest in 'nanoscience' and 'nanostructured materials'. These are materials with characteristic geometric dimensions below 10^{-7} m and new properties that result from the nanostructure[††]. By virtue of their large surface-to-volume ratios, certain nanomaterials can *adsorb* considerable amounts of hydrogen in the molecular state via weak molecular–surface interactions (so-called 'physisorption'). This is in contrast with the above-mentioned chemisorption process on metal hydrides in which the hydrogen is dissociated into atoms that chemically bond with the lattice of the storage medium. Obviously, physisorption is preferred, as it would moderate the pressure and temperature required for the respective uptake and release of hydrogen. Moreover, there is no major heat-transfer problem (of the type discussed earlier) because physisorption bonds are weak, typically with enthalpies of adsorption of –10 to –20 kJ per mole H_2[‡‡]. On the other hand, significant uptake of hydrogen is normally seen only at cryogenic temperatures, which is a major inconvenience.

Particular attention has been paid to the possibility of hydrogen storage in carbon-based materials that take the form of nanofibres or nanotubes. These are structures that derive from the fundamental carbon entity C_{60} (buckminsterfullerene) that has a spherical cage structure made up of hexagons and pentagons, as shown schematically in Figure 5.8(a). The generic term 'fullerene' is used to describe a pure carbon molecule that consists of an empty cage of 60 or more carbon atoms. Graphitic nanofibres are grown by the decomposition of hydrocarbons or carbon monoxide over metal catalysts and are made up of graphene sheets[§§] aligned in a set direction (dictated by the choice of catalyst). Three distinct structures may be produced: platelet, ribbon and herringbone; see Figure 5.8(b). The structures are flexible and can expand to accommodate the hydrogen. Carbon nanotubes are cylindrical or toroidal varieties of fullerene and have lengths of between 10 and 100 μm. 'Single-walled' nanotubes are composed of only one graphene layer and have typical inner diameters of 0.7–2 nm. 'Multi-walled' nanotubes consist of concentric rings (diameters 30–50 nm) or spirals of graphene sheets. Illustrations of different types of carbon nanotube structure are given in Figure 5.8(c). The tube ends are usually capped by hemispherical fullerene molecules. Since the aspect ratio of the tubes is often large (~ 1000), it may prove difficult for hydrogen molecules to penetrate the length of the tube, even after the end caps have been removed to allow access.

It is generally considered that the high hydrogen-storage capacities (30–60 wt.%) of carbon nanomaterials reported a few years ago are in fact impossible and were the result of measurement errors due to the presence of metal

---

[††] Nanotechnology is a broad area of applied science and technology that focuses on controlling and exploiting the structure of matter on a scale below 100 nm. It has applications as diverse as colloidal science, device physics, molecular biology and supramolecular chemistry.
[‡‡] The negative sign indicates an exothermic process.
[§§] Graphene sheets are two-dimensional analogues of the three-dimensional graphite structure with similar chemical bonding.

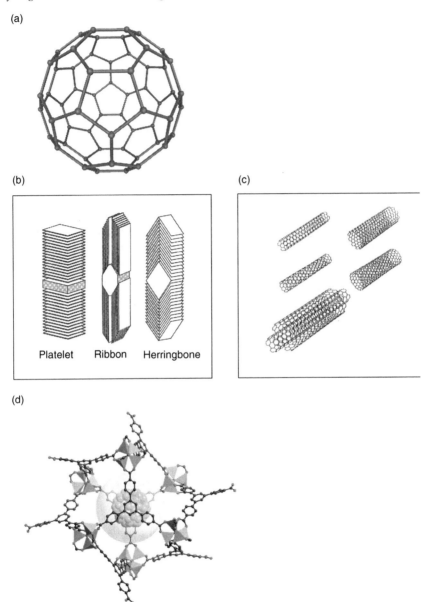

Figure 5.8 Schematic representations of: (a) buckminsterfullerene; (b) carbon nanofibres; (c) single- and multi-walled carbon nanotubes; (d) metal–organic
framework, MOF-177.

contaminants and/or water absorption. More recently, some evidence has
emerged that certain metals (*e.g.*, titanium) adsorbed on the surface of nanotubes can enhance the uptake of hydrogen. Whereas physisorption has been
clearly demonstrated, again this approach is useful only at cryogenic

temperatures (up to about 6 wt.% H_2) and carbons with extremely high surface areas are required. Room-temperature adsorption of a few wt.% H_2 is occasionally reported, but has not been reproducible. The surface and bulk properties necessary to achieve practical hydrogen storage are not completely understood and it is far from certain that carbon nanotubes can be economically and consistently synthesized in useful quantities as a homogeneous, well-defined and pure material.

In parallel with the research being undertaken on carbon, many other porous materials and composites with high surface areas are being investigated as possible storage media. These include the following.

- *Zeolites:* complex aluminosilicates with engineered pore sizes and high surface areas. The materials are used as molecular sieves and therefore the science for capturing gases other than hydrogen is well known. For an optimum hydrogen-delivery cycle, the adsorption enthalpy should be neither too low (so as not to limit storage) nor too high (to facilitate release). It is proposed that –15 kJ per mole H_2 is required, but most reported values have been rather low, namely –5 to –10 kJ per mole H_2. Recent university research, however, has indicated that the enthalpy may be modified to –15 to –20 kJ per mole H_2 by incorporating magnesium into the zeolite structure.

- *Metal–organic frameworks (MOFs):* typically zinc oxide structures bridged with benzene rings. The materials have extremely high surface areas, are highly versatile and allow for many structural modifications. By way of example, the structure of the colourless crystal designated MOF-177 is shown schematically in Figure 5.8(d). The organic component is a triangular molecule (benzene-1,3,5-tribenzoate) that is comprised of three benzoate rings which branch off a central benzene ring. These molecules are positioned at the vertices of tetrahedral zinc oxide-based complexes – an arrangement that yields a very high surface area ($4500\,m^2\,g^{-1}$) relative to the volume of the crystal. The shaded sphere represents the pore volume.

- *Clathrate hydrates:* water (ice) cage-like structures that often contain guest molecules such as methane and carbon dioxide; see Section 3.3, Chapter 3. The cage size and structure can often be controlled by the inclusion of organic molecules. Clathrates would probably be handled as slurries or solids to deliver hydrogen. As yet, however, there is little evidence that useful amounts of hydrogen can be accommodated.

For the above three classes of material, the main challenge is to find ways by which they can be engineered to store high levels of hydrogen, reversibly, at close to room temperature. The metal–organic frameworks and clathrate hydrates represent new storage ideas that should be studied to determine their respective capabilities, although from a wealth of experience with physisorption on conventional materials of high surface area (carbon blacks, zeolites) it appears unlikely that the required storage targets will be met at ambient temperatures.

To summarize, the major benefits and barriers associated with the main categories of hydrogen storage are as outlined in Table 5.4.

Table 5.4 Hydrogen storage technologies: advantages and limitations/challenges.

Technology	Advantages	Limitations/challenges
Compressed gas cylinders	• Steel cylinders widely used up to 20 MPa • Lightweight composite cylinders now becoming available for use up to 70 MPa	• Volumetric energy density of compressed hydrogen is poor at 20 MPa pressure, average at 70 MPa; possibly suitable for buses and trucks, but probably not for cars • Refuelling vehicles at high pressure requires special compressors and presents hazards • Energy required for compression is a significant fraction of the hydrogen energy
Liquid hydrogen	• Established, but specialized, low-temperature technology • Storage density comparable with gas at 70 MPa	• Expensive technology, best conducted on a very large scale • Transfer from store to user is complex and requires specialized equipment • Boil-off leads to losses and presents hazards
Metal hydrides	• Established technology for small-scale uses • Good volumetric energy density of hydrogen in some hydrides	• Low wt.% H_2 • Many hydrides only liberate hydrogen at elevated temperatures • Difficulty in controlling kinetics of hydrogen release • Slow rate of recharge with hydrogen • Problems of dissipating heat released during recharge
Chemical storage – Simple hydrogen-bearing chemicals	• High wt.% H_2 • Cheap industrial chemicals • Fast refuelling	• High temperature required for decomposition • Some candidates are toxic
– Complex chemical hydrides	• High wt.% H_2, especially borohydride when reacted with water	• Factory regeneration required • High cost of borohydride
– Nanostructured materials	• Considerable scope for research	• Academic interest only as yet • Early promising results now dubious

5.6   Hydrogen Storage on Road Vehicles

At present, the only truly practical means of storing hydrogen on a road vehicle is as high-pressure gas in a cylinder. As discussed above, modern lightweight composite cylinders are a distinct improvement on conventional steel vessels. Other options such as cryogenic liquid hydrogen and metal hydrides are likely to be uneconomic, at least in the short term. Irreversible chemical hydrides that react with water (*e.g.*, $NaBH_4$) may find some success if their manufacturing cost can be reduced dramatically. Further investigation is necessary both at the fundamental scientific level and in terms of vehicle engineering, hydride recovery plant and overall logistics.

The task of displacing liquid hydrocarbon fuels by hydrogen is formidable. This is illustrated pictorially in Figure 5.9 and numerically in Table 5.5. In the latter, a comparison is made of the probable masses and volumes of systems to accommodate the amount of hydrogen (11.75 kg) that is equivalent in energy content (1.4 GJ) to 45 L of petrol, which is typically the capacity of the fuel tank in a car. Columns 3 and 5 relate these masses and volumes, respectively, to petrol as unity. The data may be only approximate, because of the complexity of making due allowance for the containers, but they do give some idea of the magnitude of the problem. On a mass basis, only LH_2 comes anywhere close to competing with petrol. On a volume basis, none of the candidates appears to be practical, with compressed gas at 70 MPa (10 000 psi) being the least bulky. For cars and light-duty vehicles, the use of this latter option will result in vehicle owners having to accept shorter driving ranges between refuelling stops, but this may be the price that has to be paid for hydrogen-fuelled vehicles.

We conclude that the storage mode for hydrogen on road vehicles is still very uncertain. In the short term, for demonstration fleets of fuel cell vehicles, compressed gas will continue to be used. Although LH_2 has been employed by BMW for experimental cars equipped with hydrogen-fuelled internal combustion engines, considerations of cost, energy efficiency, boil-off and safety may delay the progress of this technology. Reversible metal hydrides have serious

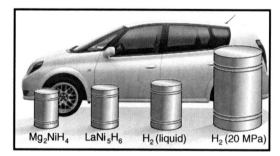

Figure 5.9   Schematic representation of the relative volumes of two hydrides, liquid hydrogen and compressed gas (20 MPa pressure) to contain 4 kg of hydrogen for a vehicle driving range of 400 km.[7]
(Courtesy of Nature Publishing Group).

Table 5.5 Approximate comparison of masses and volumes of petrol and hydrogen for equivalent energy content (1.4 GJ).

	Mass of store and fuel/kg	Index	Volume of store/L	Index
Petrol (45 L)	41	1	45	1
Compressed H_2 (20 MPa), conventional steel cylinder	~1150	28	~1080	24
Compressed H_2 (70 MPa), composite cylinder	~200	4.9	~170	3.8
LH_2 in cryostat	~100	2.4	~350	7.8
Ti–Fe hydride bed	~1050	25.6	~275	6.1

problems of mass, volume and cost, in addition to engineering difficulties with heat management. Thermal transfer and control of the kinetics of hydrogen uptake and release will undoubtedly present difficulties with all hydrides. Long times of on-board recharge will not be acceptable to drivers. The simple hydrogen-bearing chemicals (cyclohexane, methanol, ammonia, *etc.*) all require elaborate thermal processing plant, which would be difficult to accommodate on most road vehicles. Complex chemical hydrides (such as $NaBH_4$) that react with water to liberate hydrogen are at an early stage of development. The logistics of discharging an oxidized salt such as sodium borate and returning it to a factory for re-processing to sodium borohydride have yet to be demonstrated. Nevertheless, the 'simple' and 'complex' classes of chemicals are attractive as hydrogen carriers because of their large gravimetric hydrogen densities compared with metal hydrides. Hydrogen sorption on porous materials of high surface area may prove to be impractical on account of the need for cryogenic temperatures. These more recent options are highly speculative and much further work has to be done on all possible storage modes if fuel cell vehicles are to become successful.

In the USA, the technical targets for on-board hydrogen-storage systems given in Table 5.2 have been established through the FreedomCAR Partnership between the US Department of Energy and the US Council for Automotive Research (USCAR). There are also stringent cost targets to be met. Progress towards meeting both sets of targets has been slow and it has to be questioned whether these goals are realistic. The twin assumptions behind them are that (i) motorists will never accept any vehicle that is inferior in performance (or costs more) compared with those to which they are accustomed and (ii) given sufficient research funding, scientists and engineers will always produce a viable solution. Both of these assumptions have to be questioned. The challenges of on-board hydrogen storage, other than as compressed gas, are formidable. When compounded with the costs of producing pure hydrogen and with those of manufacturing fuel cells, it seems unlikely that fuel cell vehicles will be able to compete with combustion engines fuelled with synthetic fuels (*e.g.*, petroleum from coal) unless, and until, all carbon emissions are heavily taxed or outlawed. Even then, there will be competition from biofuels such as gasohol and bio-diesel. We therefore look to the long term – possibly the next century – before hydrogen-fuelled vehicles may become commonplace. Despite all the difficulties and challenges associated with realization of a practical means of hydrogen storage, nobody can say with any conviction that future research and innovation will not yield a solution and therefore efforts will continue towards this end.

References

1. *The Hydrogen Economy: Opportunities, Costs, Barriers and R & D Needs*, National Research Council and National Academy of Engineering of the National Academies, National Academies Press, Washington, DC, 2004.
2. U. Bossel, B. Eliasson and G. Taylor, *The Future of the Hydrogen Economy: Bright or Bleak?* Updated version distributed at the Lucerne Fuel Cell Forum 2003, 30 June–4 July 2003.
3. R.S. Irani, *MRS Bull.*, 2002, **27**, 680–682.
4. J.M. Ogden, *Prospects for Building a Hydrogen Energy Infrastructure*, Princeton University report, Princeton, NJ, USA, 11 June 1999.
5. E. Akiba and M. Okada, *MRS Bull.*, 2002, **27**, 699–703.
6. R.M. Dell and D.A.J. Rand, *Understanding Batteries*, Royal Society of Chemistry, Cambridge, 2001.
7. L. Schlapbach and A. Zuttel, *Nature*, 2001, **414**, 353–358.

CHAPTER 6
Fuel Cells

As discussed in Chapter 1, energy companies throughout the world are anticipating serious problems of demand growth, fuel shortages, atmospheric pollution and restrictions on greenhouse gas emissions. To meet these challenges, conventional ways of delivering energy and power from fossil fuels must be radically upgraded or replaced by alternatives. Foremost among the latter are nuclear energy and renewables for electricity generation. Fuel cells are a key enabling technology for a future Hydrogen Economy – particularly for electric vehicle propulsion – and therefore are being intensively investigated and developed in many countries.

The fuel cell is an electrochemical device for the conversion of chemicals into direct-current electricity. To this extent, it resembles a primary battery. There are, however, some important differences. In a battery, all the chemicals necessary for its operation are normally confined within a sealed container. Thus, the capacity of a battery, measured in ampere-hours, is determined by the quantity of chemicals that it holds. With a fuel cell, the chemicals are supplied from external reservoirs so that the capacity of the device is limited only by the available supply of reactants. For this reason, fuel cells are rated by their power output, measured in watts, rather than by their capacity, which is indeterminate. In brief, fuel cells may be viewed as energy-conversion devices, in contrast to rechargeable batteries that are energy-storage devices.

Another important distinction between batteries and fuel cells lies in the nature of the chemicals employed. Batteries generally are based on inorganic compounds, *i.e.*, metals and oxides, whereas fuel cells employ hydrogen, methanol or a hydrocarbon as the reactant at the negative electrode and oxygen or air at the positive electrode. This distinction, together with the need to introduce the reactants from outside and to remove the product water (and carbon dioxide if methanol or a hydrocarbon is used) necessitates a different design of cell compared with that of a battery. Provision has also to be made for the removal of heat liberated during the electrochemical reaction.

6.1 Fuel Cell History

The fuel cell has a long history. Sir William Grove invented the device in 1839, some 50 years before the advent of the internal combustion engine. The fuel cell

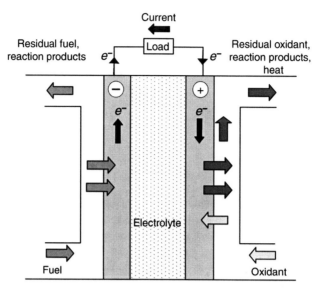

Figure 6.1   Schematic of fuel cell operation.

was not, however, widely adopted as an electricity generator, despite its many advantages. This was a consequence of seemingly intractable materials science and engineering problems. Also, the use of platinum electrodes tended to make the technology uneconomic for most applications. To demonstrate the principle of the fuel cell at low power output and without regard to cost is remarkably simple and can be done in any school science laboratory. The difficulty lies in developing an engineered, high-power system at acceptable cost that will operate untended for long periods. The fundamental operation of a fuel cell is illustrated in Figure 6.1; proper management of the water and heat produced by the reaction are key issues.

Because a fuel cell functions at a low voltage (*i.e.*, well below 1 V), it is customary to build up the voltage to the desired level by electrically connecting cells in series to form a 'stack'. This is achieved by means of a bipolar 'plate-and-frame' arrangement similar to that employed for electrolysers; see Section 4.2, Chapter 4. There are a number of different designs of fuel cell, but in each case the unit cell has certain components in common. These are as follows.

- An electrolyte medium that conducts ions. This may be a porous solid that contains a liquid electrolyte (acid, alkali or fused salt) or a thin membrane that may be a polymer or a ceramic. The membrane must be an electronic insulator as well as a good ionic conductor and must be stable under both strong oxidizing and strong reducing conditions.
- A negative fuel electrode that incorporates an electrocatalyst, which is dispersed on an electronically conducting material. The electrode is fabricated so that the electrocatalyst, the electrolyte and the fuel come

into simultaneous contact at a so-called 'triple-point junction' or 'three-phase boundary'.

- A positive electrode, also with a triple-point electrocatalyst, at which the incoming oxygen (either alone or in air) is reduced by uptake of electrons from the external circuit.
- A bipolar plate that serves to connect individual cells together. Gas channels are usually machined, or moulded, into both sides of the plate to introduce fuel and oxygen/air to the respective electrodes and to remove the reaction products, *i.e.*, pure water in the case of hydrogen fuel. This component is commonly known as a 'flow-field plate' because it serves to smooth the current across the area of the cell stack.
- Seals that keep the gases apart and also prevent cell-to-cell seepage of liquid electrolyte, which otherwise would give rise to partial short-circuits.
- Current collectors that are located at the two ends of the stack and are connected by end-plate assemblies.

There is also a requirement for air or liquid cooling of the stack.

After the pioneering work of Sir William Grove, little more was done towards furthering the fuel-cell concept until the studies of F.T. (Tom) Bacon – first at the University of Cambridge (1946–55) and then at Marshall of Cambridge Limited (1956–61). Bacon was an engineer by profession, with an interest in fuel cells that dated back to 1932. He ploughed a lone furrow, with little support or backing, but showed enormous dedication to the challenge of developing practical, working cells. As an engineer, he appreciated the many potential advantages of the fuel cell over both the internal combustion engine and the steam turbine as a source of electrical power. Early in his career, Bacon elected to study the alkaline electrolyte fuel cell with nickel-based electrodes, in the belief that platinum-group electrocatalysts would never become commercially viable. In addition, it was known that the oxygen electrode is more readily reversible in alkaline solution than in acid. This choice of electrolyte and electrodes necessitated operating at moderate temperatures (100–200 °C) and high gas pressures. Bacon restricted himself to the use of pure hydrogen and oxygen as reactants. In recognition of his outstanding contribution to the advancement of fuel cells, Bacon was elected to Fellowship of the Royal Society and, in 1991, was the first recipient of the Grove Memorial Medal awarded by the Grove Fuel Cell Symposium.

A major opportunity to apply fuel cells arose in the early 1960s with the advent of space exploration. In the USA, fuel cells were first used to provide spacecraft power during the fifth mission of Project Gemini. Batteries had been employed for this purpose in the four earlier flights and in those conducted in the preceding Project Mercury. This switch in technology was undertaken because payload mass is a critical parameter for rocket-launched satellites and it was judged that fuel cells, complete with gas supplies, would weigh less than batteries. Moreover, the objective of Project Gemini was to evolve techniques for advanced space travel – notably, the extra-vehicular activity and the orbital manoeuvres (rendezvous, docking, *etc.*) required for the Moon landing planned in the following Project Apollo. Also, lunar flights demand a power source of longer duration.

A proton-exchange membrane fuel cell (PEMFC) manufactured by General Electric was adopted for the Gemini missions (two modules each capable of providing a maximum power of about 1 kW), but this was replaced in Project Apollo by an alkaline fuel cell (AFC) of circulating-electrolyte design, as pioneered by Bacon and developed by Pratt and Whitney Aircraft (later United Technologies); see Figure 6.2(a). Both types of system were fuelled by hydrogen and oxygen from cryogenic tanks. The AFC could supply 1.5 kW of continuous power and the in-flight performance during all 18 Apollo missions was exemplary. In the 1970s, International Fuel Cells (a division of United Technologies) produced an improved AFC for the Space Shuttle Orbiter that delivered eight times more power than the Apollo version and weighed 18 kg less; see Figure 6.2(b). The system used an immobilized electrolyte and consisted of three modules that could each supply 12 kW continuously and up to 16 kW for short periods. It provided all of the electricity, and also drinking water, when the Space Shuttle was in flight.

The success of fuel cells in the space programme stimulated their deployment in terrestrial applications. Convenience and economics dictate that on Earth it is generally necessary to use air rather than pure oxygen for the positive electrode reactant, and this introduces some technical challenges.

There are six broad categories of fuel cell (see Sections 6.4 and 6.5) and academic laboratories and companies around the world have been engaged in the research and development of these for the past 30 years. Large corporations in Germany, Japan and the USA have undertaken most of the engineering and commercialization work, often with governmental support. With respect to the medium (kW) size required for residential and transportation applications, Ballard Power Systems in Canada has been foremost in advancing the capability of PEMFCs, and Siemens in Europe (Germany) has also been active in the field.

6.2 Why Fuel Cells?

The advantages of fuel cells over other generators of electrical energy such as gas turbines, steam turbines or internal combustion engines with alternators are as follows:

- potentially higher energy efficiency, especially when using the waste heat from the cell in combined heat and power schemes or to power a steam turbine;
- greater efficiency when operated at part load, in contrast to engines;
- no moving components apart from auxiliary fans and blowers; therefore, little lubrication is required;
- almost silent in operation;

(a)

(b)

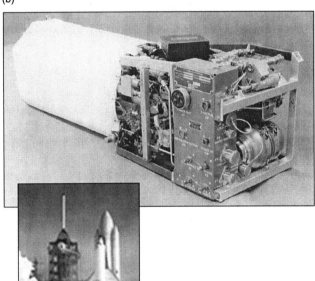

Figure 6.2   (a) Assembly of alkaline fuel-cell modules for Project Apollo missions; (b) alkaline fuel-cell module used in Space Shuttle Orbiter.

- flexibility of fuel supply (hydrogen, for example, may be derived from many primary fuels);
- virtually no exhaust pollutants (low environmental impact, suitable for indoor use);
- pure water emitted when using hydrogen as the fuel;
- rapid load-following response times (after pre-heating);
- good power-to-weight ratio;
- modular construction and ease of installation.

This impressive list of attributes has provided the incentive for much of the research that has taken place on fuel cells in recent years. Naturally, the extent to which each of these characteristics is desirable depends on the application under consideration. Against these advantages should be set the many difficulties that have been encountered in the refinement of fuel cells towards practical power devices that are commercially viable, particularly when fuelled by natural gas, oil or coal. These primary fuels must first be reformed to hydrogen or methanol and then purified, which are tasks that present obstacles when chemical engineers seek to integrate the reformer to the fuel-cell stack and balance the characteristics of the two systems in terms of reaction kinetics and thermal management.

The potential uses for fuel cells are extremely diverse and vary greatly in their power demands, from watts to megawatts. Broadly, the applications fall into the three following categories.

- Stationary

    - dispersed electricity generation at the local level (up to several megawatts);
    - combined heat and power (CHP, for residential or commercial premises, hotels, hospitals, *etc.*)[†];
    - stand-by (emergency) power supplies;
    - remote-area power supplies.

- Transportation

    - electric road vehicles (cars, trucks, buses);
    - railway locomotives;
    - submarines and sub-sea vehicles.

- Portable

    - consumer electronics, (*e.g.*, laptop computers, mobile phones, personal digital assistants);

[†] Also known, particularly in the USA, as 'co-generation'.

    – military equipment (*e.g.*, communications devices, global position-
    ing system receivers, portable radio transmitters, remote surveil-
    lance, field computing, command and control devices, chemical
    agent monitors).

Stationary fuel cells tend to lie in the 1 kW to multi-MW power range,
transportation applications may stretch from a few hundred watts for, say,
an electric scooter, to multi-MW for submarine traction, and units for portable
electronics devices are obviously very much smaller (1–100 W). Clearly, with
such a diversity of sizes (and quite apart from other considerations), the most
suitable type of fuel cell will vary from one application to another. For the
foreseeable future, most fuel cells will operate on hydrogen derived from
natural gas by steam reforming. Large modules, from 100 kW upwards, are
likely to have their own dedicated steam reformer. By contrast, smaller units
will generally draw hydrogen from a central reformer although, as mentioned
earlier (Section 2.3, Chapter 2), 'micro' steam reformers suitable for use with
CHP plant have been developed. Given the six categories of fuel cell under
investigation, the wide range of possible sizes and the variety of fossil fuels that
can be used, it is clear that there is a multi-dimensional matrix from which to
select the best system for a specific purpose. This versatility is one of the
principal attractions of fuel cells.

6.3 Fuel Cell Operation

Fuel cells are the converse of electrolyzers – they convert chemical free energy
into electricity. Fuel is oxidized at the negative electrode with the release
of electrons that pass through the external circuit and reduce oxygen at the
positive electrode. The flow of electrons is balanced by the ionic current
(*i.e.*, the electrical charge carried by ions) in the electrolyte. The reversible
voltage, V_r, produced by the cell (*i.e.*, when there is no current flow, also known
as the 'open-circuit' voltage) is given by:

$$V_r = -\Delta G/nF \tag{6.1}$$

where ΔG is the free energy of the cell reaction (joules per mole), n is the
number of electrons involved in the reaction and F is the Faraday constant
(96 485 coulombs per mole). This relationship applies also to the reverse
(electrolysis) reaction; see Section 4.1, Chapter 4.

    When the fuel is hydrogen, the reversible voltage under standard conditions
is 1.229 V at 25 °C, just as for an electrolyzer. On drawing current from the cell,
however, the voltage developed is much lower than this, typically 0.6–0.8 V.
This is illustrated in Figure 6.3(a), in which the cell voltage is plotted against the
current density. As current starts to be drawn, the voltage falls. The factors that
contribute to this loss in performance are reflected in different regions of the
voltage–current curve. The initial fall in voltage is due to electrokinetic

(a)

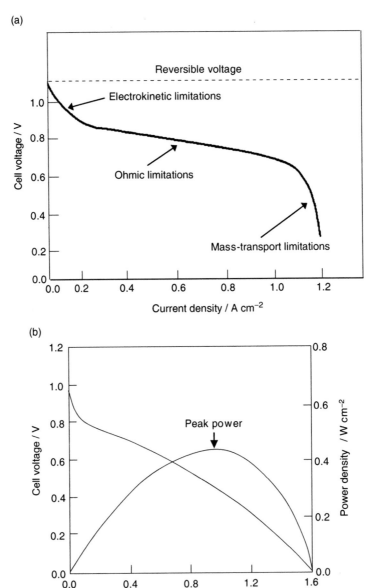

Figure 6.3 (a) Schematic of a voltage–current curve of a fuel cell; (b) power output of a cell as a function of current density.
(Courtesy of the California Institute of Technology).

('activation') limitations at the electrodes. In the central linear portion of the curve, 'ohmic' losses play a more significant role; these arise from resistance to the flow of ions in the electrolyte. Finally, mass-transport ('concentration') limitations, which result from gas consumption outstripping supply, become dominant at high current densities[‡]. There are also electrical resistance losses in the collectors and conductors of the current – and, possibly, side-reactions and leakages of ionic current between series-connected cells – all of which may contribute to the voltage drop. Because each unit cell is prone to these effects, the working cell voltage is well under 1 V, as noted earlier. An example of the power performance of a cell is given in Figure 6.3(b).

Referring back to the voltage versus temperature relationship for water electrolysis shown in Figure 4.2, Chapter 4, it will be apparent that only the free energy of the cell reaction (ΔG) is available for conversion to electrical energy, whereas the entropy term ($T\Delta S$) represents energy that is lost to the surroundings in the form of heat. The higher the temperature, the smaller will be the voltage developed and the larger will be the quantity of heat evolved. Therefore, for optimum output of electrical energy, fuel cells are best operated at low temperatures, but kinetics and electrocatalytic effects then determine the lowest practical temperature. Fuel cells that run at high temperatures are best deployed for CHP schemes or combined with gas turbines so as to make maximum use of the liberated heat. At high temperatures, the heat is both greater in quantity and of higher quality.

The half-cell reactions of the six major types of fuel cell are summarized in Figure 6.4. The systems can be broadly classified in terms of their temperature of operation:

- low-temperature (50–150 °C): alkaline electrolyte (AFC), direct borohydride (DBFC), proton-exchange membrane (PEMFC) and direct methanol (DMFC) fuel cells;
- medium-temperature (around 200 °C): phosphoric acid (PAFC) fuel cell;
- high-temperature (600–1000 °C): molten carbonate (MCFC), direct carbon (DCFC) and solid oxide (SOFC) fuel cells.

Some operational data are given in Table 6.1.

The DMFC differs from the other five types in using a liquid (methanol) rather than hydrogen as the fuel. The former is far more convenient to transport and handle than a gas. Unfortunately, however, the methanol molecule is much more difficult to activate than hydrogen and this necessitates a higher loading of platinum-based electrocatalyst in the negative electrode. Other alcohols, notably ethanol, have also been explored as potential fuels, but comparatively little progress has yet been made with these alternatives. Nevertheless, the generic term 'direct alcohol fuel cell (DAFC)' is often used to describe this technology.

With hydrogen fuel cells, the allowable level of gas purity becomes progressively less stringent with the increasing operational temperature of the respective systems. Both AFC and PEMFC units are poisoned by fairly small

---

[‡]The terms 'activation polarization' and 'concentration polarization' are also frequently used. This practice is to be discouraged because 'polarization' is an ill-defined and misleading concept, as exemplified by the many different definitions to be found in dictionaries. Indeed, Sir William Grove himself stated, in 1874, that the word is 'sadly inaccurate'.

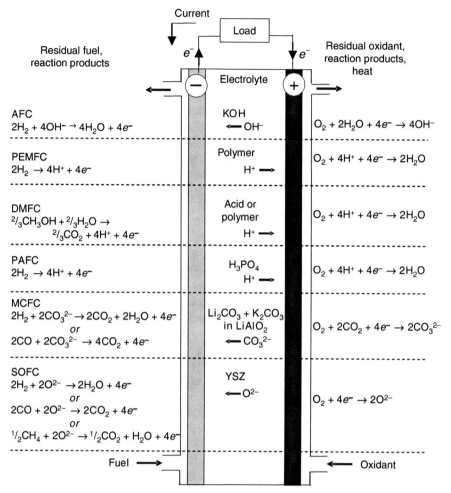

Figure 6.4   Electrochemical reactions occurring in different types of fuel cell.[1]

quantities of sulfur and carbon monoxide (*e.g.*, ~ 10 ppmv CO) and therefore demand hydrogen of the highest purity. By contrast, the PAFC is somewhat more tolerant to carbon monoxide, and the high-temperature cells (MCFC and SOFC) will accept this impurity, and also a variety of hydrocarbon fuels provided that they are free of sulfur-containing compounds. Liquid fuels that are comparatively 'clean' (*e.g.*, liquefied petroleum gas, naphtha, kerosene) may either be reformed to hydrogen externally or treated internally within a high-temperature fuel cell; see Section 6.5. The latter option considerably simplifies the overall plant design and thus reduces its cost. Solid fuels such as coal, however, must be processed externally to yield hydrogen.

A variety of sub-systems and components are required for a fuel-cell stack to function effectively. The exact composition of this so-called 'balance-of-plant'

Table 6.1 Principal types of fuel cell.

Fuel-cell technology	Electrolyte	Temperature range/°C	Electrocatalyst		Fuel	Efficiency^a/ % HHV	Start-up time/h
			Positive electrode	Negative electrode			
PAFC	H_3PO_4	150–220	Pt supported on C	Pt supported on C	H_2 (low S, low CO, tolerant to CO_2)	35–45	1–4
AFC	KOH	50–150	NiO, Ag, or Au–Pt	Ni, steel, or Pt–Pd	Pure H_2	45–60	<0.1
PEMFC	Polymer^b	80–90	Pt supported on C	Pt supported on C	Pure H_2	40–60	<0.1
DMFC	H_2SO_4 Polymer^b	60–90	Pt supported on C	Pt supported on C Pt–Ru	CH_3OH	35–40	<0.1
MCFC	Li_2CO_3	600–700	Lithiated NiO	Sintered Ni–Cr and Ni–Al alloys	H_2, variety of hydrocarbon fuels (no S)	45–60	5–10
SOFC	Oxygen-ion conductor	700–1000	Sr-doped $LaMnO_3$	Ni- or Co-doped YSZ cermet	Impure H_2, variety of hydrocarbon fuels	45–55	1–5

^{a} The reported efficiency of a given type of fuel cell varies widely and often no information is provided on whether the higher heating value (HHV) or the lower heating value (LHV) of the fuel is used (for explanation of these terms, see Box 6.1). The efficiencies that are taken here from the literature should be treated with caution as to their exact meaning and are simply included to provide an approximate comparison of the performance of the respective systems.
^{b} Proton-conducting polymer: perfluorosulfonic acid polymer.

depends upon the type of fuel cell, the available fuel and its purity, and the desired outputs of electricity and heat. Typical auxiliary sub-systems are: (i) fuel clean-up processor, *e.g.*, for sulfur removal; (ii) steam reformer and shift reactor for the fuel; (iii) carbon dioxide separator; (iv) humidifier; (v) fuel and air delivery units; (vi) power-conditioning equipment, *e.g.*, for inverting direct current (d.c.) to alternating current (a.c.) and then transforming to line voltage; (vii) facilities for the management of heat and water; (viii) overall control and safety systems; (ix) thermal insulation and packaging. Individual components include fuel storage tanks and pumps, compressors, pressure regulators and control valves, fuel and/or air pre-heaters, heat-exchangers and radiators, voltage regulators, motors and batteries (to provide power for pumps on start-up).

Sub-systems (i)–(iii) above involve the preparation of pure hydrogen and may, of course, be remote from the fuel cell (or not needed if electrolytic hydrogen is used). Alternatively, they may be integral with the fuel cell, but this adds considerably to the complexity, maintenance and cost of the installation, and also to its size and mass. Heat and water management are particular problems that have to be addressed in the engineering design. Often the balance-of-plant may account for 60–80% of the total capital cost. For this reason, the entire fuel-cell system should have a long operational life in order to prove a sound investment. For the modules themselves, the US Department of Energy has stipulated 40 000–50 000 h service in stationary facilities and 3000–5000 h in transportation applications, although other agencies have set even longer targets. In the event of premature stack failure, the modules might be replaced while retaining much of the balance-of-plant for further use. In the case of a small vehicle, such as a car, it is impractical on the grounds of both size and cost to accommodate on-board reformers and therefore the fuel supplied must be pure hydrogen. For larger vehicles (*e.g.*, buses or trucks), on-board reforming of primary fuels is a possibility, but appears unlikely to be implemented.

6.4   Types of Fuel Cell: Low-to-Medium Temperature

6.4.1   Phosphoric Acid Fuel Cell (PAFC)

The PAFC was the first technology to be commercialized. Units with power outputs of up to 250 kW have been employed for large applications, although interest in this system appears now to be waning. The phosphoric acid electrolyte is held in a porous matrix of silicon carbide, which is bonded with polytetrafluoroethylene (PTFE, trade name Teflon$^{\circledR}$) to provide a structure that is stable at the operating temperature (150–220 °C). The porous carbon electrodes are also PTFE bonded and each supports a platinum metal electrocatalyst. The platinum loadings have been decreased progressively to an average of 0.1 and 0.5 mg cm^{-2} for the negative and positive electrode, respectively. This has been achieved by reducing the platinum crystallite size to around 2 nm and dispersing the metal on carbon black of high surface area (~ 100 m^2 g^{-1}). The PTFE cements the carbon black particles together to form a porous, but integral, network that is supported on a carbon paper substrate.

Stacks of the required voltage are assembled by using bipolar plates that connect the positive of one cell to the negative of the next. These are ribbed so as to provide access for the reactant gases and to distribute them over the area of the cell. Accordingly, as mentioned above, they are often called 'flow-field' plates. There are two configurations. The simpler design is formed from graphite, with channels machined orthogonally on either face: one set of channels to admit fuel to the negative electrode and the other to admit air to the positive; see Figure 6.5(a). The drawbacks to this arrangement are the cost

(a)

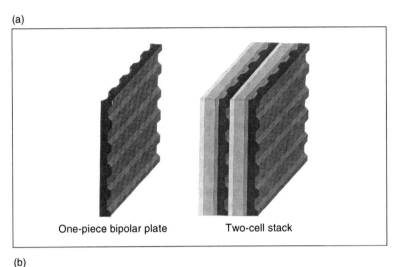

One-piece bipolar plate   Two-cell stack

(b)

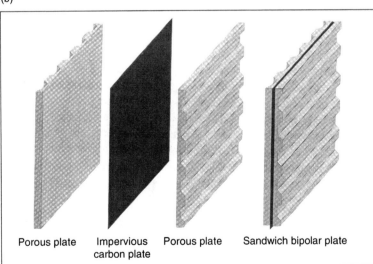

Porous plate   Impervious   Porous plate   Sandwich bipolar plate
               carbon plate

Figure 6.5 (a) One-piece, ribbed, bipolar plate and its use in a two-cell stack; (b) sandwich-type bipolar plate and its use in a two-cell stack.[2] (Courtesy of John Wiley & Sons Ltd.).

of graphite and the difficulty in machining the channels, which results in slow production rates. The alternative construction employs a thin, impervious carbon plate that is sandwiched between two porous members; see Figure 6.5(b). The latter have ribbed channels to admit the gases, can be manufactured without machining (*e.g.*, by the pressing and sintering of suitable powders) and can store phosphoric acid so that the electrolyte matrix can be made thinner and hence present a lower electrical resistance.

It is also necessary to remove the heat generated in the stack. Air cooling is employed in small stacks for which the am15ount of heat to be dissipated is comparatively small, but water cooling is mandatory for larger units with power outputs greater than about 100 kW. Although air cooling is simpler in practice, the ability of air to extract heat is much inferior to that of water. Because the operating temperature is up to 220 °C, the cooling water is usually run at high pressure. A variant is to permit it to boil in the stack, so as to take advantage of the latent heat of vaporization (and also the heat capacity of the liquid) to remove heat. Water of high purity must be employed in order to avoid corrosion of the pipework and hence the build-up of solid deposits. Cooling is accomplished via the use of additional plates through which the air or water is circulated. These are introduced after every few cells in the stack. Liquid cooling requires more complex manifolding, not least on account of the high operating pressure. The manifolds are generally external to the cell stack, as presented schematically in Figure 6.6.

Manufacturers in the USA (Electrochem, United Technologies) and Japan (Fuji Electric, Mitsubishi Electric, Toshiba) have developed PAFC plants in

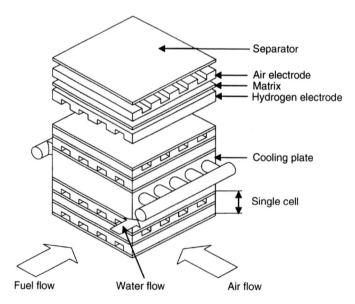

Figure 6.6   Model of a water-cooled PAFC stack.
(Courtesy of The Royal Society).

sizes from 50 kW to 11 MW. Several hundred of these plants have operated around the world to supply combined heat and power to major building complexes such as airports, hotels, hospitals, office buildings and schools. A 200 kW PAFC system offered by United Technologies is shown in Figure 6.7. This also provides up to 270 kW of heat.

Under normal operating conditions, PAFC modules at 0.1 MPa pressure deliver a current density of 150–350 mA cm^{-2}, which corresponds to a cell voltage of 0.7–0.6 V. The electrical efficiency is between 40 and 50%. Note: this efficiency is defined as the ratio of the electrical energy generated by the cell to the hydrogen energy consumed; the latter is based on the higher heating value of the hydrogen fuel (HHV), see Box 6.1 and, later, Box 6.2. When allowance is made for the efficiency of the natural-gas reformer used to generate the hydrogen (80–85%) and for that of the inverter–transformer unit (about 90%), the overall efficiency from primary fuel to electrical energy is 30–40%. This is just about the efficiency range of a conventional coal-fired electricity station and is well below that of a combined-cycle gas turbine plant. If the waste heat generated (at about 200 °C) can be put to good use, *e.g.*, in a CHP scheme, then the overall efficiency will be much improved (possibly up to almost 80%). Such comparison of efficiencies is not, however, especially relevant as a traditional generating station, at 1000–2000 MW capacity, is some 500–5000 times larger than a fuel-cell plant. A better comparison might be with the efficiency of a gas engine, which would be a competitor of broadly similar electrical output. Here, the fuel cell comes into its own on the grounds of quiet operation, low or zero emissions and a particular suitability for part-load running (the efficiency is almost independent of load, whereas that of an engine falls off sharply at low loads or during idling; see Figure 7.5, Chapter 7).

Figure 6.7   200 kW PAFC power plant (the PC25™).
(Courtesy of United Technologies Corporation).

Box 6.1 Higher and Lower Heating Values of Fuels.

All chemical compounds have a 'heat (enthalpy) of formation', ΔH_f, which equates to the heat liberated or absorbed when 1 mole of the compound is formed from its constituent elements. An element in its standard state is defined as having zero heat of formation. The standard molar heat of formation, ΔH_f°, of a compound is then the change in enthalpy, positive or negative, when 1 mole of the compound is formed at standard conditions (298.15 K and 101.325 kPa) from the elements in their most stable physical forms (gas, liquid or solid).

For hydrogen, the heat of combustion equates to the heat of formation of the product water, *i.e.*,

$$H_2(gas) + 1/2 O_2(gas) \rightarrow H_2O(liquid) \qquad (6.2)$$

ΔH_f° (for liquid water) = -285.83 kJ per mole of water. The negative sign indicates that heat is evolved in the process.

When a fossil fuel is burnt completely to carbon dioxide and water, the heat of reaction (or heat of combustion) per mole equates to the difference in enthalpy of the products and the reactants. Thus, for the combustion of methane:

$$\begin{array}{ccccccc} CH_4 & + & 2O_2 & \rightarrow & CO_2 & + & 2H_2O \\ \Delta H_f^\circ : -74.85 & & 0 & & -393.5 & & 2 \times -285.83 \end{array} \qquad (6.3)$$

$$\text{Heat of combustion} = -393.5 - 571.66 + 74.85$$

$$= -890.3 \text{ kJ per mole methane}$$

$$= -55.6 \text{ MJ per kg methane}$$

These values for the heats of combustion of hydrogen and methane are fundamental physical quantities and represent the maximum quantities of heat that can be evolved in the respective combustion processes. They are termed the 'higher heating values' (HHVs) for the fuels.

In many practical appliances, which range from steam engines to internal combustion engines, the carbon dioxide and water products are released to the atmosphere at comparatively high temperatures and the heat that these gases hold above the reference temperature of 298.15 K is lost. This heat consists of both the 'sensible heat' (*i.e.*, the heat carried by the substance – its 'heat capacity') and the latent heat of condensation of the steam. For practical engineering purposes, it has become customary, therefore, to define a 'lower heating value' (LHV), which corresponds, arbitrarily, to the maximum heat recoverable when the reaction products are emitted at 423.15 K.

When comparing the efficiencies of various appliances that are using the same fuel, it is convenient to take the LHV since this is usually the maximum amount of heat that can be recovered in the appliance itself. This does not hold, however, for modern appliances with waste-heat recovery systems (*e.g.*, condensing gas boilers). In such cases, the fundamental HHV rather than the

Box 6.1 Continued.

practical LHV should be used. Similarly, the LHV is not applicable when using different fuels as these will have different sensible and latent heats. The difference between the two values varies with the fuel. Generally, the sensible heat is small and it is the heat of condensation of steam that predominates. It follows that the richer the fossil fuel is in hydrogen, the greater is the deviation between the LHV and the HHV. For example, the ratio of LHV to HHV is almost 1.0 for carbon monoxide (no hydrogen), 0.98 for coal (a little hydrogen), 0.91 for petrol, 0.90 for methane, down to 0.85 for hydrogen.

In absolute numbers, the HHV and LHV for hydrogen are 142 and 120 MJ kg^{-1}, respectively, *i.e.*, the former is 18% greater than the latter. This is important when considering the efficiency of fuel cells, which generally burn hydrogen. Given that the heat of the exhaust gases can be recovered (*e.g.*, in a CHP scheme), the HHV should be adopted when calculating the efficiency. Some authors have erroneously employed the LHV, which flatters the fuel-cell efficiency unduly.

The useful commercial life of fuel cells depends on many factors, among which are the working temperature, the current density and voltage, the operating pressure and the frequency of start-up and shut-down. For PAFCs, modes of deterioration with time include (i) corrosion of the carbon electrodes, particularly the positive at potentials above 0.8 V, (ii) electrode flooding, which impedes the transport of reactants to electroactive sites, and (iii) progressive agglomeration of the platinum electrocatalyst particles, which decreases the active surface area. Considerable research has been conducted into minimizing these limitations with the result that lifetimes of 40 000 h are now anticipated for PAFCs.

6.4.2 Alkaline Fuel Cell (AFC)

As mentioned earlier, the AFC dates back to pioneering work of F.T. Bacon at Cambridge and the subsequent adoption of the technology by NASA. Pure oxygen, rather than air, serves as the oxidant in AFCs that are designed for spacecraft in order to minimize the mass of the gas cylinders and to simplify the system. For most terrestrial applications, considerations of cost and portability rule out the use of oxygen; air is freely available and does not have to be carried. Unfortunately, however, the AFC is susceptible to poisoning by carbon dioxide in the atmosphere. This gas reacts with the potassium hydroxide electrolyte solution to form potassium carbonate, with resultant degradation in cell performance through a variety of effects, namely: (i) decrease in the concentration of hydroxide ions, which lowers the rate of reaction at the negative electrode; (ii) reduction in the solubility of oxygen, which intensifies activation losses at the positive electrode; (iii) increase in the viscosity of the electrolyte, which results in lower diffusion rates of reactants; (iv) a fall in

electrolyte conductivity, which increases ohmic losses; and (v) eventual precip-
itation of carbonate, which is capable of deactivating the electrodes and
blocking the pores of an immobilized electrolyte or the pathways in a circulating system. It is therefore essential to pre-treat the air to remove the carbon
dioxide. The usual procedure is to pass the gas through a soda-lime trap that
has to be replenished from time to time. This is an operational inconvenience
rather than a fundamental obstacle. For a similar reason, the hydrogen fuel
must be completely free of carbon dioxide (in addition to carbon monoxide),
which necessitates rigorous cleaning and separation of the gas and thus imposes
an added cost penalty.

The nominal operating temperature of the AFC is 60–90 °C, although it will
perform over a wider range of temperature (50–150 °C) according to the choice
of gas pressure, electrolyte concentration and electrocatalyst. It is possible to
operate at even higher temperatures. For example, the stacks in the Apollo
spacecraft (see above) were run at 230 °C, but then it was necessary to increase
the concentration of potassium hydroxide in the electrolyte solution from 30 to
75 wt.% to prevent boiling. In the Space Shuttle Orbiter [see Figure 6.2(b)], the
concentration of potassium hydroxide was reduced back to 32 wt.% and the
temperature to 93 °C, but these modifications entailed the development of a
better electrocatalyst.

Given the relatively low operating temperatures of AFCs, little heating is
required at start-up, which is rapid with respect to some other types of fuel cell;
see Table 6.1. This feature provides AFCs with a distinct advantage. The
electrolyte is held in a porous matrix (traditionally, asbestos), which also serves
to separate the electrodes. Working with hot, concentrated potassium hydroxide
solution is no easy matter; in particular, this electrolyte is corrosive towards
seals and tends to 'creep' so that its containment is made difficult. In order to
separate out the product water and extract the waste heat, the electrolyte is
circulated through evaporators. The procedure has to be controlled so as to
maintain constant temperature and constant electrolyte concentration. At the
same time, the design of the stack has to ensure that shunt currents between
adjacent cells are held to a minimum. These are self-discharge currents that, as a
consequence of cells developing different voltages, travel through the external
electrolyte manifold and pumps in series-connected (bipolar) stacks. Even at
open-circuit, a residual shunt can still flow and consume reactants without doing
any useful external work. Hence shunt currents reduce efficiency. Introducing
breaks or gaps in the flowing electrolyte system can lower these currents, but this
inevitably introduces some form of flow resistance and thereby increases the
energy required for pumping.

In addition to the work undertaken by United Technologies (and later by its
subsidiary, International Fuel Cells) for the US space programme, as discussed
above in Section 6.1, major developers of AFCs have been Siemens in Germany
and Elenco in Belgium. In the 1970s, Siemens built a 7 kW, circulating-
electrolyte AFC module, which operated on hydrogen and oxygen, together
with the appropriate manufacturing technology for the bipolar stacks. The cells
employed porous nickel negative electrodes, asbestos separators and catalytic

silver positives. The gas pressure was 0.2 MPa. The efficiency of the modules was claimed to be 61–63% at full load and 71–72% at 20% load. In the early 1990s, Siemens adopted the prototype as a basis for the construction of much larger AFCs to power submarines. Oxygen rather than air modules are ideal for this application since the gas has to be generated on-board electrolytically or supplied in cylinders. These units permit longer underwater range compared with lead–acid batteries, which are the customary choice of power source for submarines.

Elenco produced AFCs in multi-kW sizes for powering electric vehicles. The stacks ran on air rather than on oxygen and were monopolar in configuration. The nominal operating temperature was 65–70 °C. A circulating electrolyte of 6.6 M KOH served as both a coolant when the cell was operational and as a pre-heating medium. The electrodes were mounted on injection-moulded frames and 24 cells were assembled in series. Due to the moderate temperature and the very small amount of platinum electrocatalyst that was used (0.15–0.3 mg cm^{-2}), the current densities were low (100 mA cm^{-2} at 0.7 V). Systems with outputs of 1.5, 15, 50 and 80 kW were field-tested in different applications that included small-scale generators and power sources for a Volkswagen electric van and an 80-passenger demonstration bus. Because the poisoning problem of carbon dioxide could not be resolved satisfactorily, AFCs employing air rather than oxygen fell out of favour and Elenco discontinued its fuel cell business in the mid-1990s.

In summary, AFCs based on pure oxygen have been proved to be both powerful and successful, whereas those operating on air appear not to have a commercial future at present. In fact, it is generally considered that the sensitivity of alkaline electrolyte solutions to carbon dioxide is the principal reason why PAFCs have made such an inroad into fuel-cell technology since the 1970s and have been followed by PEMFC systems in recent years. If it were not for this poisoning problem, it is likely that air-based AFCs would also be serious competitors.

6.4.3   Direct Borohydride Fuel Cell (DBFC)

The possible application of sodium borohydride (NaBH$_4$) as a convenient chemical carrier for hydrogen was discussed in Section 5.5, Chapter 5; reaction with water releases hydrogen. In principle, it should be possible to utilize this compound more directly in a fuel cell by oxidizing it electrochemically. Sodium borohydride is unstable in acidic media, but dissolves in strong alkali to yield a stable electrolyte solution. This, then, is a variant of the AFC, with the added attraction that it is not necessary to provide cylinders of compressed hydrogen. Rather, the 'fuel' is already present in the electrolyte solution. The discharge product is sodium borate (NaBO$_2$), which, if desired, could be factory recycled as described in Section 5.5, Chapter 5. Limited research on this system has shown some promise, while at the same time throwing up technical problems that have yet to be solved. At least one Japanese company is developing the system as an alternative to the direct methanol fuel cell (DMFC) as a power source for laptop computers and, possibly, mobile phones. Thus, the 'direct borohydride fuel cell' (DBFC) has certain characteristics of the AFC (strongly alkaline electrolyte) and others of the DMFC (chemically bound hydrogen, suitability to power portable electronic devices). The prospects for use in much larger applications are as yet unknown.

6.4.4 Proton-exchange Membrane Fuel Cell (PEMFC)

The PEMFC (also known as the 'polymer electrolyte membrane fuel cell' and as the 'solid polymer electrolyte fuel cell', SPEFC) is an acid-electrolyte fuel cell that operates on the same principle as the PAFC (see Figure 6.4), although with some important differences. The electrolyte is a thin membrane of a copolymer that is highly conductive to hydrated protons. The standard membrane material is that sold under the trade name Nafion®, which was produced by the DuPont Corporation in the USA, originally for application in the chlor-alkali industry. The backbone consists of PTFE to which side-chains of perfluorinated vinyl polyether are bonded via oxygen atoms. Sulfonic acid groups, $-SO_3H$, at the ends of the side-chains can exchange with protons to facilitate their passage through the membrane. Nafion® has two notable characteristics: (i) the PTFE backbone is water repellent and thereby prevents flooding by product water; and (ii) the sulfonic acid group attracts water molecules that form aqueous micelles within the polymer and these, in turn, serve to conduct the protons across the membrane. The perfluorosulfonic acid membranes, as a class, have inherent chemical and oxidative stability. They are, however, difficult to manufacture and are therefore relatively expensive to purchase. Accordingly, other companies, notably in the USA and Japan, have been searching for significantly cheaper alternatives. Efforts have focused mainly on the development of thinner membranes that, if attainable, would also offer the added performance benefits of lower electrical resistance (and hence higher power density) and improved hydration.

For PEMFCs to achieve widespread commercial success, membranes should also have greater durability, better water management and the ability to function at higher temperatures – all at reduced cost. In particular, operation at higher temperatures would yield significant energy benefits. For example, heat rejection would be easier and would warrant the use of smaller heat-exchangers in fuel-cell systems. Honda in Japan has recently claimed just such a breakthrough; see Section 7.4, Chapter 7. This was achieved by replacing the PTFE backbone of the usual membrane structure with a new aromatic material. As in the case of Nafion®, proton transfer is facilitated by attaching sulfonic acid groups to the side-chains. The increased conductivity of this new membrane is said to permit start-up at temperatures as low as $-20\,°C$, which is a significant advantage in cold climates, and to allow PEMFCs to be run at higher temperatures than formerly. The new 'Honda FC Stack' features 50% fewer components than an earlier prototype and thus makes for easier manufacturability and lower cost.

The central component of the PEMFC is the 'membrane–electrode assembly' (MEA), which consists of the polymer membrane sandwiched between a positive and a negative electrode, as illustrated in Figure 6.8. Typically, the MEA has a thickness of about 0.5 mm. The key objective in electrode manufacture is to produce a structure in which the electrocatalyst is finely divided and highly dispersed on a conductive substrate, so that maximum surface area is presented to the membrane electrolyte and, in turn, the electrocatalyst loading (and hence its attendant cost) is kept to a minimum. In one method of electrode preparation, an emulsion of fine carbon powder (particle size 50–100 nm) and isopropyl alcohol is mixed with platinum particles (just a few nm in size) and deionized water to form a slurry. Carbon paper (or thin carbon cloth or felt) is impregnated with an appropriate amount of the slurry and then dried. The two processes are repeated until the electrocatalyst loading reaches the desired level for the respective positive and negative electrodes. The electrolyte membrane is then placed between a pair of these electrodes and the assembly is hot pressed. Alternatively, the slurry can be applied directly to both sides of the electrolyte membrane and the carbon paper/cloth added separately. It is customary to mix PTFE into the carbon because this polymer is hydrophobic and will assist in expelling the product water to the electrode surface, where it can evaporate. In

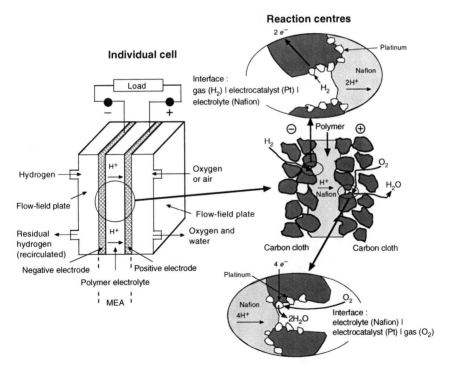

Figure 6.8 Schematic of a PEMFC showing pore structure and distribution of electrocatalyst at a three-phase boundary.
(Courtesy of Heliocentris Energiesysteme GmbH).

addition to providing a mechanical support, the porous carbon substrate allows the diffusion of reactant gas to the electrocatalyst and is therefore often called the 'gas-diffusion layer'. Since the carbon paper or cloth is electrically conducting, it serves also to carry electrons to and from the electrocatalyst. It is important that product water is removed from the stack so that the electrodes do not become flooded, otherwise access of gas to the electrocatalyst sites would be impeded and hence reaction at the three-phase boundary would be seriously impaired. The MEA microstructure and the electrode reaction mechanism are shown schematically in Figure 6.8. The membrane must be kept sufficiently humidified during cell operation to permit effective conduction of protons. Accordingly, the stack may also contain a humidification unit to maintain the gases at close to their saturation levels and thereby prevent dehydration of the membrane. Reactant gas distribution and humidity control are critical aspects of PEMFC design.

The MEA is held together by flow-field plates that serve to distribute the current uniformly and also act as bipoles to connect adjacent cells in series. A diagram of a single cell within a stack – separated into its components – is given in Figure 6.9. In this example, the electrocatalyst has been applied directly to the membrane. As with the PAFC, the flow-field plates are made from graphite, graphite–polymer composites or solid metal and contain channels to distribute the reactant gases so that each gas is in contact with the whole surface of its respective electrode via the porous gas-diffusion layer. In addition to this role, the plates have to distribute a cooling fluid throughout the stack, while at the same time ensuring that the reactant gases and cooling fluids do not mix. Several versions of flow-field plate have been investigated, *e.g.*, parallel-channel, serpentine and inter-digitated structures; major developers of fuel-cell stacks tend not to disclose their respective preferences. One of the problems commonly encountered is that of gas escape from the perimeter of the cell. This is solved by making the membrane larger than the gas-diffusion backing plates and inserting Teflon masks in front of the graphite blocks. The entire assembly is then sealed around its perimeter.

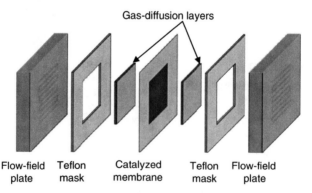

Gas-diffusion layers

Flow-field   Teflon   Catalyzed   Teflon   Flow-field
plate   mask   membrane   mask   plate

Figure 6.9   Schematic of a single-cell construction of a PEMFC.
(Courtesy of the US Department of Energy).

As mentioned earlier in Section 6.3, any carbon monoxide in the fuel gas would readily poison the platinum electrocatalyst and, to avoid deterioration in negative-plate performance, its concentration should be kept to below 10 ppmv by means of a gas-cleaning process. Thus there is a strong incentive to develop electrocatalysts that are both cheaper and less susceptible to poisoning. Nevertheless, remarkable progress has been made with PEMFCs that employ platinum. In the late 1980s, cell stacks with a platinum loading of $28 \, mg \, cm^{-2}$ delivered a power density of about 100 W per litre. Today, stacks using less than $0.2 \, mg \, cm^{-2}$ of platinum achieve over 1 kW per litre, *i.e.*, a 10-fold increase in power with only around one-fifteenth as much electrocatalyst.

Despite these advances, one of the aims of continuing research on PEMFCs is to reduce the loading of the expensive platinum electrocatalyst still further, particularly for transportation applications (electric vehicles). Modern cells can operate with a total loading (both electrodes) of $0.4 \, mg \, cm^{-2}$. The target set by the US FreedomCAR Partnership (see Section 7.7, Chapter 7) is to reduce this combined amount to $0.2 \, mg \, cm^{-2}$ by 2015 – this is considered to be a very demanding proposition. There are, however, other uses for PEMFCs, both in stationary power generation and in the transportation sector, that may accept a less exacting limitation to the quantity of electrocatalyst. Even at the present level, the cost of platinum is no longer a major obstacle for many applications of these fuel cells. It should also be appreciated that the precious metal catalyst can be recovered from spent fuel cells and recycled. This would be important if, in a future scenario, the majority of road vehicles were powered by PEMFCs since the world inventory of accessible platinum might then prove to be inadequate.

The Nexa® power module, manufactured in Canada by Ballard Power Systems, is shown in Figure 6.10(a). The unit has an output of 1.2 kW. Introduced in 2001, it became the world's first volume-produced PEMFC for stationary and portable power-generation applications. A subsequent Ballard product, the AirGen®, has been designed for indoor duty; see Figure 6.10(b). This 1 kW PEMFC can act as a stand-by or uninterruptible power supply and has a built-in suppression system that protects sensitive electronic equipment from high-voltage surges.

Intelligent Energy Holdings, a UK manufacturer, has produced PEMFC units in sizes from 1 to 75 kW for a range of applications that include consumer electronics, remote-area power supplies, military communications, industrial uses and battery charging. The hydrogen is supplied as small canisters of metal hydride that are simply plugged into the fuel-cell pack as required. A particular application is to provide motive power for electric motorcycles; see Section 7.5, Chapter 7. The units are said to be capable of cold starting from $-20 \, °C$ in less than 2 min.

In the field of much larger units, Ballard has conducted trials on 250 kW PEMFC plants to evaluate their suitability and performance as distributed power generators for commercial use. An installation at Mingolsheim in Germany logged 6000 h of operation at 75% efficiency with utilization of waste heat. Stacks manufactured by Ballard are also being used to power electric buses; see Section 7.5, Chapter 7.

The attraction of PEMFCs for propulsion applications lies in their relatively high power density and compactness. Furthermore, their present low

(a)

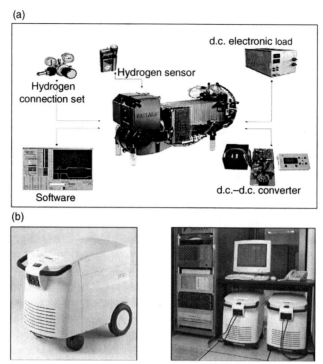

(b)

Figure 6.10   (a) NEXA® and (b) AirGen® power modules.
(Courtesy of Ballard Power Systems Inc.).

operational temperature (80–90 °C) permits rapid start-up from cold, *i.e.*, <0.1 h; see Table 6.1 Adverse features are the cost of the stacks, which arises mainly from manufacture of the MEAs, and the requirement for high-purity hydrogen and its on-board storage. For most vehicles, the production of hydrogen *in situ* by the reforming of methanol is not thought to be a practical proposition. Moreover, there remains the difficult problem of integrating the functions of the reformer and the fuel-cell stack so as to meet the varying electrical power demands dictated by the driving schedule.

Methanol is a liquid fuel that is readily dispensed and stored but, compared with conventional liquid fuels, is by no means ideal; it is both toxic and water-miscible and has an energy content per litre that is well below that of petrol; see Table 1.4, Chapter 1. A further drawback of incorporating a reformer into the vehicle is the need to supply heat for its operation. At 80–90 °C, the waste heat from the fuel cell would be inadequate and it would be necessary to burn some of the methanol to provide the heat or to use electrical heating. Either of these two options would obviously reduce the effective efficiency of the fuel-cell system. Altogether, on-board methanol reformers pose as many engineering problems as on-board hydrogen storage. Furthermore, the combined cost of a reformer and a fuel cell is likely to prove prohibitive, at least for small vehicles such as cars.

Comparable difficulties with the fuel supply do not arise for stationary applications. Space, and especially mass, are usually less critical considerations and it would be possible to integrate a fuel reformer with the fuel cell, particularly if a steady supply of electricity is required. Assuming that PEMFCs continue to perform reliably for an acceptably long life, the main barrier to the widespread deployment of such technology for stationary power generation is the cost target imposed by competitive systems. Plug Power in the USA have introduced a residential CHP system based on a PEMFC (the GenSys™) that runs on natural gas; see Figure 6.11(a). An installation of 13 such units on Long Island, USA, is being used to feed electricity back into the local grid; see Figure 6.11(b). Similar domestic CHP units are also under development by Intelligent Energy in the UK.

6.4.5 Direct Methanol Fuel Cell (DMFC)

A liquid fuel would be highly attractive for mobile or portable applications if only it could be utilized directly and did not first have to be reformed to hydrogen. As mentioned above, such a fuel is methanol (CH_3OH), which may be regarded in strictly non-chemical terms as 'two molecules of hydrogen made liquid by one molecule of carbon monoxide'. Also, it has the attractions of being plentiful and cheap, and the only products of combustion are carbon dioxide and water. Of course, carbon dioxide is a greenhouse gas, but emissions from mobile/portable uses would be small compared with those emanating from the employment of heat engines. In principle, it should be possible to oxidize methanol electrochemically in a single step, given the availability of a suitable electrocatalyst. This has long been a dream of electrochemists engaged in advancing the science of fuel cells.

Early experiments explored the direct oxidation of methanol in a sulfuric acid electrolyte, but attention gradually turned to the use of the emerging technology of proton-exchange membranes. This change in design has extended the operational temperature of DMFCs and has resulted in significant improvements in performance. The following electrochemical reactions take place in the cell[§]:

at the negative electrode:

$$CH_3OH + H_2O \rightarrow CO_2 + 6H^+ + 6e^- \tag{6.4}$$

at the positive electrode:

$$3/2\,O_2 + 6H^+ + 6e^- \rightarrow 3H_2O \tag{6.5}$$

Note: it is the protons that migrate from the negative to the positive electrode and not the methanol molecules, and it is this feature that permits the use of a proton-exchange membrane.

[§] These reactions are identical with those given in Figure 6.4, except that the latter have all been normalized to four electrons.

(a)

(b)

Figure 6.11   (a) Residential fuel-cell system (the GenSys™) for domestic generation of combined heat $(3-9\,kW_{th})$ and power $(2.5-5\,kW_e)$. System components: (1) fuel processor; (2) power generation (fuel-cell stack); (3) power electronics; (4) energy storage; (5) thermal management. (b) Series of GenSys™ units installed on Long Island, USA. (Courtesy of Plug Power Inc.).

A number of challenges have been identified with the DMFC. In brief, these are as follows.

- *Improved kinetics at the negative electrode.* Research to date has found that platinum-based electrocatalysts are the only materials that are able to activate methanol. Even then, the overall reaction at the negative electrode is sluggish, particularly at low temperatures. This is because the electrochemical oxidation of methanol to carbon dioxide is substantially more complex than that given by reaction (6.4) – the transfer of six electrons does not take place simultaneously. Partial electron transfer results in the formation of surface-adsorbed species on the electrocatalyst that, in turn, lower the activity for methanol oxidation. The main 'poison' is linearly bonded carbon monoxide (Pt–CO). Subsequent reactions are thought to involve oxygen transfer (from water) to the Pt–CO species to produce carbon dioxide, which then desorbs from the platinum surface, *i.e.*,

$$Pt + H_2O \rightarrow Pt - OH_{ads} + H^+ + e^- \tag{6.6}$$

$$Pt - OH_{ads} + Pt - CO_{ads} \rightarrow 2Pt + CO_2 + H^+ + e^- \tag{6.7}$$

These reactions occur at a very slow rate on pure platinum and this has resulted in a large research effort to discover more active electrocatalysts. At present, platinum–ruthenium offers the best performance. Ruthenium adsorbs water more readily than platinum and the resulting species, $Ru-OH_{ads}$, assists the removal of carbon monoxide from neighbouring platinum sites. Despite this beneficial effect of ruthenium, still more efficient electrocatalysts are required to enhance the power delivered by DMFCs, especially if the system is to compete favourably with hydrogen–air PEMFCs.

- *Higher temperature operation.* An improvement in the reaction kinetics at the negative electrode is also being sought through finding a means to allow cells to operate at higher temperatures (*e.g.*, 150–180 °C). This essentially rests on the development of new membrane materials that have a higher resistance to dehydration and can therefore operate with less humidification or do not require water to maintain their proton conductivity. Other desirable features include low permeability of methanol (see the following discussion), long-term thermal and chemical stability, high flexibility and good mechanical strength.
- *Mass-transport effects.* Proton-exchange membranes are permeable to both methanol and water. The 'crossover' of methanol, which is caused by protonic drag and is similar to the electro-osmotic drag of water, will result in a loss of cell performance and efficiency due to deactivation of the positive electrode and unproductive fuel consumption by direct reaction with oxygen. This problem, together with the above-mentioned poor kinetics of electro-oxidation at the negative electrode, require a DMFC to have a loading of at least 10 times more platinum-based electrocatalyst than a PEMFC to achieve a comparable power output. Methanol crossover can be reduced to more manageable levels by using dilute aqueous solutions of less than 10 vol.% methanol, but this will cause a decrease in power density. A second debilitating factor is the migration of water across the membrane; this floods the positive electrode and thus inhibits the access of oxygen. New membrane

materials and better MEA designs are being investigated as more effective means to moderate the movement of both methanol and water.

- *Improved kinetics at the positive electrode.* The electrochemical reduction of oxygen is also a complex process; each molecule requires the transfer of four electrons for complete reduction. Surface intermediates are formed and these lower the kinetic performance of the electrode. Platinum-based electrocatalysts are necessary to allow the reaction to proceed at a useful rate. A more effective and cheaper alternative electrocatalyst is an ongoing research target, as is equally the case for positive electrodes in PEMFCs.

Despite these limitations, there is still considerable enthusiasm for the DMFC because of its obvious advantages, namely: (i) the ready availability and low cost of methanol; (ii) its ease of distribution and storage; and (iii) the high specific energy $(Wh\,kg^{-1})$ of the cell compared with that of batteries. Present interest is focused more on portable than on mobile power applications since, in the former, much smaller units are required, the power demands are lower and electrocatalyst cost is less of an issue. With the growth of third-generation mobile phones, laptop computers and other hand-held information and communications technology, there is an ever-increasing desire for longer 'active' lifetimes that are beyond those obtainable from even the best rechargeable lithium batteries. The DMFC has the capability to fulfil these needs and its practical success in this sector is seen as a more immediate goal than its use in fuel cell vehicles.

6.4.6 Miniature Fuel Cells

Many companies are engaged in developing fuel cells with power outputs in the range 1–50 W for electronic appliances and communications equipment. In addition to their use in mobile phones and laptop computers, there are demands from the military for superior power sources to provide more extensive 'silent watch' operation of surveillance and also communications facilities in the battlefield. Fuel cells are attractive for all these purposes, not only because they are more portable than batteries and have a longer life, but also because refuelling is much faster than recharging a battery. For example, replacing an empty ampoule of methanol would be akin to refuelling a cigarette lighter and would take just a few seconds. It is even possible that small DMFCs will be employed for recharging batteries rather than for powering devices directly. This would offer two operational advantages: batteries are more capable of meeting fluctuating power demands than fuel cells and a mains electricity supply would not be required for recharging. One current problem is that international airline regulations do not permit the carriage of methanol on civil aircraft, which would be a limitation for laptop computers. It may prove possible to rescind this restriction for appropriately packaged, small-scale, methanol ampoules.

Because these 'miniature' fuel cells are devices for use in high-value markets, the cost of the polymer membranes and the platinum-based electrocatalyst is less significant than for larger-scale applications. There are, however, exacting specifications for the design of small fuel cells: (i) use at or near, ambient temperature; (ii) a minimum number of miniaturized balance-of-plant components; in particular, no pumps for forced air supply (so-called 'air-breathing' cells) and no membrane re-humidification facility; and (iii) automatic thermal and water management. With the advent of 'microelectromechanical systems' (MEMS) technology, it may prove possible to relax some of these provisions. For instance, miniature pumps could be used to introduce fuel and air into the cell stack.

A typical unit would comprise a fuel-cell pack, a fuel cartridge, a fuel flow-regulator and a d.c.–d.c. converter. Fujitsu, Hitachi, Motorola, Nippon Electric and Toshiba are among the many companies that are presently conducting research and development on DMFCs for portable consumer electronics. Recent examples of prototype devices are shown in Figure 6.12. Clearly, the cells produced to date are still too large for use with mobile phones. Here, the provision is for a system that can be accommodated within the handset itself. At the same time, with the advent of new communications technology (terrestrial digital television broadcasting, miniature hard-disc drives for downloading, *etc.*), there will be an ongoing desire for more power and more energy storage in mobile phones. An illustration of the evolution of power sources for electronic applications over the past two decades is presented in Figure 6.13.

Better batteries may supply the required power, but only fuel cells will provide the energy capacity needed for long run times. It has been calculated that if mobile phones are to remain at their present size, the methanol fuel cell will have to be shrunk to a volume of around $5 \, cm^3$ and still provide an output of $1 \, W$ – quite a challenge in miniaturization! Meanwhile, several Japanese companies have already produced prototypes in which the fuel cell and methanol cartridge are built into the back of the handset. This results in integrated units that are still two to three times more bulky and heavy than conventional cellular phones. Nippon Telegraph and Telephone has opted to miniaturize a PEMFC running on hydrogen gas supplied from a small built-in canister of metal hydride. Again, the outcome is a rather large device.

The widespread introduction of miniature fuel cells would obviously have little direct impact on the international energy scene, although they might affect it indirectly. For example, the technology could be used to power (i) remote sensors for measuring natural and anthropogenic emissions of greenhouse gases and atmospheric pollutants and (ii) safety systems for monitoring seismic activity in the vicinity of sites used for the underground storage of carbon dioxide (see Section 3.3, Chapter 3) and for raising the alarm in the advent of any leakage of gas. Their real significance, however, is in encouraging developers to address the generic and basic problems associated with the DMFC and in increasing public awareness and acceptance of fuel cells in general. If miniature fuel cells can be brought to a successful conclusion, then extension of DMFC technology to larger sizes may prove possible.

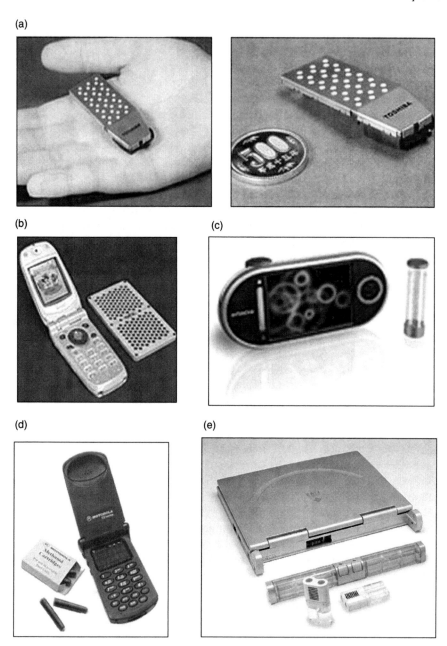

Figure 6.12 Developmental miniature methanol fuel cells for electronic applications: (a) Toshiba, for digital audio players and wireless headsets for mobile phones; (b) NEC, for mobile phones; (c) Hitachi, for personal digital assistants; (d) Motorola, with methanol capsules, for mobile phones; (e) Casio, for laptop computers.

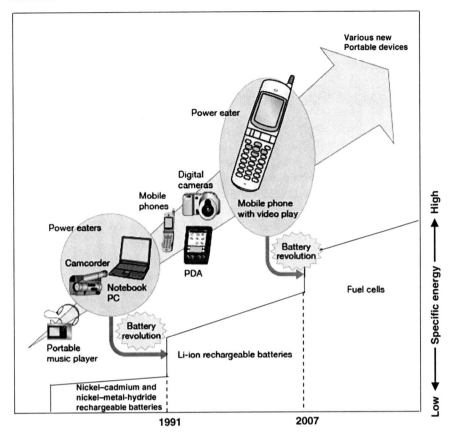

Figure 6.13   Evolution of portable power sources.
(Courtesy of Nikkei Business Publications Asia Ltd).

There is also considerable industrial interest in 'micro' fuel cells with power outputs in the sub-watt range. Potential applications for these power sources include wireless sensor networks, intelligent chip cards and autonomous systems for data collection and control. In the last-mentioned context, many possibilities are foreseen in the automation and control of distributed energy-generation systems joined together in a local network. The drive for these micro fuel cells is based on the availability of higher power, both per unit mass and per unit volume, compared with button cells and other small batteries. The fabrication techniques being explored are based on planar technology inherited from the semiconductor industry, *e.g.*, lithography and patterning of free-standing grid structures. It is likely that micro fuel cells, if they come to fruition, will be PEMFCs fuelled by methanol, either directly or via reforming to hydrogen. Accordingly, scaled-down fuel processors are also being developed. Battelle in the USA has demonstrated such a unit. It is composed of two vaporizers, a heat-exchanger, a catalytic combustor and a methanol reformer. Together, these components occupy only $0.2\,cm^3$ and weigh less than 1 g. An alternative and probably simpler option is to use a small canister of a chemical hydride, such as sodium borohydride, that reacts with water to liberate hydrogen; see Section 5.5, Chapter 5. This would offer a useful bonus by providing a route for recycling the product water from the fuel cell.

6.5   Types of Fuel Cell: High Temperature

There are two broad types of high-temperature fuel cell, namely those with a molten carbonate electrolyte (MCFCs) and those with a solid oxide electrolyte (SOFCs). For large-scale power generation (*i.e.*, $> 250\,kW$), these systems are expected to be the preferred technology on account of their high electrical efficiency. Also, with operational temperatures of $600-1000\,°C$, the quality of the waste heat produced is ideal for industrial CHP applications. When providing heat for such a purpose, the overall efficiency can be above 80%. The further attractions of high temperature are as follows:

- the achievable power densities are relatively high;
- platinum-based electrocatalysts are not required;
- hydrogen of lower purity may be used as fuel;
- when running on natural gas, the fuel reformer may be integral with the fuel-cell stack ('internal reforming', see below); this saves on capital cost and also uses the heat from the electrochemical reaction to support the endothermic reforming process.

The technical challenges posed by these systems are different from those facing low- to medium-temperature cells. For instance, there are no severe kinetic limitations at the electrodes or poisoning of electrocatalysts by impurities (other than sulfur) in the fuel gas. Instead, material science issues arise with (i) sintering of the electrodes and the electrolyte matrix, (ii) corrosion of cell components in molten salt electrolytes (MCFC), (iii) electrolyte migration in the external manifolds of MCFCs and (iv) differential expansion coefficients of the materials of construction in all-solid-state systems (SOFCs).

When internal reforming is being used, some of the heat liberated in the electrochemical reaction would be employed to support the endothermic reforming reaction and thus less would be available for CHP operations. By way of compensation, however, no additional fuel is burnt solely to provide heat for the reforming reaction. Moreover, less steam is required, since that liberated by the fuel cell itself is used in the reforming reaction and this improves the overall efficiency. Generally, the heat released by the electrochemical reaction exceeds that required for the reforming reaction, so there is still a need for some cooling, although at a much reduced level, with the excess heat available for external use.

6.5.1   Molten Carbonate Fuel Cell (MCFC)

Amongst the six main types of fuel cell, molten carbonate systems offer the best prospect for providing the largest units (in the MW range). They operate at

600–700 °C and are being developed as natural gas/coal-based power plants for electrical utility, industrial and military applications. The electrolyte is a eutectic mixture of molten lithium and potassium carbonates that is held in a porous ceramic matrix (or 'tile') of lithium aluminate, $LiAlO_2$. It should be noted that the individual carbonates have melting points that are higher than the operating temperature of the cell (*i.e.*, Li_2CO_3, 723 °C; K_2CO_3, 891 °C) and therefore any departure from the eutectic composition would result in the deleterious formation of a paste of solid carbonate in the eutectic melt. The ionic-conducting species is the carbonate anion, CO_3^{2-}, which moves through the electrolyte from the positive to the negative electrode. During cell operation, carbon dioxide is formed at the negative electrode and consumed at the positive; see Figure 6.4. Consequently, the carbon dioxide contained in the exhaust fuel gas from the negative electrode is extracted and recycled to the positive electrode, as shown schematically in Figure 6.14. Similarly, any fuel gas that is unused is recycled back to the negative electrode. Cell stacks are built up in the same way as for other fuel cells, but the high operating temperature and the aggressiveness of the molten carbonate electrolyte necessitate the use of special materials for gaskets and cell sealing.

By operating at about 650 °C, the kinetics of the oxygen reduction reaction at the positive electrode are sufficiently accelerated that it is no longer necessary to use platinum-based electrocatalysts. Usually, the positive electrode is fabricated from lithiated nickel oxide ($Li_xNi_{1-x}O$, where $0.02 < x < 0.04$) and the negative

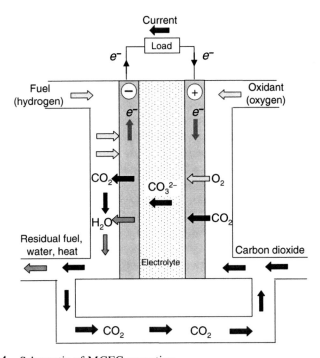

Figure 6.14 Schematic of MCFC operation.

electrode from a porous Ni−10 wt.% Cr alloy. In the latter, the role of the chromium is to reduce sintering of the nickel during cell operation. The negative electrode is able to catalyze the water-gas shift reaction (see Section 2.3, Chapter 2) and therefore the fuel cell will accept hydrogen that contains significant quantities of carbon monoxide. In practice, however, problems arise from the slow kinetics of natural gas reforming at 650 °C and from the vapour of the molten carbonates that tends to deactivate the electrocatalyst.

The downside of running a fuel cell at high temperature is that the reversible cell voltage is much reduced compared with its value at ambient temperature. Moreover, the cell voltage decreases almost linearly with increasing current density. For example, a typical cell has output voltages of around 0.9 and 0.8 V at 100 and 200 mA cm^{-2}, respectively. Elevated pressures can raise the performance somewhat, although at the expense of a higher investment for pressurized plant and an increased operating cost for compressing the fuel gas and air. There is also some evidence that high-pressure cells have a shorter service life. Thus, in designing a MCFC stack for a given application, a compromise has to be made between size, voltage and current density (power output), operating pressure, capital and operating costs, and possibly service life.

Other difficulties encountered with the MCFC are as follows.

- The oxygen reduction reaction is not straightforward and proceeds via either the peroxide ion (O_2^{2-}) or the superoxide ion (O_2^-), as dictated by the electrolyte composition. These ions can dissolve in the molten carbonate and corrode the counter electrode. The electrolyte itself is also an aggressive medium.
- The nickel oxide of the positive electrode tends to dissolve, possibly through being attacked by carbon dioxide. The resulting Ni^{2+} ions can diffuse to the negative electrode and deposit there as metal, to form short-circuits, and hence reduce the power of the cell. Such deposition will also occur in the pores of the LiAlO$_2$ electrolyte matrix and this will cause blockages and further short-circuiting. Additions of alkaline earth metal oxides to the positive electrode are claimed to reduce this problem.
- The LiAlO$_2$ matrix may experience progressive pore-coarsening due to sintering under the combined influence of the high operating temperature and the molten carbonate electrolyte.
- During extended cell operation, electrolyte may be lost from the LiAlO$_2$ matrix and this will lead to gradual decay in cell performance and, ultimately, to fuel leakage through to the positive electrode. It is therefore essential that the porous matrix should remain full of electrolyte at all times. Typically, 60 wt.% of carbonate is constrained within 40 wt.% of matrix.
- Carbon monoxide present in the cell may undergo disproportionation to form carbon via the Boudouard reaction, *i.e.*,

$$2CO \rightarrow C + CO_2 \qquad (6.8)$$

This should be avoided because it can lead to plugging of the gas passages in the negative electrode by carbon. Carbon monoxide may also react with hydrogen to produce methane ('methanation', the reverse of reforming), *i.e.*,

$$CO + 3H_2 \rightarrow CH_4 + H_2O \tag{6.9}$$

The process is detrimental to cell performance because it involves a considerable loss of reactant and thereby reduces power-plant efficiency. Moreover, the methane can, in turn, be decomposed to carbon, *i.e.*,

$$CH_4 \rightarrow C + 2H_2 \tag{6.10}$$

Carbon deposition can be reduced by ensuring that there is some water in the fuel gas, so that the formation of methane is not favoured; see reaction (6.9).

- Finally, the long start-up and shut-down times for MCFCs (5–10 h; see Table 6.1) limits their utility, as also does their restricted ability to withstand thermal cycling given the concomitant expansion and contraction that take place, which can eventually lead to cracking of the $LiAlO_2$ matrix.

Given all of the possible degradation modes of MCFCs cited above, reliability issues have yet to be fully resolved and thus the ultimate service life remains to be demonstrated.

Despite all these difficulties, significant progress has been made with MCFC technology, notably in Italy, Japan, The Netherlands and the USA. The first tests of prototype MCFC power plants (250 kW or larger) were conducted during the period 1996–2000. Conversion efficiencies of 55–60% were claimed. One US company, FuelCell Energy, is now commercializing MCFCs under the brand name Direct FuelCell®; this activity has received financial support from the government. Their systems have power outputs of 250 kW, 1 MW and 2 MW and are intended for stationary use in hospitals, schools, hotels, *etc.*, and also for commercial and industrial applications. In some situations, these will be pure electrical power devices, whereas in others they will operate in a CHP mode. Because internal reforming of natural gas is adopted, there is no need to await the creation of a hydrogen infrastructure. An example of a 1 MW unit is shown in Figure 6.15. Advantages claimed for such power plants, compared with conventional facilities, are higher fuel efficiency (especially in CHP mode), lower emissions, quieter operation, flexibility in siting, modular construction and scalability. The exhaust heat can be used for cogeneration applications such as high-pressure steam, district heating and air conditioning. An alternative is to direct the hot, high-pressure exhaust gas to a turbine for the generation of more electricity. The fuel cell produces the larger share of the electrical power ($> 80\%$) while the gas turbine serves to generate additional power by recovering the by-product heat from the fuel cell and the balance-of-plant. This hybrid system is capable of electrical efficiencies approaching 70%.

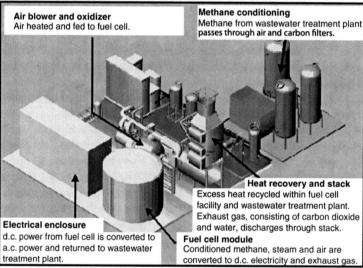

Air blower and oxidizer
Air heated and fed to fuel cell.

Methane conditioning
Methane from wastewater treatment plant passes through air and carbon filters.

Heat recovery and stack
Excess heat recycled within fuel cell facility and wastewater treatment plant. Exhaust gas, consisting of carbon dioxide and water, discharges through stack.

Electrical enclosure
d.c. power from fuel cell is converted to a.c. power and returned to wastewater treatment plant.

Fuel cell module
Conditioned methane, steam and air are converted to d.c. electricity and exhaust gas.

Figure 6.15   1 MW Direct FuelCell® MCFC power plant.
(Courtesy of FuelCell Energy Inc.).

Other advantages are as follows:

- minimal emissions including ultra-low NO_x and reduced carbon dioxide release to the environment;
- simplicity in design;
- direct reforming internal to the fuel cell;
- potential cost-competitiveness with existing combined-cycle power plants.

These advantages have to be set against the problem areas listed above. Although significant advances have been made with MCFC technology, commercial availability is still limited.

6.5.1.1 Internal Reforming

There are two variants of internal reforming: 'indirect' and 'direct'. The indirect option employs a plate-type reformer that is external to the cell stack, but in good thermal contact with it. Direct internal reforming, as illustrated in Figure 6.16 for an MCFC, incorporates the reformer and shift catalysts into the cell stack, specifically at the negative electrode. With this approach, the heat arising from the electrochemical reaction is produced in close proximity to the reformer. In practice, there are significant engineering and performance differences between indirect and direct internal reforming that have to be carefully evaluated and weighed, but it is widely held that both methods hold many advantages over having a reformer that is separated from the fuel cell. These include:

- greater energy efficiencies;
- reduced system costs, since no independent reforming apparatus is required;
- more uniform hydrogen formation, which results in a more even distribution of temperature;
- higher levels of methane conversion.

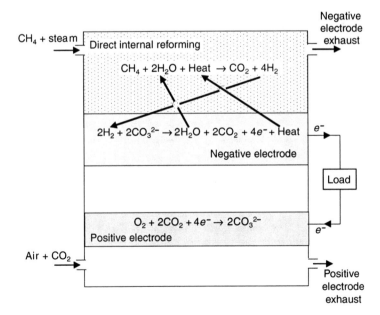

Figure 6.16 Schematic of direct internal reforming in an MCFC.

Against these benefits must be set the need to modify the hardware of the cell stack, the possibility of electrocatalyst deactivation through impurities in the fuel gas or through sintering, and the less flexible operation of an integrated system. On balance, however, it is usually considered that the advantages of internal reforming outweigh the disadvantages.

Internal reforming may also be utilized with an SOFC (see below). When operating with higher-order hydrocarbon fuels, such as propane, butane or naphtha, the addition of a pre-reformer may be necessary to prevent the build-up of unacceptable levels of carbon deposits within the cell structure.

6.5.2 Direct Carbon Fuel Cell (DCFC)

A new version of MCFC technology – the direct carbon fuel cell (DCFC) – is under development at the Lawrence Livermore National Laboratory in the USA. Instead of using gaseous fuel, a slurry of finely divided carbon particles dispersed in molten alkali metal carbonates is fed to the cell. The carbon is made by the pyrolysis of almost any waste hydrocarbon (*e.g.*, petroleum coke), a process that is already carried out industrially on a large scale to produce carbon black for use in the manufacture of tyres, inks, plastic fillers, *etc.* The pyrolysis reaction yields hydrogen that can itself be utilized in another fuel cell:

$$C_xH_y \rightarrow xC + y/2H_2 \qquad (6.11)$$

The power output of the DCFC depends critically on the structure of the carbon black that serves as the fuel, and this relationship is under study. Potentially, the cell has many advantages over other methods of generating electricity, especially as a prospective technology for producing power directly from low-grade fuels or from coal. Unlike conventional power plants, it does not require combustion in air and raising steam in boilers. The DCFC also promises a much greater overall efficiency and the off-gas is carbon dioxide that can be used industrially or sequestered without the difficulty and expense of separation from flue gases. Similarly, when making a comparison with a MCFC operating on hydrogen (from reformed natural gas), a higher efficiency is again projected for the DCFC. To date, however, this process has only been examined at the laboratory scale and it is therefore far too early to predict the likelihood of commercial success.

6.5.3 Solid Oxide Fuel Cell (SOFC)

The SOFC is an all-solid-state system and, as such, does not encounter the problems associated with the containment and corrosivity of molten salt electrolytes. Likewise, electrolyte flooding or starvation is avoided. On the other hand, the higher operating temperature ($700-1000\,°C$) gives rise to a different set of materials science challenges, notably those of fabricating the cells and stacks, matching the expansion coefficients of the various solid components and ensuring a gas-tight seal between the fuel and air chambers.

The unit cell consists of the following parts:

- a ceramic electrolyte, which conducts oxygen ions and may have either a planar or tubular geometry; it is in the form of a supported layer with a thickness of about 40 μm;
- an air electrode (positive), which is usually $La_{0.8}Sr_{0.2}MnO_{3+x}$;
- a fuel electrode (negative), which is a sintered mixture of the electrolyte and nickel, a so-called 'cermet';
- a cell interconnect of doped lanthanum chromite $LaCr_{0.8}Mg_{0.2}O_3$ with a thickness of 80–100 μm, which serves to join the positive electrode of one cell to the negative electrode of the next.

These components are fabricated and assembled into a cell of either tubular or planar geometry.

The ceramic electrolyte is based on zirconia (ZrO_2) or ceria (CeO_2), each of which is an electronic insulator but is conductive towards oxide ions (O^{2-}) at high temperatures. This conductivity is achieved by doping the chosen oxide with ions of lower valency. For instance, when zirconia is doped with yttrium ions (Y^{3+}), oxide ion vacancies are formed in the parent lattice to compensate for the charge difference, *i.e.*,

$$(1 - x)ZrO_2 + x/2Y_2O_3 \rightarrow Zr_{1-x}Y_xO_{2-x/2} \tag{6.12}$$

Conduction then takes place at high temperature by the diffusion of O^{2-} ions through these lattice vacancies. A plot of conductivity versus reciprocal of the absolute temperature for a number of doped zirconia, ceria and thorium oxides is given in Figure 6.17. To achieve a usable conductivity of $10^{-2}-10^{-1}\,S\,cm^{-1}$,

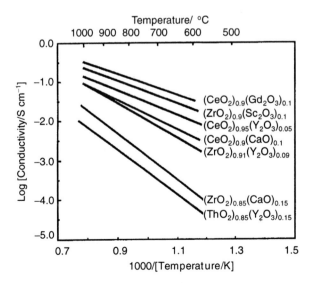

Figure 6.17 Ionic conductivity of solid electrolytes.

the data show that it is necessary to employ temperatures in the range 600–1000 °C, as dictated by the choice of electrolyte. The most widely investigated and adopted composition is that of yttria-stabilized zirconia (YSZ) with around 10 mol% Y_2O_3; the electrolyte is operated at 900–1000 °C. Whereas high temperatures favour the electrolysis of water by supplying the entropy term via thermal energy (see Figure 4.2, Chapter 4), they are unfavourable for fuel cells and result in a reduced output voltage, *i.e.*, typically around 0.5 V at 1000 °C. This is a consequence of thermodynamic considerations and is a penalty that is inevitably associated with SOFCs, to be set against their many advantages.

The electrolyte layer has to be both thin (to provide sufficient conductivity) and impermeable to gases (both hydrogen and oxygen). This implies that the material must be fully dense. In the tubular design of cell, shown in Figure 6.18(a), either the air electrode or the fuel electrode is in the form of a thick (mm)-walled, porous inner tube that serves as a support for the respective thin layers of electrolyte, counter electrode and interconnect. By being porous, the electrode allows gas transport to all the electroactive sites. Alternatively, an inert porous support tube may be used on which all four components are laid down as thin layers. The interconnect joins adjacent cells with the aid of a soft and compliant nickel felt. Other nickel felts are used to provide parallel connections between the fuel electrodes of adjacent cells. In Figure 6.18(a), the air electrode acts as the support and the electrolyte covers 300° of its outer surface, This leaves an axial strip of 60° width down the length of the cell for placement of the interconnect. Finally, a fuel electrode covers the solid electrolyte over about 280°. The cells are assembled in a series–parallel array. The series arrangement is essential for the generation of a practical stack voltage, whereas the parallel connections provide paths by which the current can bypass any defective (open-circuit) cells. Fuel flows axially through the spaces between the cells.

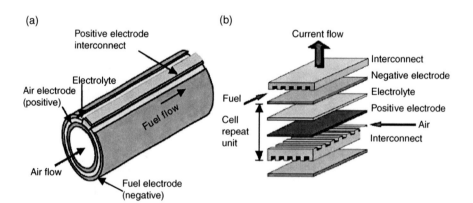

Figure 6.18   Schematics of (a) tubular and (b) planar SOFCs. The tubular design is that of Siemens Westinghouse.

In the planar cell, the components are assembled in flat stacks, with air and fuel passing through channels that are built into the interconnect; see Figure 6.18(b). Clearly, the configuration is similar to that employed in PAFCs and PEMFCs. Both rectangular and circular shapes of cells have been demonstrated. A relatively thick interconnect provides the mechanical support, while the typical thickness of each electrode is 50 μm and that of the electrolyte is 5–15 μm. Cells have also been constructed with one of the electrodes providing the mechanical support.

The tubular SOFC was the first to be studied and is the more advanced in terms of scale-up and commercialization. In the longer term, however, the planar version appears to be more amenable to mass production and cost reduction, once its outstanding technical problems are fully resolved. The advantages of each design have to be weighed against the difficulties encountered in its fabrication.

The fabrication of durable cells is, in fact, the main problem area that is hindering the progress of SOFCs. Given the high operating temperatures, the various cell components must be compatible with each other: not only chemically, but also physically in terms of expansion coefficients. If the latter requirement is not met, the stresses encountered in repeated heating and cooling will lead to cell fracture. Finding a suitable interconnect is especially difficult as only certain high-grade alloys (*e.g.*, Inconel[®], an alloy based on nickel and chromium) can be used at 800–1000 °C and these tend to have a mismatch in expansion coefficient with respect to the solid electrolyte. The alternative is the electronically conducting ceramic ($LaCr_{0.8}Mg_{0.2}O_3$) mentioned above, but this is more costly and also is brittle. Each of the various components has to be deposited as a thin layer by any one of several possible techniques (*e.g.*, vapour-phase deposition, sputtering, screen printing). Different system developers have employed proprietary materials and production techniques. The planar configuration of SOFC has a lower ohmic resistance than its tubular counterpart and, therefore, a higher power density. Nevertheless, it is more difficult to build on account of the much larger area of the cells, which leads to greater mechanical stresses in the stack. Also, the cells have to be hermetically sealed around a far larger perimeter than in the tubular design in order to ensure gas tightness. These technical issues have, however, been addressed with some success.

Another approach to easing the manufacturing challenges, particularly with respect to planar cells, is to lower the operating temperature of the SOFC by using an electrolyte of higher oxide ion conductivity. Cerium gadolinium oxide, $(CeO_2)_{0.9}(Gd_2O_3)_{0.1}$ (see Figure 6.17), has the potential to reduce the operating temperature to 500–600 °C. In 2001, Ceres Power was set up in the UK to exploit this 'intermediate-temperature' technology for planar units in the range 1–25 kW. A target market is domestic CHP. The electrodes and electrolyte are present as very thin layers that are deposited on the surface of a stainless-steel structural component. This simplifies the sealing of the cell by means of conventional welding techniques.

A potential drawback to the use of chromium-containing stainless steel is the volatility of the material, which can result in contamination of the stack components. This has an increased significance for future reclamation of materials and components from used stacks where the presence of a toxic ion such as Cr^{6+} would require special disposal procedures. Nevertheless, there are significant advantages for intermediate-temperature SOFCs based on stainless steel, namely, operation at < 700 °C:

- offers the possibility of more rapid start-up and shut-down procedures;
- simplifies the design and materials requirements of the balance-of-plant;
- significantly lowers corrosion rates.

In addition, an appreciable reduction in the manufacturing cost of the total system is expected.

One of the attractions of the SOFC, a consequence of the high temperature of operation, is that it can accept a variety of feedstocks – not just hydrogen (of most purities), but also natural gas, LPG, petrol and diesel. Fuels containing significant amounts of sulfur or halides are to be avoided, but carbon monoxide is acceptable. In effect, this cell (like the MCFC) may serve as its own internal reformer. Designs of SOFC based on either indirect or direct internal reforming have in fact evolved. External reformers will be required for low-grade feedstocks, such as residual oils, tars and coal. Generally, not all of the fuel is utilized in a single pass: part of the depleted supply is recycled and part is combusted to heat up the incoming gases. In deciding what percentage conversion is to be targeted in a single pass (typically, 50–90%), it is necessary to balance the power output of the device against operational efficiency.

Several companies around the world have been active in the development of SOFCs, *e.g.*, Acumentrics, General Electric and Westinghouse in the USA, Ceres Power, Rolls Royce, Siemens Power and Sulzer Hexis in Europe, Mitsubishi Heavy Industries in Japan and Ceramic Fuel Cells in Australia. Since the early 1980s, Westinghouse (now Siemens Westinghouse) has constructed and demonstrated successively larger units of tubular cells. A 100 kW system has operated successfully on natural gas for over 20 000 h in Germany with a reported efficiency of 46% and with no detectable degradation in performance. By 2004, Siemens Westinghouse had built and was testing 250 kW power plants, as shown in Figure 6.19(a). These are to be marketed for local generation in CHP mode and it is claimed that overall efficiencies of at least 70% should be achievable.

Several of the other manufacturers have reached the stage of offering SOFCs for distributed electricity generation and for CHP applications, both in industry and in large commercial undertakings. At the opposite end of the size range, developers are working on compact 5 kW units, based on planar technology, for residential CHP (gas-fuelled) and even auxiliary power (petrol-fuelled) markets.

Other options are being investigated. One is to configure units that can run at elevated pressures as this results in a higher voltage and hence an improvement

(a)

(b)

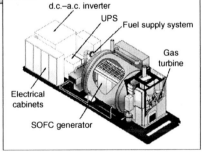

Figure 6.19 (a) 250 kW SOFC power plant and (b) 330 kW pressurized SOFC–turbine hybrid unit, both developed by Siemens Westinghouse. (Courtesy of Siemens Westinghouse).

in performance. For example, operation at 0.3 MPa increases the power output by about 10%. On integration with a gas turbine, the pressurized air required by the SOFC can be provided by the turbine's compressor and the cell itself can act as the system combustor with the exhaust being used to drive the compressor and a separate generator. This is a dry (no steam) hybrid-cycle power system, in which the auxiliary generator augments the direct electrical output from the fuel cell and thus makes for unprecedented electrical efficiency. This prospect has encouraged the pursuit of such hybrid technology for enhanced power production. A 330 kW hybrid plant with tubular SOFC technology has been built and undergone field trials; see Figure 6.19(b). Analysis indicates that an *electrical* efficiency of 55% can be achieved from hybrid power plants of this size; this rises to 60% at the 1 MW level and, possibly, 70% at 2–3 MW. The last-mentioned efficiency far exceeds the performance of the best combined-cycle steam plant, although it relates to a much smaller unit than is found in conventional, gas-fired, power stations. The SOFC–turbine hybrid is therefore well suited to distributed generation of electricity.

Clearly, great progress has been made with tubular SOFCs in recent years and many of the materials science problems appear to have been solved. The remaining issues to be addressed are the usual commercial ones of reducing the initial capital cost, optimizing performance and ensuring reliability and lifetime in service. In the USA, the target is to lower the capital cost to US$400 per kilowatt by 2010. With planar cells, there is scope to obtain higher power densities and to permit the use of low-cost, high-throughput fabrication techniques such as tape casting, calendaring or screen-printing. Nevertheless, planar geometry based on high-temperature electrolytes is still beset with difficult technical problems, although the use of intermediate-temperature electrolytes deposited on stainless steel structural supports may be more promising.

The balance-of-plant components that are required for the operation of both MCFCs and SOFCs all embrace conventional technology, but they must be tailored to the power unit under development. Together, they are likely to constitute well over 50% of the overall system cost. Therefore, ways must be found to reduce this expense if the two fuel-cell systems are to compete successfully with gas engines. On the other hand, because high-temperature fuel cells can accept a wide range of feedstocks and produce heat at a temperature that is industrially useful, power units running on natural gas may be seen as a bridge, or stepping-stone, to hydrogen energy at a future date. When resources of natural gas become scarce, MCFCs and SOFCs may provide an effective means to utilize impure hydrogen produced by the gasification of coal and other low-grade fuels.

6.6   Fuel Cell Efficiencies

The efficiency of a fuel cell – the fraction of the energy in the fuel that is converted into useful output – is a critical issue. Much is made of the fact that fuel cells are not heat engines, so their efficiency is not limited by the Carnot cycle[¶] and therefore should be high. This reasoning has driven much of the interest and investment in the technology. The thermodynamic 'theoretical' efficiency, defined as the ratio of electrical energy output to the enthalpy of the fuel combustion reaction, can be above 80% for low-temperature fuel cells. Nevertheless, electrochemical kinetic theory says that this ratio is an upper limit that is only reached at equilibrium when the current is zero. For further explanation see Box 6.2. In practice, the efficiency is lower, particularly at high power outputs (a schematic representation of the cell voltage, and therefore the efficiency, as a function of current density has been given earlier in Figure 6.3). Just how much smaller is difficult to calculate and depends on numerous kinetic and other parameters, such as the overpotentials at the electrodes, the ohmic losses in the electrolyte, the occurrence of side-reactions, fuel loss by leakage across the electrolyte, partial fuel usage and parasitic energy consumption by the auxiliary components (fans, blowers, inverters, transformers, *etc.*). These factors cannot be dismissed as temporary practical impediments that are simply waiting to be overcome by further research and development, although they may be reduced through advances in system design. Since the various losses in performance depend on the duty cycle of the fuel cell and are therefore difficult to determine, reliance must be placed on practical measurements of the electrical output as a fraction of the hydrogen energy input.

By way of example, consider the PEMFC when acting as a power source in an electric vehicle. The stack runs at around 100 °C, at which temperature the open-circuit cell voltage is 1.19 V. During the powering of a vehicle, however, the voltage typically falls to between 0.7 and 0.6 V, as dictated by the current that is being drawn. This corresponds to efficiencies of 59 and 50% for the upper and lower working voltage, respectively. Such a calculation, however, relates only to the free energy of the hydrogen combustion process and does not take into account the entropy lost as heat. When the calculation is based upon the *total enthalpy*, ΔH, of the reaction (see Figure 4.2, Chapter 4), which is a measure of the *overall* efficiency of the process, the corresponding efficiencies are 47 and 40%. In addition, the losses in pumps, heaters, blowers, *etc.*, together with those in the electrical system (inverter, transformer and traction motor), must be deducted. In round numbers, the collective losses in each system can be taken as 10%. Thus, the total energy efficiency of the fuel-cell system, from hydrogen to useful electrical output, lies between $0.47 \times 0.9 \times 0.9 = 38\%$ and $0.4 \times 0.9 \times 0.9 = 32\%$. This performance is still well above that of the internal combustion engine, but nowhere near the theoretical efficiency of around 80% for a fuel cell. Moreover, no account has been taken of the energy consumed in transforming the primary fuel (*e.g.*, natural gas) to hydrogen. The significance of the resulting 'well-to-wheels' efficiency in comparing the performance of fuel cell vehicles with that of internal combustion engine vehicles is discussed in detail in Section 7.7, Chapter 7.

[¶] The Carnot cycle states that only a fraction of the heat produced by an engine can perform work and that the remainder dissipates into the engine and the environment.

Box 6.2 Fuel Cell Efficiency.

As noted in Section 4.1, Chapter 4, the thermodynamically reversible voltage for the electrolysis of water is $V° = 1.229$ V at 298.15 K and 101.325 kPa. Similarly, this is the reversible voltage of a hydrogen fuel cell operating under the same conditions of temperature and pressure. On increasing the temperature, the free energy of formation of water decreases and so, therefore, does the reversible cell voltage; see Figure 4.2, Chapter 4, and reaction (6.1). Thus, at 200 °C the reversible voltage of a hydrogen fuel cell is only 1.14 V. The remaining enthalpy of reaction is liberated as heat (the entropy term $T\Delta S$). From this fact, it is apparent that high temperatures favour electrolyzers (since the entropy term can be supplied by heat rather than by electricity), but penalize fuel cells – unless the heat liberated by the cell reaction can be gainfully employed (as for instance in a CHP scheme). In practice, the output voltage of a fuel cell will be considerably less that the reversible voltage on account of losses within the cell stack. Just as in electrolysis, these losses arise from activation and concentration losses at the electrodes, ohmic losses in the electrolyte and non-uniform current distribution across the surface of the electrodes. For fuel cells, inadvertent crossover of fuel from the negative to the positive electrode and incomplete utilization of the hydrogen as it passes through the stack are further problem areas. All these factors conspire to reduce the practical cell voltage to well below the reversible value, typically to between 0.5 and 0.7 V.

Fuel cell efficiency is a topic that has given rise to much confusion in the literature. One measure of efficiency is simply the practical cell output voltage divided by the thermodynamically reversible voltage at the stated temperature and pressure of operation. This indicates how much free energy is lost by inefficiencies in operating the fuel-cell stack, without regard to how much of the enthalpy of the reaction is liberated as heat (the $T\Delta S$ term). For purposes of calculating *overall* energy efficiencies, it is necessary to compare the electrical output of the cell stack (in joules) with the *enthalpy* of the cell reaction, which for hydrogen fuel equates to the heat of formation of water. As explained in Box 6.1, there are two values of this function: (i) a higher heating value (HHV), which corresponds to the product water being present as liquid; and (ii) a lower heating value (LHV), which applies when the product water is present as uncondensed vapour or steam. The difference between the two values represents the latent heat of evaporation of water. For hydrogen, the HHV is approximately 18% greater than the LHV, as discussed in Box 6.1.

In a low-temperature fuel cell, where the product is liquid water, the HHV should be employed in efficiency calculations, while for high-temperature fuel cells it may be permissible to use the LHV if the product steam is put to good use. Alternatively, the work done by the steam, for instance in driving a turbine, might be added to the electrical output and the combined figure compared with the HHV to calculate the overall efficiency. The resulting value will be 18% lower than that obtained when using the LHV for hydrogen.

Box 6.2 Continued.

In addition to the losses that originate in the cell stack, there are also external inefficiencies to be taken into account. These include electrical losses in compressing the incoming hydrogen and air, and in converting the low-voltage d.c. output to high-voltage a.c. The total effect is a significant reduction in overall system efficiency. Finally, if the fuel cells are to be used to propel electric vehicles, for example, there are also inefficiencies in the electric motors and the drive-train to be considered.

Electrical efficiencies approaching 50% have been claimed for large MCFCs, *i.e.*, those of MW output, shown in Figure 6.15. These figures, however, relate to the LHV for hydrogen. The corresponding efficiency in terms of the HHV for hydrogen would be about 43%. It should be reiterated that there is confusion in the literature over fuel-cell efficiencies because of (i) the varied use of HHV and LHV for hydrogen, (ii) the reporting of values obtained at different current densities or different temperatures, (iii) the use of different allowances for parasitic losses in balance-of-plant and (iv) the failure sometimes to include the efficiency of generating hydrogen from primary fuels.

6.7   Applications for Fuel Cells

As outlined in Section 6.2, there are many potential applications for fuel cells. Some of these are specialized and therefore do not command large markets, *e.g.*, spacecraft, submarine traction. Prospective mass markets fall into four broad categories, as follows.

6.7.1   Large Stationary Power Generation

Among the fuel cells with outputs of 10 kW to several MW that are being developed for the stationary generation of power, the PAFC has until now been regarded as the prime candidate, at least for units up to 250 kW. Concerns over cost and durability of this system are, however, causing a reconsideration of this choice. At the upper end of the size range, it is possible that MCFCs or SOFCs will become more attractive propositions for both commercial and industrial purposes, on account of their internal-reforming function and their CHP capability. Again, cost and durability targets present a real challenge. For stationary applications, the US Department of Energy has set 40 000 h (5 years) as the minimum target for operational life. In the case of larger units, *i.e.*, greater than 1 MW in output, there is growing interest in hybridization with gas turbines to take advantage of the high-quality heat for the production of additional electricity.

6.7.2   Small Stationary Power Generation

There are numerous future mass markets for fuel cells with power outputs between 0.1 and 10 kW. These include electronic communications equipment, small-scale distributed electricity generation, CHP for individual homes, small apartment blocks or offices, uninterruptible power supplies and remote-area power supplies. In recent times, less emphasis has been placed on the residential applications as the costs are still too high and the lifetimes of the systems too short. A life of at least 5 years with little maintenance is required. When such reliable service has been demonstrated, and provided that the cost is acceptable, the fuel cell may well find widespread use in homes – both for the on-site generation of electricity and, in CHP mode, as a replacement for the conventional gas boiler. The PEMFC has emerged as the best candidate system, although a remaining major objective is to raise the temperature of operation so as to enhance performance and provide more useful heat. The Japanese government and industry are committed to this technology and are investing heavily. To date, 1 kW domestic co-generation systems have achieved lifetimes of 15 000 h towards the target of 40 000 h. The programme calls for 1.2 million units to be installed in Japanese homes by 2010.

6.7.3 Mobile Power

A consensus appears to be emerging among automotive companies that the *all-electric* vehicle, if it comes into widespread use, will be powered by a fuel cell rather than by a battery. This is primarily because of the longer driving range and faster refuelling that would be made possible. Nevertheless, a battery-powered vehicle is still an option for purely urban operation where range limitation may not be such an impediment. The focus for general transportation applications is the PEMFC, although the specifications that must be met to compete with the internal combustion engine are stringent, particularly as regards the size, performance and cost of the overall propulsion system. The progress being made with fuel cell vehicles is discussed in more detail in Chapter 7.

6.7.4 Portable Power

Most portable electronic devices would require a fuel cell of less than 100 W output, and often considerably less. Here, effort is concentrated on perfecting the DMFC, because of the convenience of methanol as a liquid fuel. Clearly, portable power is a very promising high-value opportunity for small fuel cells, but the outcome would have little impact on overall energy consumption and is not strictly part of the hydrogen energy scene. Nevertheless, many observers believe that volume production of micro fuel cells would be a key technical and economic driver for the entire fuel-cell market.

6.8 Prognosis for Fuel Cells

Almost 170 years after their invention, fuel cells are approaching commercial reality. As we have seen, the space programme that started in the early 1960s provided the initial stimulus for their practical development. Efforts into advancing the technology have intensified in recent years and have been motivated by mounting concerns over atmospheric pollution, greenhouse gas emissions and the search for new fuel options to meet the rapidly growing demand for energy world-wide. Looking ahead a few decades, it is expected that depletion of the reserves of petroleum and natural gas will drive a move towards the increasing consumption of lower grade fossil fuels such as oil sands, shales and coal. At the same time, there will be growing pressure for environmental sustainability and the control of carbon emissions. These conflicting requirements will necessitate the conversion of low-grade fossil fuels to clean forms of energy (electricity, low-sulfur liquid fuels, hydrogen), with accompanying sequestration of carbon dioxide. The same considerations will lead to growth in renewable sources of energy and, with them, a projected increase in the production, storage and utilization of hydrogen. A key to the effective adoption of hydrogen as an energy vector lies in fuel cells, once they become reliable, affordable and widely available.

The six principal types of fuel cell have their own distinct characteristics and are at different stages in their development. Some of their individual features are given in Table 6.2. Of necessity, this is an oversimplified list that could be greatly refined in detail, especially with respect to the proposed applications, the scale of power generation and the nature of the fuel available. A concise, visual summary of the fuels that might be used in different fuel cells and the various possible future applications for these devices is given in Figure 6.20.

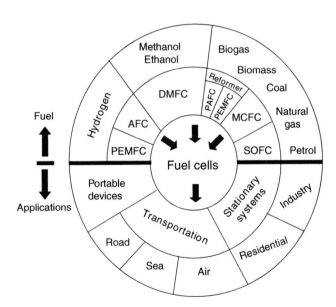

Figure 6.20 Fuel cells: their fuels and applications.
(Courtesy of the European Commission).

Table 6.2 Overview of fuel-cell systems: technical issues and applications.

PAFC (150–220 °C)	*AFC (50–150 °C)*
• Fuelled by hydrogen or reformed natural gas • Specifications for hydrogen purity less stringent than for PEMFCs • Uses Pt-based electrocatalyst • Corrosion and sintering of positive electrode are problems • Most commercially advanced technology • Medium-to large-scale stationary applications • Unsuitable for automotive applications	• Fuelled by pure hydrogen • Removal of carbon dioxide adds to system complexity • Inexpensive (not Pt-based) electrocatalyst may be used, although the addition of Pt-group metals improves kinetics • Corrosion problems • Used in space missions (high power-to-weight ratio)

PEMFC (80–90 °C)	*DMFC (60–90 °C)*
• Fuelled by pure hydrogen • Uses Pt-based electrocatalyst; lower loadings and higher tolerance of carbon monoxide being sought • Electrolyte membrane is expensive • Improved membranes required for cell to operate at higher temperatures with greater efficiency • Thinner, lighter, cheaper bipolar plates required • First developed for space missions • Portable or stationary applications • Attracting major investment for automotive applications	• Fuelled by methanol • No need for external reformer • Advantages over compressed gas, *e.g.*, ease of handling, distribution (methanol can integrate with existing distribution systems for liquid fuels), no need for compressors or gas cylinders, better safety • Requires high loading of Pt-based electrocatalyst • Improved membranes are needed for cell to operate at higher temperatures with greater efficiency • Methanol crossover and toxicity are issues of concern • Portable electronics applications (micro fuel cells)

MCFC (600–700 °C)	*SOFC (700–1000 °C)*
• Fuelled by hydrogen, natural gas, petroleum, propane, landfill gas, biogas, diesel, *etc.* • No need for external reformer • Uses less expensive (not Pt-based) electrocatalyst • Slow start-up • Slow response to changes in load • Corrosion and durability problems • Requires carbon dioxide management system • Limited thermal cycling capability • Medium- to large-scale stationary applications, possibly industrial multi-megawatt plants • Suitable for cogeneration operations	• Fuelled by hydrogen or almost any hydrocarbon fuel including coal gas • No need for external reformer • Uses less expensive (not Pt-based) electrocatalyst • Slow start-up • Slow response to changes in electrical and thermal loads • Difficult materials science problems • Requires significant thermal shielding to avoid heat losses • Limited thermal cycling capability • Medium- to large-scale stationary applications • Suitable for cogeneration operations • Small systems being developed for auxiliary power units in automotive applications

Many of the challenges in realizing practical fuel-cell systems centre on the requirement for low-cost materials that have a long operational life in the aggressive environments of the respective cells. Some types of fuel cell may fall by the wayside as, variously, their technical problems prove to be intractable, their reliability and lifespan inadequate, and their costs unacceptable. Some systems may find niche markets that are of key importance to society, even if not contributing much to the overall energy/environment scene. Others may become influential in determining future energy policy throughout the world. There is still much to be done before all the technical issues are settled, mass production lines are established and competitive products become available. Good progress is being made, however, and the international enthusiasm and backing for fuel cells augur well for ultimate success.

References

1. R.M. Dell and D.A.J. Rand, *Clean Energy*, Royal Society of Chemistry, Cambridge, 2004.
2. J. Larminie and A. Dicks, *Fuel Cell Systems Explained*, 2nd Edition, Wiley, Chichester, 2003.

CHAPTER 7
Hydrogen-fuelled Transportation

During the past 20 years, under the stimulus of rising fuel prices and growing concerns over atmospheric pollution, there have been important advances in the technology of road vehicles. The efficiency of petrol engines has been raised substantially and the problem of exhaust pollution has been ameliorated by the introduction of catalytic converters. To reduce their fuel bills, many people have moved to smaller cars or to diesel engines that are more efficient and consume less fuel. Others, less concerned with issues of cost or energy conservation and influenced by advertising campaigns, have gone in the opposite direction and have demonstrated their increasing affluence by purchasing high-performance cars, large multi-purpose vehicles (MPVs), or sports utility vehicles (SUVs). This latter trend has been especially pronounced in the USA, where fuel prices are still comparatively low, whereas in Europe (where taxation leads to fuels that are two to three times as expensive) more people have elected for economical vehicles. It is of interest to note that, despite the difference in fuel costs, the average annual distance driven in Europe is not appreciably different from that in the USA. Clearly, owners value the mobility and freedom that is provided by their cars and are prepared to pay for these benefits. The growth in road transportation world-wide has been such that, in terms of total emissions, it outweighs all savings from cleaner fuels and lower engine capacities. Consequently, the automobile industry is showing increasing interest in the development of vehicles powered by hydrogen.

In assessing the prospects for hydrogen-fuelled transportation, it is important first to review the recent advances that have been made in conventional drive systems and the possibility of further significant improvements. This will provide a moving benchmark against which the hydrogen vehicle has to compete in terms of both performance and purchase price. With respect to running expenses, a comparison has to be made between the probable costs of hydrogen production, distribution and on-board storage and the future price of liquid fuels. This is not an easy task – as oil reserves decline and the demand for personal mobility grows (especially in populous countries such as China and India), it is difficult to predict the level to which petroleum prices might rise, particularly when account is taken of arbitrary taxation imposed by national authorities.

7.1 Conventional Vehicles and Fuels

Most road vehicles today are fuelled by petrol or diesel, although small numbers run on compressed natural gas (CNG) or liquid petroleum gas (LPG). The last-mentioned fuel consists primarily of propane (C_3H_8, boiling point $-42\,°C$), butane (C_4H_{10}, boiling point $-0.5\,°C$), or a mixture of the two. At normal atmospheric pressure and temperature, these two hydrocarbons are gases, but they may be liquefied under moderate pressures (less than $1.5\,MPa$) and stored in cylindrical steel containers. Compared with petrol or diesel, LPG is clean burning and emits almost negligible amounts of nitrogen oxides, particulates (soot) or non-combusted hydrocarbons. The container weighs more than a conventional petrol tank, but occupies only slightly more space for the same amount of stored energy. Often, LPG vehicles are dual-fuelled and have the facility to switch to petrol if liquefied gas is not available.

Compressed natural gas is composed mainly of methane (90–95 vol.%). In order that a CNG vehicle shall have an equivalent driving range to that of petrol between refuelling stops, it is necessary either to have a very large fuel tank or to compress the gas to exceedingly high pressure. In reality, neither of these alternatives is totally practical and a compromise is adopted in which such vehicles generally have a shorter range than their petrol or diesel counterparts. Thus, in terms of energy content and convenience, LPG is more acceptable for road vehicles than CNG. A further disadvantage of CNG, supplied by the national gas grid at near atmospheric pressure, is that a compressor is required at the refuelling point; LPG, on the other hand, is delivered to the service station in pressurized containers. On the positive side, natural gas liberates around 30% less carbon dioxide per unit of energy consumed than petrol. Nevertheless, for many drivers this appears not to provide adequate compensation for the operational drawbacks encountered.

Efforts are also being made to reduce the emissions from internal combustion-engined vehicles (ICEVs) through the use of CNG blended with 20 vol.% hydrogen. This fuel, with the registered trade mark Hythane®, is being promoted as a transitional stage between the fossil-fuelled present and a hydrogen future; see Section 5.1, Chapter 5. It is claimed that Hythane® can be used in any CNG vehicle, but engine modifications are required to obtain the maximum benefit of 20% less carbon emissions.

The options available for future modes of transportation are listed in Table 7.1. For road vehicles in particular, Figure 7.1 illustrates the complexity of possible interactions between the various energy sources (feedstocks), the derived fuels and the types of power-train, together with the many feasible combinations to be evaluated when oil becomes scarce.

It is now widely accepted that reserves of native petroleum will be depleted before those of natural gas. Consequently, despite the drawbacks mentioned above, increasing interest is being shown in the greater use of natural gas as a transportation fuel. There are at least four possible ways to

Table 7.1 Road transportation options: power units and energy sources.

Power unit	*Road*	*Rail*	*Aircraft*	*Shipping*
Spark ignition (Otto) engine	Petrol LPG CNG Bio-ethanol			
Compression (diesel) engine	Diesel Biofuel	Diesel		Diesel
Electric motor	Batteries Fuel cells Electrochemical capacitors	Mains electricity Diesel–electric		Diesel–electric
Turbines			Kerosene	Bunker oil

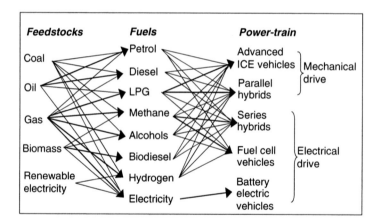

Figure 7.1 Candidate fuels and power-train technologies for road vehicles. (Courtesy of the University of Queensland).

accomplish this:

  (i) CNG-fuelled spark ignition engines;
 (ii) reforming of natural gas and conversion of the resulting syngas to liquid fuels via the Fischer–Tropsch reaction, *i.e.*, 'gas-to-liquids' technology; see Section 2.3, Chapter 3;
(iii) using natural gas to generate electricity, which will then charge the batteries of electric vehicles;
 (iv) conversion of natural gas to hydrogen and subsequent utilization in hydrogen combustion engines and/or in fuel cells.

Each of the four routes has its merits and demerits, as summarized in Table 7.2. Before making a choice, it is therefore essential to undertake a careful assessment of comparative energy efficiencies, capital and fuel costs, convenience, and emissions. The decision will vary according to the relative weightings placed on the different attributes, and also on the situation in the country in question, the type of vehicle and the required duty. Whether or not natural gas will find widespread acceptance while supplies of conventional oil are still available is an open question. In this chapter, we are primarily concerned with option (iv) above, that is, the conversion of natural gas to hydrogen. First, however, we should introduce the concept of 'hybrid electric vehicles'.

Table 7.2 Natural gas as a primary fuel for road transportation.

Technology (based on natural gas as primary)	Advantages	Disadvantages
Spark ignition engine	• Established technology • Low pollution • Less carbon dioxide emissions • Distribution networks for natural gas already in place	• Need for compressor at service station • Bulky, heavy on-board storage tank and/or reduced vehicle range
Gas-to-liquid technology	• Established technology • Refinery operation and conventional distribution of liquid fuels	• More expensive than conventional refining of petroleum
Battery electric (electricity generated from natural gas)	• High efficiency (55%) of combined-cycle gas turbine generation of electricity • In principle, carbon dioxide can be captured at the power station	• Limited vehicle range – suitable only for city and suburban use
Hydrogen (steam reforming of natural gas)	• Zero pollution from vehicle[a] • With centralized reforming, carbon dioxide can be captured at the reformer	• Need for hydrogen distribution system • Problem of on-board hydrogen storage • Reduced efficiency compared with direct use of natural gas in internal combustion engine • Present high cost of fuel cells for electric drive

[a] Ignoring loss of hydrogen (a possible atmospheric pollutant) and water vapour.

7.2   Hybrid Electric Vehicles (HEVs)

Internal combustion-engined vehicles (ICEVs) are frequently designed for power rather than for economy. In order to have power in reserve for acceleration, hill climbing and overtaking, these vehicles are fitted with engines that are excessively large and inefficient for steady driving. The way to avoid this profligate waste of petroleum and thereby reduce vehicle emissions is to divorce steady-state performance ('cruising') from acceleration by having two separate power sources, *i.e.*, one for each of these two functions. Accordingly, many automotive companies are putting sizeable efforts into the development of 'hybrid electric vehicles' (HEVs) that have electrical or electromechanical drive-trains.

In one of its manifestations, the hybrid concept employs a small internal combustion engine for steady driving on the flat and downhill, with an auxiliary electric drive to provide the extra power required for more demanding duties. Such a combination should reduce both fuel consumption and exhaust emissions. In some hybrid designs, a computer switches off the engine when the vehicle is stationary in neutral gear. This saves fuel and minimizes both pollution and noise.

Naturally, the use of two power sources rather than one brings with it additional engineering complexity and cost. For HEVs to succeed, the savings in fuel over the life of the vehicle should outweigh the expected higher initial cost. Moreover, the driving experience must be no less favourable. The viability of HEVs will depend very much on future movements in the price of petroleum, on advances in hybrid technology itself and possibly on emission regulations – especially as applied to diesel engines, which are the main competitors to hybrid drive-trains.

7.2.1   Classification of Hybrid Electric Vehicles

There are two basic architectures for ICE–battery hybrid vehicles, namely, 'series-hybrids' and 'parallel-hybrids'. In a series-hybrid [Figure 7.2(a)], the drive is all-electric and a small heat engine coupled to a generator serves to recharge the battery which, in turn, powers the electric motor. The engine–generator combination provides sufficient power to maintain the vehicle at a steady cruising speed, but is insufficient for acceleration or hill climbing. In the latter two operations, the battery reserve provides the necessary boost in power. When the vehicle is used in a purely electric mode (*e.g.*, in cities, with the engine switched off), the battery may, if desired, be re-charged overnight from the mains – a so-called 'plug-in hybrid'. An alternative approach is to have a larger engine and a smaller traction battery, such that the engine provides most of the power and is only switched off for limited urban operation.

In a parallel-hybrid [Figure 7.2(b)], there are dual transmission systems: a mechanical one driven directly by the engine and an electrical one driven by the electric motor. A parallel-hybrid corresponds to a conventional vehicle with a smaller engine and a larger battery. To date, most automotive manufacturers

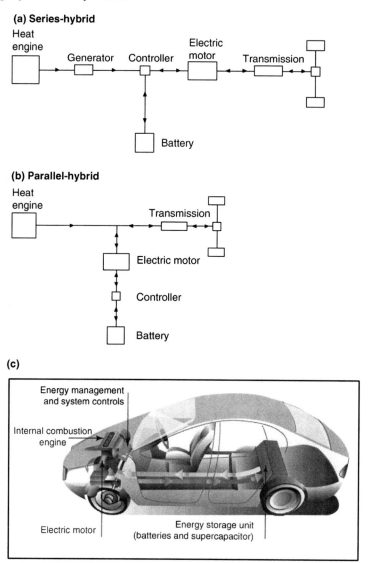

Figure 7.2 Schematics of (a) series-hybrid; (b) parallel-hybrid; (c) layout of power-train components.

have opted for this HEV design as it makes use of existing technology for engines, gearboxes and induction motors. There are several possible configurations. In one, the mechanical and electrical drive-trains are separate with the engine acting through a gearbox and drive shaft, as usual, and the battery supplying electric motors, for example wheel motors. In another version, the electric motor is mounted on the mechanical drive shaft and serves to augment the power transmitted. Although the two energy sources operate through a common shaft, they may be activated independently and hence the vehicle is still a parallel-hybrid. It is even possible to incorporate the rotor of the electric motor in the flywheel of the ICE and then surround the arrangement with the stator coils. A typical arrangement of the power-train components in a HEV is shown schematically in Figure 7.2(c).

During braking, the electric motor in both series- and parallel-hybrids becomes a generator so that the energy, which otherwise would be lost as heat, can be recovered as electricity and directed to the battery. This technique is known as 'regenerative braking'. Alternatively, the energy can be recuperated by fitting the vehicle with an auxiliary electrochemical capacitor (a so-called 'supercapacitor' or 'ultracapacitor')[†], which serves to meet peak-power demands of short duration, *e.g.*, during vehicle acceleration. Capacitors can accept very high rates of charge, as in regenerative braking, and also can be charged–discharged many thousands of times before failure, *i.e.*, well in excess of the life performance obtained from batteries.

Hybrid vehicle technology permits many different types of operation. At one extreme, the vehicle behaves essentially as an ICEV and the battery/capacitor acts as an auxiliary power source. This is termed a 'mild hybrid'. At the other extreme, when a relatively large traction battery is fitted, the HEV may operate as a pure battery electric vehicle (BEV) when in an urban environment; this is then designated a 'full (or strong) hybrid'. In this mode, the engine serves as a 'range extender' and may only be brought into use when the desired trip length exceeds the capability of the battery. Between these two extremes, there is scope for flexibility in design and operational practice, and in the relative proportions of liquid fuel and electricity that are employed. As noted above, full hybrids may include the option of overnight charging from the electricity mains (plug-in hybrid) rather than from the engine. This would allow for daily driving of up to 60–80 km powered solely by electricity, which is a range that is sufficient for most urban commuting. For longer journeys, the petrol engine would be called into service.

7.2.1.1 Cars

The Japanese companies Toyota and Honda have assumed the lead in the introduction of hybrid electric cars to the world market. Both companies have chosen the parallel architecture with a conventional, but smaller, petrol engine than is fitted to the equivalent ICEV and an auxiliary battery to provide boost power. The first such cars were the Toyota *Prius* and the Honda *Insight*, which were released in 1997 and 1999, respectively. Subsequently, in 2003, both companies brought out improved models, namely the Toyota *Prius THS 11* and the Honda *Civic* hybrid. These are both mid-sized family saloon (sedan) cars. The Toyota *Highlander* (a seven-passenger SUV) and the Honda *Accord*

[†] An electrochemical capacitor differs from a conventional electrostatic capacitor in that the energy is stored in the form of ions rather than electrons. This results in an increase in the amount of energy that is stored at the expense of reduced power output per unit mass/volume. Thus, an electrochemical capacitor begins to assume some of the characteristics of a rechargeable battery. Such devices are variously referred to as 'supercapacitors' and 'ultracapacitors'. Since there is some confusion over the distinction, we prefer to use the generic term 'electrochemical capacitor'.

(a full-size family car) are now also available as hybrids. The vehicles show much improved fuel economy on an urban driving schedule, but their fuel consumption is only comparable with that of a conventional diesel-engined car on long motorway journeys. Automotive companies in several other countries are also busy developing hybrid electric cars. In 2006, world sales of 350 000 hybrid cars and light vehicles were reported; of these, 60% were in the USA. The vehicles represented around 0.7% of total world car production. There is growing public awareness of the availability of hybrid vehicles and their environmental attributes, although there may be a tendency to exaggerate the latter. Recently, General Motors (USA) announced plans to build a plug-in series-hybrid electric car, the Chevrolet *Volt*, with commuters principally in mind as customers. The car will be equipped with a 1 litre engine to recharge a lithium battery while driving, although it is envisaged that the bulk of the energy would derive from mains electricity with overnight recharging. Commercial success will be greatly dependent upon the performance and cost of the traction battery.

7.2.1.2 Buses

Hybrid buses are becoming popular in the USA for city use, with many hundreds in service across the country. Subject to the route and level of traffic congestion, these vehicles are seen as delivering a 10–50% improvement in fuel efficiency and therefore correspondingly reduced emissions. Hybrids have an advantage over CNG buses in that their introduction involves no major, and hence expensive, modifications to existing fuelling infrastructure. On the other hand, the disadvantages seen for hybrid buses are a 60–80% increase in capital cost with respect to comparable diesel buses (a penalty that should reduce with mass production) and uncertainty over the durability of the batteries. There is a view developing that, for most routes, hybrid electric buses are more practicable than their pure-battery counterparts and are more likely than fuel cell buses to be commercially available at an acceptable cost in the next 10 years.

7.2.1.3 Batteries

The battery pack is the Achilles heel of hybrid vehicles. It is required to operate mostly at a partial state-of-charge and to deliver and accept charge at very high rates, which are encountered during accelerating and braking, respectively. Traditional forms of the lead–acid automotive battery are not well suited to this mode of operation and exhibit premature failure. Intensive efforts are under way to solve this taxing problem.[1] To date, nickel–metal hydride batteries have been the favoured choice for hybrid vehicles, but these are expensive. It is estimated that the battery pack contributes anything from 25 to 75% of the additional weight, volume and cost of a hybrid vehicle compared with a conventional counterpart, according to whether the hybrid is of mild, moderate or strong design.

For buses, a better choice of battery might be the sodium−nickel chloride (ZEBRA) system, which is now being used by some operators. Because these units operate at high temperature ($>200\,^{\circ}C$) and are well insulated, their performance is not affected by the ambient temperature. This is important for buses operating in extremes of climate. Another possibility is the lithium-ion battery, which is sometimes considered to be the ultimate technology. Both sodium−nickel chloride and lithium-ion batteries are also costly and each presents potential safety issues unless the recharging is tightly controlled. Several automotive manufacturers (General Motors, Nissan, Subaru, Toyota) are evaluating lithium-ion batteries for use in their hybrid vehicles. In addition to safety, key considerations are performance, service life, manufacturing cost and tolerance to abuse.

Conventional battery electric vehicles (BEVs) have large, heavy traction batteries that, typically, are discharged daily to a considerable depth-of-discharge and then recharged. Parallel configuration HEVs, on the other hand, have much smaller batteries that are discharged–charged frequently (possibly several times per hour) to an intermediate depth of charge. The battery for a series-hybrid vehicle depends on the operational mode. If the engine is used more or less continuously to charge the battery, then only a small pack is needed. Conversely, if the vehicle is to be employed as essentially an EV in urban environments, with the engine used for extension of range, then a much larger battery pack has to be fitted, comparable in size with that of a BEV. The battery is a particular problem for the much-favoured plug-in hybrid (such as that now proposed by General Motors), where most of the electrical energy stems from the mains. It has to be capable of fully recharging overnight, but also be able to accept charge (and discharge) very rapidly when running in hybrid mode. The requirement to fulfil both duties is particularly demanding, with implications for battery life. It is probably true to say that the future of all types of HEV will be determined to a large degree on advances in battery technology.

Battery research and development are relevant not only to hybrid vehicles based on internal combustion engines, but also to fuel cell vehicles (FCVs); see Section 7.5. It is likely that the latter will always require a sizeable battery to aid start-up, to buffer the fuel cell when peak power is demanded and to recuperate energy otherwise lost in braking. In short, the FCV is an all-electric hybrid.

As mentioned above, an alternative (or supplement) to a battery is an electrochemical capacitor that is also charged by the generator. These are not equivalent options. Electrochemical capacitors are capable of considerably higher power output than batteries, although for much shorter periods of time (from seconds to a minute or so) and are therefore ideal for supplying short-duration boost power. Moreover, they can also be charged at a high rate, as in regenerative braking. High-rate charging is a problem for certain batteries and leads to a shortened service life. Some combination of battery and high-energy capacitor may therefore prove to be optimum for hybrid electric vehicles.

7.2.2 Conventional *versus* Hybrid Vehicles

Most heavy commercial vehicles and many smaller freight vehicles are diesel powered. In the case of private cars, however, competition exists between petrol and diesel engines, with the latter being rather more expensive to purchase but consuming less fuel per kilometre travelled. As discussed above, petrol–electric hybrids are now entering the market. Europeans, in particular, have shown a marked preference for diesel cars and this has stimulated a rapid growth in sales to around 50% of the car market. The reasons are (i) better fuel economy, (ii) superior engine durability, coupled with improvements in design that make for smoother, quieter operation, (iii) a modest extra purchase price and (iv) greater safety in an accident situation (diesel is less flammable than petrol).

Sales of hybrids in Europe are still small by comparison with diesels. By contrast, there are still relatively few diesel cars in Japan and the USA where, to achieve greater fuel economy, the preference is for petrol–electric hybrids, despite their higher initial cost. This choice is partly historical in origin (there is little tradition of diesel cars in these countries) and partly the result of recent concern over emissions from diesel engines. There is mounting evidence that such fumes can cause lung cancer and other respiratory diseases such as asthma, and the growth in diesel traffic is giving rise to anxiety, especially in urban environments. Particulate traps are now available for heavy diesel vehicles and are under development for cars. If the emissions from these two classes of vehicles can be reduced to the levels that have been achieved with petrol engines, then the logical next step is a diesel–electric hybrid that should have greater fuel economy than either a simple diesel or a petrol–electric hybrid. Already some diesel–electric hybrid buses have been built. In Europe, the challenge has been taken up by Ford, who introduced a diesel hybrid version of the *Transit* van for door-to-door delivery services.

So long as petroleum is still widely available, it is difficult to forecast which of the various HEV options will prove to be most successful over the next decade or two. Indeed, fuel-cell enthusiasts consider both diesel and hybrid vehicles to be merely an interim technology until the day when FCVs are mass produced. Others are more sceptical over the future for FCVs and favour hybrids. Yet others see the hybrid as an expensive option with a performance that does not fully match that of a petrol engine. At present, it is impossible to say which view is correct; the market will decide in the light of factors such as the extent to which hybrid technology progresses, the resulting vehicle prices, the future availability and cost of petroleum, and the possible imposition of carbon taxes or emission permits.

7.3 'Green' Fuels for Internal Combustion Engines

Another environmental competitor to the hybrid electric vehicle is the internal combustion engine fuelled with bio-ethanol. Since this ethanol is produced by the fermentation of maize, sugarcane or sugar beets – and not from fossil fuel – it is rated as 'carbon neutral' and can therefore be burnt in an engine with impunity. Brazil, a country rich in arable farmland, has embraced this technology and up to 25 vol.% of bio-ethanol derived from sugarcane is added to the nation's petrol. In Europe and the USA, the present limits on biofuel addition are 5 and 10 vol.%, respectively. India also has an extensive programme of introducing ethanol-blended petroleum. In some countries, the blended fuel is referred to as 'gasohol'. The advent of this commodity has given rise to a growth industry in fermentation.

At present, bio-ethanol is not fully competitive with petrol in terms of economics and often receives a government subsidy that is provided not only to keep farmers in gainful employment, but also to help the given country meet its greenhouse gas targets and ease its bill for importing petroleum. These are important national targets that have little to do with the day-to-day costs of running a car as perceived by the consumer. From a purely energy perspective, however, bio-ethanol is not as attractive as it first appears. This is because cultivating the feedstock involves energy input in the form of fertilizers, and also fuel for the machinery used in harvesting, transporting and processing of the crop. Fermentation yields a dilute solution of alcohol that has to be concentrated by distillation. The latter is an energy-intensive process, although it requires only low-grade heat that might originate from burning waste straw. Overall, the energy input, much of it from fossil fuels, can be greater than the energy content of the bio-ethanol produced.

Another approach to the manufacture of bio-ethanol is through the enzymatic breakdown of cellulosic material, notably the waste stalks from the growing of cereal crops. There is vastly more of this material than there is of the grain and it is of no value, except for ploughing back into the soil to maintain fertility or combusting to produce heat. Enzymatic breakdown (using fungi and bacteria) takes place at ambient temperature to produce glucose and other sugars that may then be fermented (using microbes, primarily yeasts) to ethanol. The preparation of bio-ethanol via this process involves less input of energy *overall* than the fermentation of carbohydrates (starches or sugars) because the yields per unit mass of feed are so much greater. A pilot plant is running in Canada and the facility is being scaled up to produce 1 million gallons of ethanol annually. If bio-ethanol is to make more than a minor contribution to the world's transportation fuel then it will be vital to utilize cellulosic feedstocks, since the supply of starches and sugars is strictly limited. As yet, however, this route to ethanol is more costly than that of direct fermentation. Biotechnology companies are busy investigating a range of new enzymes from genetically engineered fungi and bacteria that will deliver higher yields of bio-ethanol and facilitate the breakdown of a greater variety of cellulosic products, *e.g.*, grasses, wood chips, leaves and paper pulp. The ultimate goal is to develop a single species that will both convert the cellulose to sugar and ferment the latter to ethanol, while at the same time being capable of surviving and operating at a higher ethanol concentration than current yeasts can withstand. In brief, the aim is to build a custom genome almost from scratch.

Bio-diesel, an alternative biofuel, is made from a vegetable oil that is extracted from the seeds of plants such as canola (rape plant), cotton, soybean and sunflower. This oil, which is used mostly in the manufacture of margarine, is reacted chemically with methanol to produce a methyl ester that can substitute satisfactorily for diesel fuel. Glycerol is produced as a by-product. Generally, up to 20 vol.% of bio-diesel is mixed with conventional diesel. Again, there is the problem of unfavourable economics. Bio-diesel is not yet fully cost competitive and the overall energetics of production, transport and consumption require careful evaluation.

Biofuels are claimed to offer many different benefits at the national level. These include: (i) reduced dependence on the diminishing reserves of petroleum; (ii) greater energy security; (iii) a smaller bill for oil imports; (iv) less pollutants released to the atmosphere; (v) the introduction of major industries to produce bio-ethanol and bio-diesel; and (vi) a new market for farmers. On the debit side, one has to consider the additional fossil fuel that will be expended (and carbon dioxide released) in the provision of extensive and sustainable supplies of these sources of energy. There is also the competition for land to produce food and the possible impoverishment of the soil. Overall, it is probable that in the short term biofuels will assume an increasing, but still modest, role in transportation.

7.4   Hydrogen-fuelled Internal Combustion Engines

The first recorded instances of burning hydrogen in internal combustion engines were in airships in the late 1920s. At that time, dirigibles employed hydrogen to provide buoyancy and had petrol engines for propulsion. On long flights, hydrogen had to be discharged periodically to the atmosphere in order to maintain constant buoyancy as fuel was consumed and the craft became lighter. It was realized that this hydrogen was a potential fuel and that up to 30 vol.% could be added to the petrol. This change in operation was found to increase the power output of the engine and also to extend the range of the craft before refuelling became necessary. Cruises of several days non-stop duration were common at speeds comparable with those of road vehicles. Nevertheless, it appears that the practice of fuel blending was not widely adopted for commercial airship flights, which terminated after a trans-Atlantic airship (the *Hindenburg*) caught fire and was destroyed on arriving in New Jersey in 1937. The public turned towards faster, more cost-efficient (albeit less energy-efficient) aeroplanes.

7.4.1   Road Vehicles

In the 1930s, Rudolf Erren, a German engineer, carried out extensive trials on hydrogen-fuelled road vehicles. The combustion characteristics of hydrogen are different from those of the lower hydrocarbons and this necessitates some changes in engine design. In general, hydrogen engines tend to exhibit pre-ignition, backfire and knock[‡], which are phenomena attributable to the low ignition energy and high flame speed of hydrogen compared with petrol. Erren showed that these malfunctions could be overcome by feeding hydrogen at a slightly increased pressure directly into the combustion chamber, rather than by introducing it with the fuel–air mixture through the carburettor. By adding hydrogen as an adjunct to petrol, better combustion of the hydrocarbon was achieved with higher power output and lower fuel consumption. More than 1000 cars and trucks were converted to Erren's dual-fuel system in the 1930s. A German self-propelled railcar with an internal combustion engine of 75 bhp (56 kW) was also re-engineered and this raised the power output to 83 bhp (62 kW).

It soon became apparent that a pure-hydrogen engine, rather than a dual-fuel system, was highly desirable from a pollution standpoint. At the same time, it was equally clear that there was a problem storing sufficient hydrogen on-board the vehicle. For this reason, experimenters switched to liquid hydrogen (LH_2) as a fuel. This was possible in the 1970s because liquid hydrogen was required in bulk for use in space programmes; see Section 5.3, Chapter 5. Efforts to utilize hydrogen fuel in cars and trucks sprang up in several countries in Europe, and also in Japan and the USA.

The automotive company BMW has been working with hydrogen engines since 1978. After developing a series of one-off prototype cars based on the large (*Series 7*) saloon chassis, BMW produced an experimental fleet of 15 dual-fuel cars in 2000. The 12-cylinder, 5.4 litre ICEVs were each fitted with a 140 litre storage tank that was capable of holding $\sim$8 kg of LH_2 to give the vehicle a cruising range of around 200 km. Because LH_2 supply points are likely to be few and far between for many years yet, the car also had a petrol tank that allowed a further 500 km of travel. These vehicles have been driven on public roads and between them have covered over 170 000 km. In 2001, BMW introduced the 4.4 litre *745h* dual-fuel car; the design is illustrated in Figure 7.3(a). This vehicle is equipped with an auxiliary power unit in the form of a proton exchange-membrane fuel cell (PEMFC). The latter functions independently of the engine via a direct hydrogen feed from the LH_2 tank. Thus, power accessories such as air conditioning can be operated when the engine is shut off and this leads to greater overall fuel economy.

In collaboration with various industrial partners, BMW has been engaged in the development of superior cryogenic storage tanks to reduce the 'boil-off' of hydrogen to acceptable levels and also in the establishment of bulk LH_2 storage and dispensing facilities. Refuelling at a service station is a critical issue. First, there is a requirement for large, cryogenic storage reservoirs to hold bulk quantities of LH_2. These might be surface mounted or located underground as for existing petrol tanks. Then, a special type of pump is required to dispense LH_2. Finally, there is the cryogenic linkage [Figure 7.3(b)] for coupling the pump to the vehicle's storage tank (shown in Figure 5.5, Chapter 5) without

---

[‡] In petrol engines, pre-ignition occurs when the fuel–air mixture ignites before the spark plug fires; backfiring results from a fuel–air mixture that is too lean in fuel; knock ('pinking') is associated with incorrect ignition timing or with the use of a poor-quality fuel.

(a)

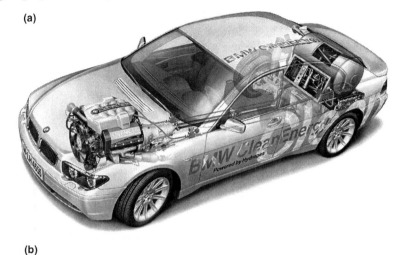

(b)

Figure 7.3 (a) Schematic of LH$_2$-fuelled BMW *7-series* car; (b) re-fuelling at a service station.
(Courtesy of Linde AG).

losing significant amounts of hydrogen through boil-off. All such aspects have safety implications. Prototypes of these components are in use in some of the hydrogen demonstration programmes. In particular, BMW and General Motors are working together to develop fully satisfactory cryogenic couplings.

Recently, Ford in the USA has announced that it is starting production of a supercharged 6.8 litre hydrogen engine. This is a V10 model intended for use in hydrogen-fuelled shuttle buses, first in Florida and then across the country.

Liquid hydrogen has a number of major disadvantages as a transportation fuel. Among these the following may be cited (see Section 5.3, Chapter 5):

- the high electricity consumption involved in liquefying hydrogen, which would negate the greenhouse gas benefits of hydrogen as a fuel;

- the logistics of supplying LH_2 to garage forecourts;
- the large size of the cryogenic storage tank, compared with a conventional fuel tank of similar energy content (a factor of 3.5 times larger internal volume, before adding the external insulation);
- the high cost of the specialized LH_2 tanks;
- the safety aspects of the refuelling operation by untrained personnel, bearing in mind the low ignition energy and wide flammability limits of hydrogen;
- the inevitable loss of hydrogen as the transfer lines cool during refuelling;
- the problem of evaporative boil-off, especially when the vehicle is not being used; this would be particularly serious in confined spaces such as a garages and tunnels, where it would present an explosive hazard.

Altogether, and despite the best efforts of BMW, some scepticism has to be expressed about the future of LH_2 as a practical and affordable fuel for motor vehicles, especially for cars.

7.4.2 Aircraft

As early as the mid-1950s, a modified American *B57* twin-jet bomber flew several experimental missions with one engine operating on LH_2. In the aerospace industry of the 1970s, some interest arose in extending this concept to civil aircraft. The attraction lay in the fact that the gravimetric energy density of LH_2 (expressed as GJ of energy per unit mass) is 2.5–2.8 times more than that of conventional aircraft fuel (kerosene). Thus, a switch to hydrogen would permit the possibility of increased aircraft range or greater payload, either of which would be commercially attractive – always assuming that the reduced fuel load was not offset unduly by the extra mass of the cryogenic storage tanks and other requisite equipment. Another advantage is that a hydrogen engine gives off no carbon dioxide, no unburnt hydrocarbons and limited emissions of nitrogen oxides (NO_x).

The downside to employing LH_2 as an aircraft fuel is its low volumetric energy density, which is only one-quarter that of kerosene. Consequently, there is a critical problem of where to accommodate the fuel aboard the aircraft. Various solutions and preliminary designs were advanced for a *Boeing 747*. These included storage in tanks along the length of the fuselage above the passenger seats [Figure 7.4(a)] or in wing-tip nacelles [Figure 7.4(b)]. The latter option was possible by virtue of the low density of LH_2, but would have entailed moving two of the four engines to the rear of the fuselage. In 1974, NASA awarded Lockheed Aircraft a 1-year contract to conduct an engineering assessment of a long-range aircraft powered by LH_2 for introduction in 1990–95. Unfortunately, nothing practical came from this venture.

Later, in 1988, a Russian *Tupolev 155* passenger plane flew near Moscow with one of its two engines adapted to run on hydrogen. The plane took off on jet fuel, but then switched to hydrogen during the cruising part of the flight.

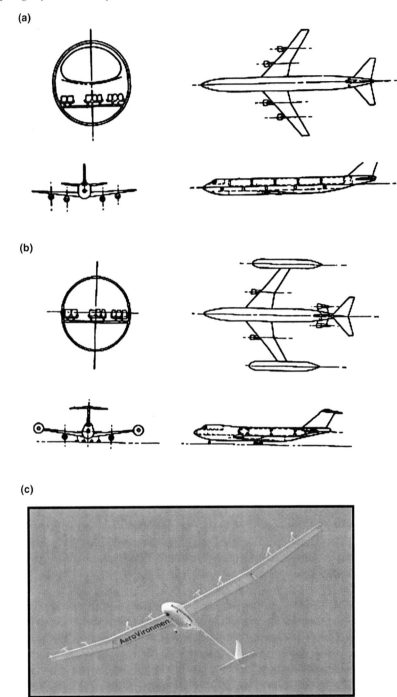

Figure 7.4 (a), (b) Two proposed layouts for LH_2 storage tanks on a Boeing 747 aircraft (1970s); (c) unmanned surveillance aircraft (the '*Global Observer*') powered by a 20 kW fuel cell.

Given their large indigenous reserves of natural gas, the Russians also showed an interest in running the aircraft on LNG. This fuel should be easier to store than LH_2 since the temperature of LNG ($-164\,°C$) is about $90\,°C$ higher. The emissions would clearly be greater, but far less than those from the customary use of kerosene. The principal motivation, however, was to ensure national energy security by introducing a new domestic supply of aviation fuel on a shorter time-scale than would be possible by switching to hydrogen.

The above demonstrations led to a Russian–German venture in the early 1990s for the realization of hydrogen-powered aircraft propulsion systems. Known as 'Project CRYOPLANE', the aim initially was to build a demonstrator aircraft based on the *Dornier 328JET*. Extensive design studies were conducted, but in 1999 the work was halted because of high costs and the plane was never built. The baton was then taken up by the European Union (EU) in its Fifth Framework Programme for Research and Technological Development which supported a multi-nation, 2-year project (2000–02) to investigate the feasibility of an LH_2-fuelled airliner based on the *Airbus A310*. The project involved 35 partners from 11 EU countries. A comprehensive systems analysis was conducted. This included aircraft and engine modifications, LH_2 on-board storage and the logistics of making the fuel available at European airports. The results and achievements of CRYOPLANE were as follows:

- identification of overall aircraft configurations to meet the requirements of efficient and safe operation in all categories, from business jets to very large long-range aircraft;
- assessment of the technical feasibility and availability of components for fuel systems for different aircraft categories;
- confirmation that hydrogen engines can be as efficient as kerosene engines in terms of energy consumed, with benefits that include substantial reduction in emissions;
- assessment of aviation safety aspects, which showed that hydrogen-fuelled aircraft will certainly be able to exceed current international airworthiness requirements;
- confirmation of outstanding environmental benefits of hydrogen engines;
- detailed quantification of global and regional scenarios for a soft transition to LH_2 fuel, by considering Northern Europe as the leading region during transition.

In the UK, the Royal Commission on Environmental Pollution published a study in 2002 that was much less sanguine about the prospects for hydrogen-fuelled aircraft. It was reasoned that when account was taken of the weight of cryogenic fuel tanks (cryostats) compared with that of kerosene containers, there would be a much reduced saving in the overall mass of the fuel system. The substantially larger volume of LH_2 would necessitate a larger aircraft with an increased drag coefficient. Also, bearing in mind that water vapour is a powerful greenhouse gas, concern was expressed over the release of so much extra water vapour in the upper atmosphere. Thus, it might transpire that there was no environmental benefit of hydrogen as an aircraft fuel over that of kerosene. For the foreseeable future, it was concluded that aircraft will continue to use conventional fuel. It remains to be established whether the EU or the UK view is the most accurate.

There is also interest in unmanned aerial vehicles (UAVs – essentially, small aircraft), powered by hydrogen fuel cells (PEMFCs), for both military and civil applications such as surveillance, meteorology, communications and road traffic control. Several companies (including Boeing) have been collaborating in the design and construction of a prototype model with a 20 kW PEMFC. AeroVironment, a US company, reports that it has successfully completed test flights of a UAV powered by a fuel cell with LH_2 storage [Figure 7.4(c)]. The electricity so generated drives eight propellers along the wings. The so-called *Global Observer* is able to operate at altitudes as high as 20 000 m, carry up to 450 kg and remain airborne for one week.

7.5 Fuel Cell Vehicles (FCVs)

The attractions of electric vehicles (EVs) with respect to ICEVs lie in the simplicity of the electric motor, the silent operation of the vehicle, the lack of polluting emissions at the point of use and, crucially, markedly different curves for energy consumed versus power output. When stationary with the engine running, ICEVs simply waste energy. Their efficiency of energy utilization increases with increasing power output. By contrast, batteries and fuel cells are at their most efficient when low currents are being drawn, much less so at high power output. This characteristic of EVs is particularly important in the urban environment where congestion and traffic hold-ups are common; when stationary, no energy is consumed. The drive-train efficiency of a typical fuel cell vehicle (FCV) as a function of power output is compared with those of ICEVs in Figure 7.5. Theoretical values for petrol engines (43%) and diesel engines (58%) are shown as horizontal lines. Except at the highest power levels, the drive-train of the fuel cell is seen to be superior to those of internal combustion engines; at low power outputs, such as are required for steady driving[§], it reaches a maximum efficiency close to the theoretical value for a petrol engine.[2]

Apart from the direct alcohol fuel cells (see Section 6.4, Chapter 6), most low-temperature fuel cells operate on pure hydrogen as fuel. For mobile applications, this raises the issue of whether the hydrogen is to be produced at a land-based reformer (or electrolyzer) and then stored on the vehicle or whether a primary liquid fuel (*e.g.*, methanol) could conceivably be reformed on-board. The latter approach is attractive in that the infrastructure for dispensing liquid fuels already exists at petrol stations and only a simple tank would be needed for storage on a vehicle. Another option, in principle, would be to move to natural gas as a primary fuel and to reform that on-board. Storage would, of course, be much less convenient although, as noted above,

---

[§]The precise significance of these quoted efficiencies and how they were measured are not known.

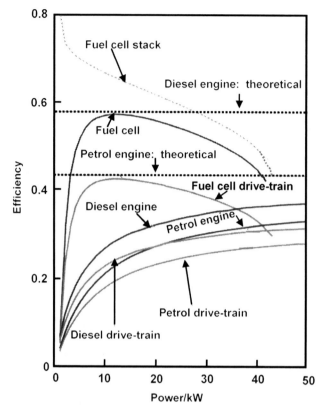

Figure 7.5 Power-train efficiencies of vehicles with internal combustion engines or fuel cells.
Note: the 'fuel cell' plot refers to fuel cell plus balance-of-plant, whereas the 'fuel cell drive-train' plot also includes the electromechanical drive to the wheels.
(Courtesy of the Government of the State of North Rhine-Westphalia).

compressed natural gas in cylinders is sometimes used as fuel for internal combustion engines.

There are, however, serious problems with in-vehicle reforming. A reformer is a complex piece of equipment that will be bulky, heavy and expensive – effectively, it is a 'mini-refinery'; see Section 2.3, Chapter 2. In addition, the product hydrogen would have to be purified so as to be essentially free of sulfur and carbon monoxide for use in a low-temperature fuel cell, if the electro-catalyst is not to be poisoned. There would be great difficulty in accommodating the total system within a private car, even if this were economically viable, although it might be possible on a heavy truck or a bus. Another engineering challenge lies in integrating the reformer with the fuel cell so that hydrogen is produced at the variable rate demanded by the fuel cell from moment to moment so as to ensure smooth 'driveability'. This would necessitate some form of short-term storage for hydrogen to meet instantaneous demand for acceleration and hill climbing. A further issue is the time taken for start-up of the reformer. Finally, carbon dioxide would still be released to the atmosphere. For these various reasons, few scientists or engineers are pursuing the idea of on-board reforming of primary fuels. An exception is Renault, which has been developing a vehicle with an embedded reformer. The rationale for this pro-gramme is the long time-scale that would be required (decades) to establish a national supply base for hydrogen and the difficulty of storing it on-board, coupled with a desire to market FCVs sooner rather than later.

A fuel cell for vehicular use is also a complicated system. In addition to the stack itself, an electronic control system must be included to match the power output to that demanded by the vehicle, as dictated by the driver, and, in turn, to balance the hydrogen supply to that called for by the fuel cell. There may also be a need to invert the d.c. output to a.c. (if a.c. motors are employed) and to transform the voltage. A battery and/or electrochemical capacitor bank are normally included for start-up and to provide peak power for acceleration and hill climbing. In the case of a PEMFC stack – the most commonly employed technology – a water-management system is mandatory in order to regulate the humidity of the incoming hydrogen and air. It is essential that the electrolyte membrane does not dry out and neither should it become over-saturated with water. Finally, provision must be made to control the temperature of the stack itself. This requires a cooling system that comprises a water pump, a condenser and radiator and an air fan, *i.e.*, a similar arrangement to that in a conventional ICEV. Despite these design difficulties, remarkable progress has been made with PEMFC-driven buses and vans/cars, as outlined below.

One of the first recorded instances of an electric vehicle powered by a fuel cell was a 20 hp tractor built in 1959 by Allis Chalmers in the USA. This machine utilized a 15 kW, hydrogen–oxygen alkaline fuel cell, which was a spin-off from space technology. During the 1960s, there were various experiments with cells fuelled by hydrazine (N_2H_4), which is a rather expensive and toxic liquid fuel but has the merit of being portable. Around 1970, Shell Research in the UK converted a car to run on a hydrazine fuel cell. In 1968–69, a hydrogen-fuelled electric car was privately built by Karl Kordesch in the USA. This was an *Austin A40* (a vehicle mass-produced in the UK) converted to electric traction with a 20 kW d.c. motor; it was equipped with a 6 kW alkaline fuel cell, which operated at 60–70 °C and with an auxiliary pack of lead–acid batteries. Up to 25 m^3 of hydrogen was compressed into six lightweight tanks mounted on the roof and gave a driving range of about 300 km. The car completed 21 000 km in four years.

In the years following these early attempts, companies in Europe, Japan and the USA have been endeavouring to develop more practical FCVs. Almost exclusively, these have used hydrogen due to the aforementioned difficulty of incorporating on-board reformers and integrating them successfully with the fuel-cell stack. The emphasis has been on urban buses and private cars for the reasons discussed below. Apart from a few concept vehicles that employed alkaline fuel cells, almost all of the serious programmes have used PEMFCs, which are seen as the most practical system for motive power use. A major attraction of the PEMFC for transportation applications is its low temperature of operation, which permits moderately rapid start-up from cold.

7.5.1 Buses

Urban (transit) buses constitute a prime market for the introduction of fuel cells. In addition to the fact that there is adequate internal space and roof area to accommodate the fuel-cell stacks and the hydrogen storage cylinders, buses have several other positive features with respect to utilizing fuel cell power. The vehicles usually run on well-defined routes and schedules, are maintained by qualified technicians and operate from a home depot so that only a single hydrogen-refuelling pump is required to service a fleet. Furthermore, buses are frequently delayed by city congestion and electric designs would neither emit pollutants nor waste energy while stationary in dense traffic. By contrast with traditional diesel buses, which are noisy and polluting, fuel cell buses offer a new form of transportation that is attractive to both passengers and the general public alike. Moreover, the vehicles are highly visible in the community and constitute an excellent showcase for electric propulsion.

A notable demonstration programme of fuel cell buses is being conducted in California. Between 2002 and 2004, SunLine Transit Agency in Thousand Palms, Chula Vista Transit in San Diego and AC Transit in San Francisco evaluated the performance on public roads of a 30-foot *ThunderPower* bus, which was developed by ISE Corporation in partnership with Thor Industries [Figure 7.6(a)]. The vehicle had a hybrid design and was primarily powered by a 60 kW PEMFC manufactured by UTC Fuel Cells. To economize on energy usage, the bus was also fitted with a pack of 48 12 V lead–acid batteries for back-up power and storage of electrical energy captured via regenerative braking. The fuel cell operated on 25 kg of compressed hydrogen that was contained in nine roof-mounted cylinders and gave the bus an approximate range of 300 km. From the field data collected, it was claimed that such hybrid electric buses nearly doubled the energy efficiency of a standard diesel-engined counterpart and produced no nitrogen oxides (NO_x) or particulates [Figure 7.6(b)].

Encouraged by the success of the above study, AC Transit formed a partnership for the modification of a 40 foot, Belgian-built *Van Hool A330* bus to accommodate a 120 kW PEMFC power system produced by UTC Power and a hybrid-electric drive system developed by ISE. A maximum of 50 kg of hydrogen is stored at 35 MPa in 10 cylinders and offers a driving range of 400–480 km. The bus has three sodium–nickel chloride (ZEBRA) battery modules to provide 95 kW of auxiliary power and to accept regenerative-braking energy. Three of these buses officially started service with AC Transit in March 2006 and a fourth is being operated by SunLine Transit Agency.

In Europe, a consortium of nine cities, backed by the EU, has conducted a large-scale field trial of fuel cell buses, together with an evaluation of pathways for the production, storage, distribution and dispensing of high-pressure

(a)

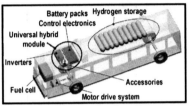

ThunderPower fuel cell bus

Van Hool A330 fuel cell bus

(b)

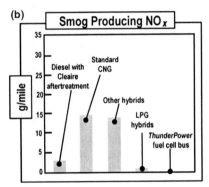

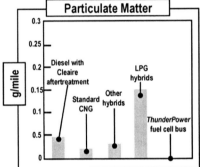

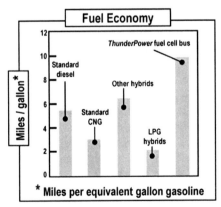

Figure 7.6 (a) Fuel cell buses in California; (b) comparison of emissions and fuel economy of transit buses with diverse motive-power sources. Cleaire technology is a retrofit catalytic device for reducing emissions of NO_x and particulates.
(Courtesy of AC Transit).

hydrogen. The initiative, known as the Clean Urban Transport for Europe (CUTE) programme, began in November 2001. Twenty-seven buses were built and integrated into the public fleets of Amsterdam, Barcelona, Hamburg, London, Luxembourg, Madrid, Porto, Stockholm and Stuttgart. Each city received three buses [examples are shown in Figure 7.7(a)] and these were serviced through the construction of a hydrogen-refuelling station at each location. The various operators adopted different sources of hydrogen and also different procedures for compression and delivery to the pumps. Some opted for electrolysis of water, using electricity from renewable sources, others for small-scale steam reforming of natural gas (50–100 N-m^3 H$_2$ per hour) or for refinery hydrogen. All the sites using electrolyzers installed pressurized units. Just one of the cities (London) chose liquid hydrogen, which was produced externally and conveyed by tanker. The others had cylinder storage facilities for compressed gas; the Hamburg facility is shown in Figure 7.7(b). When the storage pressure was lower than that on the bus (35 MPa), a booster pump was required to transfer gas to the vehicle. Refuelling typically took 10–30 min. In future designs, for a wider variety of hydrogen vehicles, it will be necessary to reduce this time and also to cater for on-board cylinders of differing sizes and pressures up to 70–80 MPa.

The buses were built by DaimlerChrysler and were based on the Mercedes-Benz *Citaro* design – a full-size, low-floor vehicle that can accommodate up to 70 passengers. The first of these was sent to Madrid in May 2003. A 250 kW PEMFC developed by Ballard Power Systems provides power to a 200 kW electric motor to give a performance comparable with that of normal diesel engines; a schematic of the system layout is presented in Figure 7.7(c). The hydrogen, stored at 35 MPa in nine roof-mounted cylinders, provides a vehicle range of around 200 km. During the course of the CUTE programme, data were collected to measure the performance, reliability, economy, safety and public acceptance of the buses. This information was exchanged freely between the participating cities. Traction motors and auxiliary components were reported to show no critical failures and the fuel-cell stacks gave lifetimes of over 3000 h. The drivers were satisfied with the vehicles, passengers enjoyed travelling on a new form of transportation and technicians were trained to operate filling stations and maintain buses without any major problems.

A further fuel cell bus project – the Ecological City Transport System (ECTOS) – was undertaken in Reykjavik, Iceland, between 2001 and 2005. It was closely aligned with the CUTE programme, was also funded by the EU and involved three *Citaro* buses that were identical with those operated in the above nine European cities. The project is of particular interest in that it was based entirely on the application of renewable energy to produce the hydrogen. Iceland is in an unusual position in having available renewable energy in excess of that required for its normal electricity generation. Consequently, the hydrogen fuel for the buses was obtained electrolytically through the use of surplus geothermal and hydroelectric power. An electrolyzer from Hydro Hydrogen Technologies (a subsidiary of Norsk Hydro) was employed for this purpose and was installed in a facility built in collaboration with Shell. Open to the public,

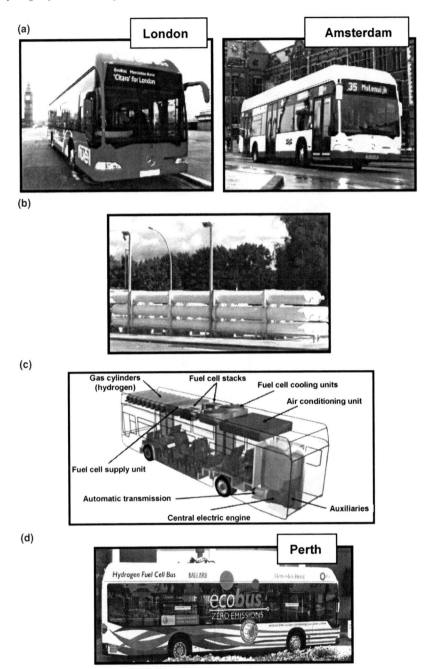

Figure 7.7 (a) Fuel cell buses operating under the CUTE programme in Europe; (b) hydrogen storage facility at Hamburg; (c) arrangement of fuel-cell system and on-board gas cylinders; (d) fuel cell bus operating under the STEP programme in Australia.

this was claimed to be the world's first commercial hydrogen service station for cars and buses. In the long term, it is conceivable that Iceland, which has no indigenous fossil fuels and is therefore eager to minimize its fuel import bill, may become totally energy secure with a sustainable future. This is a compelling national vision, even if the overall use of the energy is not particularly efficient and the cost of hydrogen is appreciable in present-day fuel terms.

In September 2004, three more *Citaro* fuel cell buses were subjected to a similar evaluation as part of the public transportation system in Perth, Western Australia [Figure 7.7(d)]. This Sustainable Transport Energy Project (STEP) continued for 2 years. The hydrogen fuel for the trial was produced by British Petroleum in its nearby oil refinery at Kwinana. A further three *Citaro* buses were delivered to Beijing in November 2005 to enter duty on a busy 19 km route past the Summer Palace and the site for the 2008 Olympic Games. The characteristic of all these bus programmes is that, at present, they are not economically viable without substantial government grants or subsidies.

The CUTE, ECTOS and STEP projects were judged to be highly successful. By October 2005, 17 of the 33 buses in operation in Europe, Iceland and Australia had exceeded 2000 h of operation, 10 had exceeded 2500 h and one had exceeded 3000 h. Between them, the buses had carried more than four million passengers and covered almost 1.1 million km. The reliability of the buses was high, with an availability rate of over 90%. Nearly 9000 refuelling operations were carried out safely and collectively dispensed almost 200 t of hydrogen, of which approximately half came from renewable energy sources. Seven of the cities operating fuel cell buses in regular service – Amsterdam, Barcelona, Hamburg, London, Luxembourg, Madrid and Reykjavik – decided to continue the work for one more year, starting in January 2006. The extension, known as the HyFLEET:CUTE project, was sponsored by the EU as part of its Sixth Framework Programme for Research and Technological Development. Associated bus trials were continued for the same period in Perth and Beijing. In March 2007, Amsterdam announced a further extension of its bus trials until January 2008.

Japan also has been active in the development of fuel cell buses. For example, Toyota and Hino Motors built the *FCHV-BUS1*, which was followed by the *FCHV-BUS 2* [Figure 7.8(a)]. The latter is a 60-passenger vehicle powered by two Toyota-designed PEMFCs that have an output of 90 kW and run on gaseous hydrogen stored at 35 MPa. A nickel–metal hydride battery stores the energy regenerated during braking and, for highly efficient operation, regulates the electric supply to the motor as determined by the operational status of the vehicle. The bus was placed into regular duty by the Tokyo Metropolitan Transportation Service in August 2003. Refuelling was carried out at a hydrogen dispensing station constructed by Showa Shell Sekiyu in partnership with Iwatani [Figure 7.8(b)]. It has been claimed that the *FCHV-BUS2* is three times more energy efficient than a conventional petrol-driven counterpart.

It is notable that in virtually all of these fuel cell bus projects the hydrogen is stored as compressed gas in roof-mounted gas cylinders. This mode of storage has been chosen by virtue of its ready availability and practicality for rapid

(a)

(b)

Figure 7.8 (a) Toyota–Hino *FCHV-BUS2*; (b) Showa Shell hydrogen-refuelling station in Tokyo.

refuelling. Roof mounting has several advantages: the cylinders do not occupy valuable space in the chassis, the roof of a bus is rarely damaged in a road accident, and in the event of a leak the gas dissipates upwards rather than entering the vehicle.

7.5.2   Delivery Vehicles

DaimlerChrysler has converted one of its large delivery vans (a *Dodge Sprinter*) to fuel-cell drive and three of these vehicles have been pressed into everyday service with a national parcel delivery company (United Parcel Service of America) – two in California and one in Michigan. This venture is receiving support from the US Environmental Protection Agency. The van is said to have a range of almost 250 km and an acceleration that is similar to that of the conventional diesel-engined version. At present, the use of such vans is limited by the availability of refuelling facilities.

7.5.3   Cars (Automobiles)

Following work on prototypes commencing in 1999, Honda became the first company to produce a passenger car powered by a fuel cell – the *FCX-V4* – and have it certified for commercial use in both Japan and the USA. The vehicle was supplied to customers in both countries in December 2002. Subsequently, in 2003, Honda built its own PEMFC stack, which was capable of start-up at temperatures as low as $-20\,^\circ$C. This was a significant breakthrough and was achieved by replacing the usual polymer electrolyte, based on a perfluorinated polyethylene backbone (see Section 6.4, Chapter 6), by a material with an aromatic structure. The new membrane gave a two-fold increase in conductivity and thereby permitted operation at below-freezing temperatures. Other technological and engineering advances raised the peak power output from 60 to 86 kW, reduced both the mass and the volume of the fuel-cell stack by half and improved the fuel economy by 20%, all compared with the *FCX-V4*. The power pack was introduced into the 2005 Honda *FCX*, which duly received certification for commercial service and was made available in limited numbers for leasing to selected organizations in Japan and the USA. In particular, the car has undergone field trials on the Japanese island of Yakushima, which is a World Heritage Site with abundant supplies of hydroelectric power that are far in excess of its local electricity needs. The choice of this testing site, as in Iceland, is motivated by the idea that hydrogen may be generated electrolytically from the surplus electricity; the initiative provides a practical example of truly 'green' personal transportation.

The 2005 Honda *FCX*, together with the chassis layout, is shown in Figure 7.9(a). The vehicle is a four-seat family hatchback (somewhat higher than usual in order to accommodate the two fuel-cell stacks under the floor) with a driving range of 250–300 km and a top speed of 150 km h^{-1}. The hydrogen is stored in two high-pressure (35 MPa) cylinders: one is located under the rear seat and the

(a)

(b)

Figure 7.9 (a) 2005 Honda *FCX* car and chassis layout; (b) *FCX Concept*. (Courtesy of Honda).

other under the boot (trunk). It is notable that the car was initially equipped with a bank of electrochemical capacitors, instead of a battery, to provide boost power of up to $1750 \, W \, kg^{-1}$ during start-up and acceleration. Recently, however, Honda has switched to a lithium-ion battery pack to allow longer periods of acceleration. Motoring journalists and others who have test-driven the car have reported favourably on the experience. Refinement of the vehicle is ongoing and Honda has released a working version of the *FCX Concept*, shown in Figure 7.9(b), which is scheduled for limited marketing in Japan and the USA in 2008. It is claimed that the new design gives a 30% improvement in range, has a maximum speed of $160 \, km \, h^{-1}$ and can be driven at temperatures as low as $-30 \, °C$. The new fuel-cell stack is 20% smaller and 30% lighter than that in the 2005 *FCX*, yet its power output is 14 kW greater. Clearly, rapid advances in the technology are being made.

Honda is by no means the only automotive company to be working on fuel cell cars. Toyota began producing its product, the *FCHV*, at the same time as Honda (December 2002). It is a very similar car, with comparable performance. The principal differences are that it is a five-seat car (a modification of the *Highlander* mid-sized SUV), has four 35 MPa hydrogen cylinders and relies upon nickel–metal hydride batteries for boost power. The Toyota fuel-cell stack has a peak power of 90 kW and provides the car with a top speed of $155 \, km \, h^{-1}$ and a maximum cruising range of about 300 km. The cars have been leased to customers in Japan and the USA. More recently, in parallel with its road-testing programme of *FCHV-4* vehicles, Toyota has been investigating the possibility of incorporating an on-board reformer in the next-generation *FCHV-5* to allow the use of ultra-low sulfur liquid fuel (Clean Hydrogen Fuel®, an evolved form of petrol) that can be supplied by existing petrol pumps.

In Germany, Opel (General Motors) has retro-fitted its *Zafira* compact van to produce a fuel cell car – the *HydroGen3* [Figure 7.10(a)]. This is a somewhat larger car than the 2005 Honda *FCX* or Toyota *FCHV*, and differs from them in that it can run on either liquid or compressed hydrogen. The vehicle has a range of 400 km with liquid storage (4.6 kg H_2) or 270 km with gaseous storage (3.1 kg H_2 at 70 MPa); the top speed is $160 \, km \, h^{-1}$. In 2004, *HydroGen3* set a world distance record for FCVs by travelling from Hammerfest, northern Norway, to Cabo da Roca, Portugal, in 38 days; a total of 9696 km was said to have been driven, which suggests an indirect route.

Also in Germany, Mercedes-Benz has been developing electric vehicles for over 30 years. Many of the later vehicles utilized advanced sodium–nickel chloride (ZEBRA) batteries. By 1997, over 1 million km of road testing of these batteries had been accumulated, much of it in the electric version of the Mercedes-Benz *A-Class* car. In 1994, the company commenced its research on fuel cell vehicles – the New Electric Car (NECAR) programme. Following the evaluation of a series of prototypes, the *NECAR 5* was launched in November 2000. This, too, was based on the *A-Class* car. Power was supplied by a 75 kW fuel-cell system that was fed by an on-board methanol reformer. The car featured a cold-start facility to remove the need for the reformer to

(a)

(b)

(c)

Figure 7.10 European fuel cell cars: (a) Opel *HydroGen3*; (b) DaimlerChrysler *B-Class F-Cell*; (c) *F 600 Hygenius*.

'warm up'. In 2002, a *NECAR 5* successfully traversed the American continent from San Francisco to Washington, a journey of 5250 km, during which it withstood first the Californian heat and then the cold, snowy conditions of the Sierra Nevada and the Rocky Mountains. It also crossed several mountain passes at heights of up to 2640 m and endured the heavily congested traffic of large cities. The objective of the undertaking was to investigate the technological limits of the vehicle.

In October 2002, DaimlerChysler unveiled its sixth-generation fuel cell car – the *A-Class F-Cell*. This employed an 85 kW Mark 902 Ballard fuel-cell stack, compressed hydrogen storage at 35 MPa, and a nickel–metal hydride battery for boost power. Sixty of these cars were manufactured and delivered for testing to customers in Europe, Japan, Singapore and the USA. Many completed over 2000 h of service without any loss of performance. The latest version of the *F-Cell* (October 2005) is based on the Mercedes-Benz *B-Class Compact Sports Tourer* [Figure 7.10(b)]. A research vehicle – the *F 600 Hygenius* – is being used to develop improved drive-train technology for this new series of *F-Cell* cars [Figure 7.10(c)]. Innovations include (i) a storage facility that can hold 4 kg of H_2 at 70 MPa and thereby increases the driving range from 160 to 400 km, (ii) a more compact fuel cell with higher power output (100 kW) and cold-start capability, and (iii) a lithium-ion battery that produces 30 kW in continuous operation and 55 kW at peak loads, *i.e.*, twice the output of the nickel–metal hydride batteries previously used. Under a partial-load regime, this vehicle is said to have a fuel efficiency of 60% (presumably based on the LHV for hydrogen).

In the USA, General Motors presented its latest fuel cell concept car – the *Sequel* – in January 2005 [Figure 7.11(a)]. This sporting sedan has a 'skateboard' chassis made of aluminium (the GM *Autonomy* concept) that integrates a 73 kW fuel-cell stack, a 65 kW lithium-ion battery, three carbon-composite TriShield™ hydrogen cylinders (8 kg H_2 total at 70 MPa; see Figure 5.3, Chapter 5), electronics, wheel motors and suspension components [Figure 7.11(b)]. The vehicle is operated with revolutionary 'drive-by-wire' technology (GM *Hy-wire*®) that allows for steering, acceleration, braking and other systems to be controlled electronically, rather than mechanically, through the use of two handgrips. Another novel feature of this vehicle is that the passenger compartment is manufactured separately from the chassis and then 'snapped in place'. This permits multiple body styles to suit customer preference. The approach is possible because the drive-by-wire technology eliminates mechanical linkages between the chassis and the body. The nerve centre of the electrical system is a universal 'docking port' in the chassis that connects all the body systems – controls, power and heating – to the rolling chassis. The *Sequel* is reported to have a performance that is comparable with that of traditional cars, *viz.*, a driving range of 480 km and acceleration to 48 and 96 km h^{-1} in 3 and 9 s, respectively. As yet, this is only a concept vehicle, but it points to the direction in which car design may go.

Early in 2007, General Motors disclosed plans to build a different version of the Chevrolet *Volt* (see Section 7.2), in which the internal combustion engine

(a)

(b)

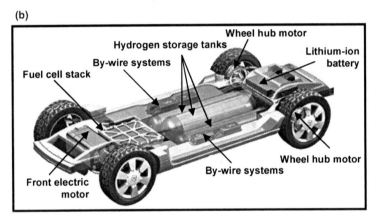

Figure 7.11 (a) General Motors *Sequel* car; (b) General Motors *Autonomy* chassis. (Courtesy of General Motors Corporation).

will be replaced by a fuel cell, so making it a fuel cell and battery (*i.e.*, all-electric) hybrid. An 80 kW fuel-cell stack will be coupled to an 8 kWh (50 kW peak power) lithium-ion battery pack. By making the former the primary electricity source, the storage battery is reduced to half the size of that in the earlier *Volt*, which was an engine–battery hybrid. The vehicle is shown in Figure 7.12(a).

Ford, like General Motors, has chosen to 'build a bridge' from hybrid electric cars to fuel cell cars by means of its hydrogen hybrid research vehicle – the *H2RV* – that was introduced in 2003. This car has a 2.3 litre, four-cylinder internal combustion engine/generator that is powered by hydrogen rather than by petrol and supplies the electric drive. A 288 V, 3.6 Ah lithium-ion battery assists the electric motor and stores regenerative braking energy. The electric

(a)

(b)

Figure 7.12 (a) General Motors fuel cell variant of the Chevrolet *Volt*; (b) Ford *Airstream Concept* car.

motor in the automatic transmission, together with the embedded controls, allow the *H2RV* to stop the internal combustion engine when the vehicle is at rest and then start it again quickly and smoothly with extra power for acceleration. The vehicle is a full hybrid (see Section 7.2), but has no carbon dioxide emissions because of the hydrogen-burning engine. It also offers enhanced fuel economy and has a range of about 200 km.

In January 2007, Ford announced the *Airstream Concept*, shown in Figure 7.12(b), which uses a similar design approach to that taken by General Motors with the Chevrolet *Volt*. The vehicle is powered by a plug-in hybrid drive-train that combines a 336 V lithium-ion battery pack with a compact Ballard fuel-cell system as a range extender. The latter operates in a steady state so that a significantly smaller, less-expensive unit can be employed. The sole function of this power source is to recharge the battery, as needed, and thereby allows further substantial reductions in the size, weight, complexity and cost of a conventional fuel-cell system. Moreover, the lifetime of the stack is expected to be doubled. The car can travel 40 km in battery mode, which depletes the state-of-charge of the pack to about 40% before current is drawn from the fuel cell for recharging – a strategy that increases the driving range by another 448 km. Both the General Motors and Ford concept vehicles are conventional series hybrids with the internal combustion engine replaced by a fuel cell. Obviously, there is scope for much further experimentation of hybrid vehicles based on fuel cells, particularly those with a plug-in battery recharge facility.

The US Department of Energy has set up its FreedomCAR and Vehicles Technology Program under the umbrella of the United States Council for Automobile Research (USCAR). The objective is to develop highway transportation technologies that are more energy efficient and environmentally friendly. At the same time, the initiative aims to reduce dependence on imported petroleum and so enhance national energy security. The ultimate target is no less than the realization of affordable cars and light trucks (vans) that are both emission free and petroleum free. The US government plans to achieve this goal by joining forces with companies engaged in the development of innovative technology. The overall programme has included the formation of the 21st Century Truck Partnership to produce prototype heavy-duty trucks and buses with improved fuel efficiency, reduced emissions, enhanced safety and performance, and lower operating costs. Vehicles with hybrid drives are included. The consortium involves 16 industrial companies, drawn from the vehicle and transportation sectors, and four federal agencies (Energy, Defense, Transportation and Environmental Protection). It will also seek, as appropriate, the technical expertise of 12 national laboratories and universities. Major government initiatives such as this are encouraging given the impending world oil shortage and the threat of climate change.

In a similar fashion, the three large US domestic car manufacturers (General Motors, Ford and DaimlerChrysler) formed a consortium in 2005 and signed agreements with the US Department of Energy to conduct research in the areas of lightweight materials and advanced batteries for vehicles. This also falls under the auspices of USCAR. On the world scene, eight of the major automotive companies (DaimlerChrysler, Ford, General Motors, Honda, Nissan, Renault, Toyota and Volkswagen) banded together with two major oil companies (Royal Dutch Shell and BP), an electricity producer (Norsk Hydro) and a tyre company (Michelin) to form The Sustainable Mobility Project, which was sponsored by the World Business Council for Sustainable Development. The object of the enterprise was to examine jointly the challenges facing the automotive industry in the next 25 years and the options available for meeting them. The final report of the project (Mobility 2030, published in 2004[3]) sets out a vision for sustainable mobility and the ways in which it might be achieved. This was a broad-brush approach to analyzing and addressing the road transportation problems that are faced by the entire world. It was concluded that widespread consumer acceptance of new vehicle and fuel technologies will depend either on policy instruments (pricing, voluntary agreements, regulations, subsidies, taxes) or on a fundamental change in society's attitudes and values to transportation.

7.5.4 Other Vehicles

A number of entrepreneurs and small companies have engaged in the development of novel FCV*s*. For instance, Formula Zero (a Dutch-based company) is making go-karts for competitive track racing [Figure 7.13(a)]. These vehicles employ a 10 kW fuel cell boosted by an electrochemical capacitor that can

Figure 7.13   (a) Formula Zero go-kart; (b) Intelligent Energy *ENV* motorcycle (fuel cell shown in inset).
(Courtesy of Intelligent Energy).

deliver 50 kW of power for 5 s. This gives a spectacular acceleration from 0 to 100 km h^{-1} in less than 8 s. The karts have a top speed of 130 km h^{-1} and can run for 12 min on a full tank of hydrogen. The idea behind this exercise is to demonstrate to the public, by means of exciting competitive events, that hydrogen fuel cells really do have the capability to power road transportation.

Intelligent Energy, a British company, has built the first motorcycle, the *ENV* ('Emissions Neutral Vehicle'), to be powered by a fuel cell; see Figure 7.13(b). This is an all-electric hybrid that has a small (1 kW) fuel cell that operates on compressed hydrogen (20 MPa) and is completely detachable. Electricity produced by the fuel cell is directed to a pack of batteries and an electric motor that provide the propulsive power; a maximum 6 kW is available to meet peak power demands. The motorcycle weighs 80 kg, has a top speed of 80 km h^{-1} and provides a range 160 km. Such a machine may prove popular in Asian cities, where motorcycles are used extensively and pollution is often severe, and also perhaps in London, where electric vehicles are exempt from the daily congestion charge. A German company is advertising a pedal bicycle with an auxiliary electric motor powered by an integrated fuel cell. In Canada, electric forklift trucks have been converted to fuel-cell power. These can be operated for 12 h between refills and then refuelled in ∼2 min. In Tasmania, an electric scooter is being converted to operate with a hydrogen fuel cell using a metal hydride to store the gas. Like Iceland and Norway, Tasmania has a plentiful supply of hydroelectricity, which makes for an interest in electric traction. In summary, various enterprising manufacturers around the world are addressing the market for novel personal electric transportation, driven by fuel cells, for local urban use.

The above projects have served to demonstrate the practical operation of fuel cell vehicles using PEMFC stacks. There is now little doubt that technically successful buses, cars and other vehicles can be manufactured and that the results from showcase in-service programmes have been encouraging. The remaining challenges relate to:

- the lifetime, reliability and (especially) cost of fuel cells;
- the development of a more effective on-board hydrogen storage facility;
- the introduction of an affordable infrastructure to produce, supply and dispense pure hydrogen;
- the overall energy efficiency, and the quantity of greenhouse gas emitted when the hydrogen is produced from fossil fuel; see Section 7.7 below;
- the timing and size of the emerging market;
- the establishment of internationally agreed protocols and regulations to govern the supply and utilization of hydrogen as a transportation fuel;
- public acceptance of hydrogen.

Other operational issues to be tackled include possible hydrogen release in confined spaces (garages, indoor car parks, tunnels), road accident situations and the need to train the emergency services to deal with hydrogen hazards, and the instruction of vehicle mechanics on how to handle the new technologies.

These are all factors that have to be fully resolved before FCVs will come into general use. Perhaps the greatest current problem is that of cost, but relative costs tend to change with the passage of time. The past 10 years have seen major improvements in the performance of PEMFC stacks for road transportation applications. The challenge now is to reduce their cost substantially through redesign and mass production without sacrificing performance.

7.5.5 Submarines

Fuel cells hold considerable attraction as underwater propulsion units for military submarines. By using hydrogen−oxygen (rather than hydrogen−air) systems, it is possible for a boat to remain under water for much longer periods and have greater operational range than a conventional submarine. Moreover, the tactical disadvantage of the latter in having to surface and run its diesel engine in order to recharge the propulsion batteries would no longer be an issue. A further advantage is that fuel cells are silent in operation, which makes the detection of the submerged submarine exceedingly difficult.

Germany has taken the lead in this field and two of its naval yards are now building submarines that use hydrogen−oxygen PEMFCs produced by Siemens, together with diesel engines for propulsion on the surface. The first of these boats, the 1500 tonne *U-31*, was fitted with nine fuel-cell stacks that could each deliver up to 50 kW. The power source gave the submarine a capability of travelling submerged for up to 20 days before surfacing. Although still well short of the underwater range of a nuclear submarine, the hybrid diesel−fuel cell design is a major advance on battery-powered submarines, thanks to the ability to recharge the batteries from the fuel-cell system while submerged. The German yards are now building vessels of this general type for other navies around the world.

7.6 Hydrogen Highways

At the end of 2005, there were around 115 hydrogen-refuelling stations across the world, with the majority located in the USA. This is a rapidly developing situation, with some 30 new centres opening in 2005 alone. In Europe, similar facilities were established in the nine cities that participated in the CUTE bus programme. Also, there are stations in Beijing (China), Perth (Western Australia), Reykjavik (Iceland), Tokyo (Japan) and Vancouver (Canada). The BP-branded station in Beijing is one component of a large Hydrogen Park and will service the hydrogen vehicle fleet for the 2008 Olympic Games. Interestingly, it is reported that the Chinese intend to produce the hydrogen from coal, via synthesis gas, rather than by the more usual steam reforming of natural gas.

In the USA, the main interest in FCVs is centred on California. This activity is driven by the State's strict environmental laws on the control of vehicle exhaust emissions, as well as by a desire to be independent of petroleum imports. Hydrogen is seen as both a clean fuel and one that can be multi-sourced from non-petroleum forms of primary energy. There is also the expectation that being a leader for hydrogen-powered transportation will bring jobs, investment and continued economic prosperity to California, and will progressively reduce the health hazards associated with emissions from conventional vehicles.

Activity is concentrated around the Los Angeles metropolitan area, where there is a sizeable cluster of existing or planned refuelling stations. The Governor of California has enunciated his vision for 2010 of 'tens of thousands' of FCVs being commercially available. For this to be a realistic proposition, it will be necessary to have an extensive hydrogen-dispensing network. To achieve this vision, the California Hydrogen Highway Network Action Plan encourages public/private partnerships to collaborate and invest in the early infrastructure development and to address key hydrogen commercialization challenges. Specifically, the aim is to install 150–200 refuelling sites throughout the State, *i.e.*, approximately one every 30–35 km along the major highways, in time for when FCVs arrive. This would constitute a sufficient density of such facilities to make hydrogen available to most Californians. There is, however, a potential 'chicken-and-egg' situation here: consumers will not buy FCVs unless hydrogen is widely available, while service stations will not wish to incur the expense of installing hydrogen stores, compressors and dispensers until there is an assured market for the hydrogen. Problems of this kind have been encountered before when introducing other types of new technology and have been successfully overcome by adopting a policy of building gradually from a nucleus, which in the present case would be outwards from Los Angeles. In the early stages, hydrogen-fuelled ICEVs will also require access to the gas and this may provide an initial market, before their fuel-cell brethren appear in numbers.

Canada has embarked upon a more modest demonstration and deployment programme for the implementation of a hydrogen highway. A consortium of organizations will work together to design, build, operate and evaluate a hydrogen-fuelling infrastructure at seven nodes along a corridor between Vancouver and Whistler (170 km to the north), with an extension to Victoria on Vancouver Island. The intention is not only to use this so-called 'British Columbia Hydrogen Highway' to demonstrate hydrogen-fuelled cars, be they powered by fuel cells or by internal combustion engines, but also to publicize other hydrogen technologies – from micro-fuel cells to stationary units. Hythane®, which, as discussed earlier, is a blend of hydrogen and natural gas, will also be evaluated as an interim fuel. The enterprise is scheduled for completion in time for the 2010 Winter Olympics that will be held jointly in Vancouver and Whistler. The infrastructure project will (i) create a critical mass of knowledge and experience, (ii) provide data for the formulation of international codes and standards for the technology, (iii) stimulate demand by allowing the media and general public to see at first-hand the potential benefits of a Hydrogen Economy and (iv) open doors for international partnerships. The choice of British Columbia for a Canadian hydrogen highway is apposite, given that Ballard Power Systems, a major manufacturer of PEMFCs, is located in the Province and the 2010 Winter Olympic Games, a fine international showcase for sustainable road transportation, are to be held there.

Proposals have also been made, although at a less advanced stage, for a hydrogen highway in Europe. Germany's leading hydrogen supplier, Linde, has suggested an initial circular network running from Berlin via Leipzig, Munich, Stuttgart, Cologne/Düsseldorf and back, with semi-automated refuelling points every 50 km. With a distance of around 1800 km, this would require some 35 new hydrogen stations, in addition to those already operating in Berlin and Munich. The route passes through major centres in which most of the German automotive production sites are located. This would allow carmakers, many of which already have hydrogen vehicles, to test the viability of their fuel-cell power systems in everyday use under real conditions. The highway differs from those under consideration in the USA and Canada by not focusing first on a nucleus of supply points around one city and then spreading outwards from there. If the German demonstration is enacted and proves successful, the plan would then be to implement progressively a Trans-European Hydrogen Highway. It is generally held that for hydrogen vehicles to be widely adopted, it will first be necessary for refuelling pumps to be available at 30–50% of all service stations.

These national and regional projects are ambitious, but realizable, although the rate at which they reach maturity will be determined by the progress being made towards the production of affordable FCVs. The most probable way forward is for FCV operation to spread outwards from centres such as Los Angeles, Vancouver or Tokyo, where the hydrogen supply system is initially established. As more vehicles are purchased, further hydrogen stations would be introduced along main routes over a broad area, until eventually the given nation had a full hydrogen network. The introduction of petrol stations early in the 20th century provides a precedent for this.

One interesting suggestion that has been raised is that FCVs should be seen as distributed generators of electricity. There is no reason why the fuel cell should only operate when the vehicle is in use. At other times, for example when a car is garaged at home, the fuel cell could be plugged into a domestic electricity 'outlet' to convert it to an 'inlet' and so supply the household. A typical unit might generate a maximum power of 50–100 kW (note: the latest Honda vehicle, the *FCX Concept*, has a peak output of 100 kW), which would be well in excess of that demanded by the average family home and thus would leave some electricity over for export to the grid. Similarly, in company car parks, the employees' cars collectively could generate sufficient power to supply the workplace. These possibilities could prove to be economic, since once the FCV has been purchased there is little additional capital cost apart from the inverter–transformer and the required banks of storage batteries. Moreover, this concept could have a massive impact on the requirement for centralized electricity generation in large plants and on national transmission and distribution systems. Savings here, too, would contribute to the overall economic benefit. Consider the UK as an example. If, in the next century, most of the nation's road vehicles were powered by fuel cells, the total generating capacity they could provide would be at least 10 times greater than that currently installed. Of course, if FCVs were to be used as local electricity-generating plant, correspondingly greater levels of hydrogen production would be needed, along with enhanced transport and storage facilities for the gas.

7.7 Efficiency Calculations and Fuel Consumption

As we have seen in Section 6.6, Chapter 6, the efficiency of fuel cells depends on a number of factors, not least the power output. In the case of electric road vehicles driven by PEMFCs, the reasons generally advanced for this form of transportation are:

(i) the opportunity to diversify the primary energy base of vehicles beyond petroleum;
(ii) the lack of airborne pollution from fuel cells;
(iii) the silent operation of electric vehicles;
(iv) the high theoretical efficiency of fuel cells, *i.e.*, much greater than that of internal combustion engines;
(v) the consequent reduction in carbon dioxide emissions.

The first two arguments constitute a powerful case for electric vehicles. The third point is of questionable benefit since it might lead to an increase in the number of accidents involving pedestrians. The final two points merit deeper consideration; they are closely inter-related.

When comparing different transportation fuels, different motive-power units and different vehicles, it is essential to speak of 'well-to-wheels' efficiency. This concept embraces:

(i) all of the energy consumed in extracting the fossil fuel from the ground (*e.g.*, oil and gas from wells, coal from mines) and then refining it, conveying the refined fuel to the service station and, finally, dispensing into the vehicle's storage tank (the 'well-to-tank' process);
(ii) the combined efficiency at which the fuel is combusted in the engine (or fuel cell) and the resulting heat energy (or electrical energy, respectively) is converted into mechanical energy at the wheels, with due regard to the friction losses in the drive-train (the 'tank-to-wheels' process).

It is well known that ICEVs have rather low on-the-road efficiencies (currently 20–25%, at best), but their overall performance is in fact even lower when account is taken of the energy used in the extraction, transport and refining of the oil and then in the delivery of the petrol. Together, these extra losses typically amount to around 13% (*i.e.*, a net efficiency of 87%). On multiplying the two values, the overall 'well-to-wheels' efficiency is reduced to 17–22%.

Electrochemical generators, such as fuel cells, are certainly more efficient than internal combustion engines, but there is considerable confusion over the numerical efficiency in real applications. For instance, use of the higher heating value (HHV) or the lower heating value (LHV) for hydrogen makes a difference of 18%. It has been shown in Section 6.6, Chapter 6 that the practical efficiency of a hydrogen-fuelled PEMFC (based on the HHV of hydrogen and having regard to electrical and mechanical losses) is likely to lie in the range 32–38%, depending on the current density, with the greatest efficiency at low power outputs[†]. As seen earlier in Figure 7.5, this relationship is precisely the opposite to that of the internal combustion engine, for which efficiency declines at low power outputs. The dependence of fuel-cell performance on current delivered is especially important in vehicle applications. We have already observed that for both fuel cells and electrolyzers, there is a trade-off between size and capital cost on the one hand, and efficiency and running cost on the other. The larger the unit, the higher is the capital cost of construction – but the resulting low current density leads to higher efficiency of operation. For small vehicles, such as cars, where size and capital cost are at a premium, the fuel cell will probably operate at a high current density and, therefore, at a relatively low efficiency. This will be mitigated to some degree by the use of a battery or electrochemical capacitor when peak power is required. In making the comparison between efficiencies of engines and fuel cells, it is important to remember that fuel cells consume no energy when the vehicle is held up in traffic and are therefore well suited to urban use, especially during peak hours.

This is not, however, the whole fuel cell story because hydrogen is simply an energy carrier that has to be produced from either fossil fuels or water. Each of these processes involves the input of energy, which must be included. If the hydrogen is derived from fossil fuels there are the losses in chemically reforming the feedstock (natural gas or coal), whereas if it is produced by electrolysis there are losses in generating and transmitting the electricity. The steam reforming of natural gas to hydrogen on a large scale is at best 75–80% efficient; see Section 2.3, Chapter 2. Further energy is consumed and hydrogen is lost in its separation from carbon dioxide, which reduces the efficiency to, say, 70%. It is reasonable to assume a 10% energy loss in compressing the hydrogen and another 10% in transporting it from the centralized steam reformer to the vehicle-refuelling depot. Hence the overall efficiency from natural gas to traction effort, via hydrogen, is as follows:

- natural gas to distributed hydrogen: $0.70 \times 0.9 \times 0.9 = 57\%$;
- hydrogen to low-voltage d.c. tractive effort: 32–38% (say, 35%);
- 'well-to-wheels' efficiency (natural gas to tractive effort): $0.57 \times 0.35 = 20\%$.

[†] Other authors have quoted somewhat higher efficiencies, usually based on the LHV of hydrogen, and lower power outputs. For instance, the Ballard Mark 902 module is said to deliver a maximum efficiency of 48% (LHV) at partial load. This equates to 40.6% (HHV). Thermodynamically, the HHV should be used. When electrical and mechanical losses are included, the tank-to-wheels efficiency reduces to around 33%, in line with our stated range. As noted above, DaimlerChrysler claims 60% efficiency for their *F 600 Hygenius* car, but this figure needs clarification.

This overall figure of 20% is clearly an approximation. Nevertheless, even if the efficiency of the fuel cell were a few percent more than we have assumed, there would be no change to the general conclusion that the well-to-wheels efficiency of FCVs operating on reformed natural gas is not far removed from the 17–22% of current high-performance ICEVs. As discussed above, however, FCVs would offer certain other benefits. For instance, if the natural gas is steam reformed regionally or locally, rather than centrally, there would be some savings in the energy otherwise wasted in hydrogen distribution, but this would be offset by the lower efficiency (not to mention the higher cost) of the smaller reformers. Also, the sequestering of the carbon dioxide would become more complicated. In terms of environmental pollution in urban areas, the replacement of ICEVs by FCVs would provide much better air quality and thus reduce the incidence of asthma and other chest complaints, but possibly at the expense of increased accidents involving pedestrians – unless electric vehicles were required to give an audible warning of approach. A precedent for this is the reverse warning alarms fitted to heavy-duty vehicles.

What if the hydrogen is produced by electrolysis rather than directly from natural gas? The practical cell voltage for the electrolysis of water is around 1.47 V; see Section 4.1, Chapter 4. The output voltage of a PEMFC lies in the range 0.7–0.6 V, as determined by the current density. Thus, the electrolyzer and fuel cell combination is likely to be 41–48% efficient (say, 45%). Although this is higher than for a high-performance automobile, as emphasized by advocates of FCVs, the losses incurred in producing electricity from primary fuels have yet to be included. The efficiency of a conventional power station lies in the range 30–35% (coal/nuclear-fired, say 33%) to 55% (combined-cycle gas turbine). The overall efficiency from primary fuel to traction effort is then as follows:

- coal or nuclear plant: $0.33 \times 0.45 \times 0.9 \times 0.9 = 12\%$;
- natural gas plant: $0.55 \times 0.45 \times 0.9 \times 0.9 = 20\%$.

In these calculations, the 10% energy loss in compressing the hydrogen has been retained, while the 10% loss in distributing hydrogen has been replaced with a 10% loss in the electricity-supply system that would result from transmission/distribution, voltage reduction and rectification operations.

Again, the above are only approximate calculations. Nevertheless, from the viewpoint of overall primary energy efficiency, the analyses clearly show that there is no incentive to replace ICEVs with FCVs, irrespective of whether the hydrogen stems directly from fossil fuels or indirectly via electricity. Fuel-cell vehicles will not contribute significantly to the abatement of greenhouse gas emissions unless the carbon dioxide is sequestered centrally and disposed of permanently, or unless electricity from nuclear or renewable sources is used to generate the hydrogen. On the other hand, benefits are to be gained in heavy traffic where FCVs are more efficient than ICEVs and also non-polluting. There are, however, considerations other than energy efficiency and greenhouse gases that may favour FCVs. Foremost among these, as discussed in Chapter 1, are the world's diminishing reserves of petroleum, the concentration of the major oil resources in relatively few regions and the huge import bills that are being incurred by many nations to keep their transportation running. A new infrastructure based on natural gas or coal, rather than on petroleum, holds many attractions.

Whenever (and wherever) renewable electricity becomes affordable and available on a large scale, the overall energy economy of FCVs will improve. This is because the conversion of mechanical energy (*e.g.*, wind or wave power) to electricity does not involve a Carnot cycle and the efficiency of this primary step should be 80–90%, rather than 30–55%. Solar conversion efficiencies are, for the foreseeable future, much lower. Of course, the energy losses associated with the electrolyzer, the pressurization and distribution of hydrogen and the fuel cell would remain. The extent to which such a move to renewable electricity is possible will be determined by cost considerations and political acceptability, in addition to technical aspects relating to the feeding of large amounts of such power into the grid system. So long as supplies of renewable electricity are limited, there is a strong argument for them to be utilized as such, rather than converted to hydrogen.

In parallel with the development of FCVs, advances are also being made in conventional internal combustion engines. The efficiencies of the latter are expected to improve rapidly, as demonstrated by the projections made by Massachusetts Institute of Technology (MIT) for various configurations of family-sized cars in 2020.[4] These are listed in Table 7.3. The data suggest that by widespread conversion to ICE–battery hybrids – a perfectly feasible proposition by 2020 – it should be possible to reduce fuel consumption by two-thirds compared with the 1996 model family car. On both energy efficiency and present cost grounds, this appears to be a much more realistic option than the introduction of fuel cell cars.

The MIT report also considered plausible full life-cycle figures for (i) energy use, (ii) greenhouse gas emissions and (iii) running cost per kilometre travelled. These calculations encompassed the extraction, refining and distribution of

Table 7.3   Anticipated improvements in ICE technology.[4]

	Fuel	*Petrol equivalent consumption*[a]	
		$MJ\,km^{-1}$	*Litres per 100 km*
1996 Reference car	Petrol	2.73	8.45
2020 Technology			
Advanced spark ignition engine	Petrol	1.54	4.80
Advanced compression engine	Diesel	1.35	4.20
Hybrid spark ignition engine	Petrol	1.07	3.30
Hybrid spark ignition engine	CNG	1.03	3.20
Hybrid compression engine	Diesel	0.92	2.85
Fuel cell hybrid	Hydrogen gas	0.81	2.50

[a] The data for fuel consumption take no account of the energy consumed in producing the fuel. When this is included, the FCV moves from being the most energy efficient of the 2020 options to the least efficient.

petroleum and the manufacture of the vehicle. The study investigated 10 different fuel–propulsion systems and compared them with the 1996 reference automobile. All of the systems were, as might be expected, considerably improved in terms of energy use and greenhouse gas emissions, but none of the vehicles were cheaper and many were considerably more expensive to buy and run. It was concluded that ICE–battery hybrids appear to have advantages over FCVs on all three counts listed above. It was emphasized, however, that there are considerable, but unavoidable, uncertainties in such a futuristic analysis.

Hybrids are attractive to the automotive industry since a paradigm shift in technology is not required, and also to the users because such vehicles are not overly more expensive than conventional automobiles and will deliver superior fuel economy. Nevertheless, it must be borne in mind that the improvements in ICE technology given in Table 7.3 relate only to a single 'reference car' driven under a defined set of conditions and that the scope for variability is considerable. Some authorities doubt the superiority of the diesel hybrid car over a fully developed petrol–electric plug-in hybrid, *i.e.*, one equipped with urban capability of running in an all-electric mode using mains electricity; see Section 7.2 above. There will always be a wide choice of vehicles to suit customer tastes, based upon designers' ingenuity, fuel availability, performance characteristics, capital and running costs, and government taxation policy.

The introduction of an entirely new technology for road transportation, such as that of FCVs, would represent a quantum leap both for the vehicle industry and for the energy supply and distribution industries. The authors of the above-mentioned MIT report note that there are at least six sets of stakeholders that must be consulted and satisfied before such a change can be realized:

- fuel manufacturers and refiners;
- fuel distributors;
- vehicle manufacturers;
- vehicle repair and maintenance shops;
- customers (vehicle purchasers and drivers);
- regional, national and local governments who set regulations and taxes.

The greater the shift in fuel or vehicle technology, the greater will be the difficulty of the task and the longer the time-scale for its successful implementation. This poses a much greater challenge to the hydrogen fuel cell than it does to the ICE hybrid.

References

1. A. Cooper, L.T. Lam, P.T. Moseley and D.J. Rand, The Next Great Challenge for Valve-regulated Lead–Acid Batteries: High-rate Partial-state-of-charge Duty in New-generation Road Vehicles, in *Valve-regulated Lead–Acid Batteries*, D.A.J. Rand, P.T. Moseley, J. Garche and C.D. Parker (eds.), Elsevier, Amsterdam, 2004, 549–565.

2. *Hydrogen – Sustainable Energy for Transport and Energy Utility Markets. Supported by State and Industry*, Government of the State of North Rhine-Westphalia, Düsseldorf, February 2006. www.energieland.nrw.de.
3. *Mobility 2030: Meeting the Challenges to Sustainability*, Full Report 2004, World Business Council for Sustainable Development, July 2004. www.wbcsd.org/web/mobilitypubs.htm.
4. M.A. Weiss, J.B. Heywood, E.M. Drake, A. Schafer and F.F. AuYeung, *On the Road in 2020: A Life-cycle Analysis of New Automobile Technologies*, Energy Laboratory Report MIT EL 00-003, Massachusetts Institute of Technology, Cambridge, MA, October 2000.

CHAPTER 8
Hydrogen Energy: The Future?

This book set out to explain the concept of hydrogen energy, to identify the barriers to its implementation, and to explore the prospects for success. In individual chapters, we have identified specific challenges and discussed how they might be met. The emphasis has been placed on technical matters. As scientists, the authors do not feel competent to go deeply into the sociological, political, legal or financial aspects of introducing an entirely new energy vector into society, although all these issues will need to be addressed. This closing chapter draws together the main conclusions that we have reached, with the understanding that by no means everyone will agree with our findings. Forecasting the future is not an exact science.

One of the difficulties faced in evaluating the prospects for hydrogen energy is the broad scope of the subject. Any new energy technology must be based on a sound platform of science, engineering and economics. All the various factors concerned with hydrogen production, purification, transport, storage and utilization are interrelated in a complex fashion. The reasons advanced for promoting hydrogen as a clean fuel (and the ultimate solution to many of mankind's energy problems) are diverse, while the possible applications can range from electronic devices requiring 1 W of power to electricity generation at the hundreds of MW scale – a factor of 10^8! Furthermore, prevailing circumstances vary greatly from country to country and thereby give rise to different priorities for hydrogen energy and its end use. In an overall review such as this, it is only possible to generalize and present some examples, rather than to treat exhaustively any one application or the situation in any given location. A further variable is the future date under consideration. New technologies in the energy field are generally introduced only slowly, *i.e.*, over a period of decades and some aspects of hydrogen energy may not be widely adopted until the 22nd century, if at all. It is therefore important to set a realistic target date when assessing the prospects for a particular technology.

8.1 World-wide Energy Problems

There is almost universal agreement that the two major energy issues facing the world today are:

(i) the future security of energy supplies, especially as regards oil and gas; this involves not only of the availability and affordability of fuels for the consumer, but also the impact of increasing imports on a country's wealth and balance of payments.

(ii) climate change, particularly with respect to reducing anthropogenic emissions of greenhouse gases, especially carbon dioxide.

How are these two issues to be addressed?

8.1.1 Security of Energy Supply

Already petroleum is being consumed at a greater rate than new discoveries are being made. Mounting concern over this shortfall is compounded by forecasts of a maximum in the global production of oil (so-called 'peak oil') within the next 10–20 years. The price of crude has risen sharply, so that novel sources of hydrocarbons, such as oil sands and bitumen, are becoming more cost-competitive. Unfortunately, extraction of petroleum from unconventional sources is accompanied by severe environmental problems and also consumes considerably more energy than traditional oil recovery. Moreover, some of these new sources have high sulfur contents and thus demand greater quantities of hydrogen for catalytic hydrodesulfurization. This latter process will result in a growth in hydrogen production and consumption within refineries.

Natural gas is more plentiful and more widespread geographically than oil, but already some major basins are becoming exhausted (*e.g.*, the North Sea.). This is leading to a build-up of the liquefied natural gas industry. Gas found in remote areas of the globe is liquefied and shipped in cryogenic tankers to centres of population. Such activity requires the construction of export/import terminals in many different countries.

It is generally accepted that natural gas fields result from the breakdown of organic matter deep beneath the Earth's surface under the influence of the prevailing very high pressures and temperatures. The gas can also be formed through the biogenic transformation of organic matter by tiny microorganisms. This process generally takes place close to the surface of the Earth and the methane produced is usually lost into the atmosphere. In certain circumstances, however, it can become trapped underground as, for example, in the case of landfill gas. It is thought that huge reserves of natural gas of non-biological (inorganic) origin may lie at extreme depths under the Earth's crust and that these might be tapped. So far, this is little more than a suggestion and the supporting evidence is scant, but it remains a possibility to be investigated.

Coal is the world's largest reserve of fossil fuel and, in the years ahead, will be exploited aggressively by countries that have significant deposits, but no oil or gas, in order to achieve energy independence. This raises the question of how best to utilize coal as a modern, clean energy source. Growth in demand suggests that electricity generation will be the prime market. Clean coal technology involves gasification of the solid fuel, separation of the hydrogen so produced and its combustion in gas turbines (or fuel cells), and finally permanent storage of the captured carbon dioxide. Most probably, there also will be an increasing requirement for coal as a raw material for the synthesis of liquid fuels that will be needed in ever-increasing quantities for the transportation sector.

In the near future, renewable sources of energy – especially wind and solar power – will play an expanding, but still secondary, role in the world electricity scene. Some countries will take up renewables more vigorously than others, as determined by the resource base and the economics. There is scope for increasing the extent of hydro-power, but this lies mostly in regions that are remote from centres of population. Opportunities also exist for the further harnessing of marine (tidal, wave) energy. Nevertheless, renewables will continue to represent only a very minor component of the primary energy supply in the majority of developed nations for some years to come.

Nuclear power is well established in some countries, whereas its future is hotly debated in others. Clearly, this is an unsatisfactory situation. It is probable, however, that a consensus will gradually emerge as new designs of fail-safe reactor are developed and as procedures for the secure disposal of radioactive waste are progressively improved. For the present, nuclear energy is the prime option for large-scale generation of electricity that does not involve the use of fossil fuels. As existing stations reach the end of their lives over the next 10–20 years, it will become increasingly urgent to decide whether or not they are to be replaced by alternative and superior technology. This is not an easy decision to make. First, a political battle would have to be won in many countries. Almost certainly, any proposal to build a new civil nuclear reactor will face fierce opposition and, consequently, the planning and approval process will be protracted. Then there is the issue of which reactor type to commission, given the likelihood of improved and safer designs evolving. Past experience in the UK has shown that it is a mistake to keep changing reactor design and that series construction of one model, albeit not the latest, is more economical in the long run. The construction time for a new nuclear power plant, once approval has been obtained, will be at least 5 years and much of the next 25 years would be taken up simply with replacing existing installations as they are retired, as opposed to expanding the programme to combat climate change. Finally, there is the matter of cost. Nuclear power involves high up-front investment, compared with fossil-fuel electricity plant. The subsequent operating costs are low, but those of radioactive waste disposal and end-of-life decommissioning are also high. It is unlikely that many private companies will assume such a financial burden without government assurances and assistance.

The assorted pathways to electricity generation will be adopted to varying degrees by different countries, as dictated by the associated costs, by politics and by national priorities. Where there are rich indigenous deposits of fossil fuels, governments will prefer to use these and, where not, will draw upon a diversity of primary energy sources to provide a secure future.

8.1.2 Climate Change

The effect on climate of mankind's overwhelming reliance on fossil fuels is the second major energy-related issue now facing the world. Arguably, its resolution is even more urgent than finding ways to safeguard the energy supply itself. Taken together, however, the twin problems are embodied in the search for ultimate global 'sustainability', which may be expressed as the ability to meet humanity's needs on an indefinite basis without producing irreversible environmental effects.

The burgeoning demand for energy stems from the growth in world population, together with the aspirations of developing nations to industrialize and so improve their standard of living. People everywhere want their homes and workplaces to be properly heated and cooled and also to possess their own powered means of transportation. When this requirement for more energy is coupled with an increasing dependence on oil sands, bitumen and coal in the decades ahead, it is clear that advanced technology is needed for the cleaner utilization of these fuels and for the capture and storage of unavoidable emissions of carbon dioxide. With respect to the latter, the term 'sequestration' has come into popular usage in setting the ground rules for the activity. There is a danger that politicians, in planning an overall energy strategy, will simply assume that it is a *fait accompli* on whatever scale they choose to adopt. As yet, this is far from the truth.

No matter which route is selected for carbon capture and storage, the operations will dwarf those conducted to date and will pose major technical and logistical challenges, not to mention added costs. Only large 'point sources' of carbon dioxide (power stations, oil refineries, cement works, *etc.*) will be amenable to direct carbon capture and many of these will be located far from suitable storage sites. Such situations will involve the construction of pipelines to convey the gas appreciable distances to a geological or oceanic repository. Furthermore, the development of a cost-effective method for the separation of carbon dioxide from mixed streams of hot gases (produced during either pre-combustion or post-combustion) will command a substantial research effort. Unless these carbon disposal problems can be solved in a timely and practical manner on an appropriate scale, and there is no certainty that they will be, society may be destined to witness steadily increasing levels of carbon dioxide in the atmosphere and ever-greater changes in global climate, with all the attendant consequences that have been so widely predicted.

As the climate warms, a new fear has arisen in relation to the tundra of Northern Canada and Russia. It is known that substantial quantities of natural gas are trapped in the 'permafrost' (permanent frozen subsoil) of these regions. As this melts, the methane – a far more powerful greenhouse gas than carbon dioxide – will be liberated and so enhance the global warming. This positive feedback will exacerbate the overall process and has been designated the 'super greenhouse effect'. It is almost impossible to see how to capture and exploit this methane and the only route to preventing its accelerating release is to hold atmospheric levels of greenhouse gases below those that will promote melting of the permafrost. It may already be too late for this to be accomplished.

Serious issues associated with almost all initiatives proposed for combating climate change (whether they involve building new nuclear reactors, capturing and storing carbon dioxide or exploiting renewable forms of energy) relate to the speed and costs of implementation. There is a grave mismatch between the forecast dates at which actions might be put into practice and the urgency of the situation as projected by climate scientists. Only major economies in the use of fossil fuels, which will entail changes in lifestyle, will meet the requisite time-frame.

8.2 Hydrogen Energy: The Challenges

Where does hydrogen fit into the evolving energy scene? The one certain prediction is that increasing quantities will be required, as mentioned above, for the hydrodesulfurization of low-grade fuels such as oil sands, bitumen and coal. Greater amounts of hydrogen will also be used in the manufacture of fertilizer (ammonia) as the world population expands and more land is brought into agriculture. These are important applications, but they hardly realize the vision of 'hydrogen energy' or indeed a 'Hydrogen Economy.' The latter implies large-scale production of hydrogen as a universal energy storage and transfer medium. In the short term, this would be by the reforming or gasification of fossil fuels with carbon capture and storage. In the longer term, for a sustainable energy future, other forms of primary energy (renewables and/or nuclear) would be used to decompose water into hydrogen and oxygen. The hydrogen, once separated, may be employed to fuel a boiler, combusted in a gas turbine to generate electricity and heat, or utilized in a fuel cell. The last-mentioned device may be a stationary installation for localized production of electricity, possibly as part of a combined heat and power (CHP) scheme or a unit designed for the propulsion of road vehicles. It should be pointed out, however, that the sustainable energy scenario for road transportation (*viz.*, primary renewable energy → electricity → hydrogen → electricity → traction) will be seriously inefficient and costs will escalate.

8.2.1 Production

Hydrogen manufacture via the reforming of natural gas is a mature industry with 45–50 Mt produced annually for use mainly in oil refineries and for the synthesis of ammonia. To gain acceptance as an energy vector, hydrogen must be produced far more cheaply than when it is employed as a chemical. Substantial cost savings are not easy to achieve in a mature, high-volume industry. Small-scale operations, as required for distributed local production, would be especially expensive. Because of the additional processing involved, hydrogen obtained by steam reforming can never compete economically with natural gas as an energy carrier when judged on a unit energy basis (*i.e.*, cost per MJ of energy delivered). In summary, the barriers to a major increase in the production of hydrogen by steam reforming are as follows.

- An impending shortage of natural gas (*i.e.*, within a few decades) given the present strong growth in world demand.

- The loss of energy associated with conversion of natural gas to hydrogen. The available stocks of the former ought therefore to be conserved for more energy-efficient uses, *i.e.*, for heating and electricity generation.
- The huge capital investment required to establish a hydrogen industry of sufficient size to make an impact on the world stage. Also, the time needed to execute such an enterprise.
- The development of cost-effective techniques for the capture and storage of carbon dioxide if hydrogen energy is to become the ultimate tool for repressing climate change.
- Cost-competitive production, distribution, storage and use of hydrogen, particularly when sequestration of carbon dioxide is included.

This is a daunting list of obstacles to be overcome and raises real doubts concerning the future of hydrogen energy. Of course, if the primary concern is that of security of fuel for use in road transportation and the aim of a hydrogen programme is to substitute for petroleum without having much regard for energy efficiency or greenhouse gas emissions, then many of these obstacles become more manageable.

Bulk production of hydrogen via electrolysis appears improbable until renewable or nuclear electricity becomes widely available and considerably cheaper than at present. The principal attribute of electrolytic hydrogen is its ultra-purity, which is an important requirement for proton-exchange membrane fuel cells. Nevertheless, the use of valuable electricity to electrolyze water and then feeding the resultant hydrogen to a fuel cell is intrinsically wasteful by virtue of the combined inefficiencies of the two devices involved. This really only makes sense in situations where there is more electricity than can be consumed as such, or where there are reasons for wanting hydrogen that transcend considerations of efficiency and cost.

Most of the other options for producing hydrogen are either limited in scope or at an early stage of development. It is too soon to envisage their commercial prospects. Manufacture from dry biomass by gasification is perfectly feasible. Moreover, given that the process is deemed to be 'carbon-neutral', direct combustion can be employed with no need to capture the carbon dioxide. The size of the operation is generally restricted by the quantity of material that can be harvested and collected close to the processing plant. This is because biomass is not dense and generally cannot be conveyed economically by road over distances of more than 100–200 km. The bacterial fermentation of wet biomass can be made to yield biogas that contains a moderately high percentage of hydrogen, but this line of investigation is still in its infancy.

Plasma reforming of natural gas is attractive in that solid carbon is formed rather than carbon dioxide. The drawbacks are that the process requires substantial amounts of electricity and, obviously, no advantage is taken of the energy that, otherwise, would have been obtained from the combustion of carbon. In terms of energy efficiency, therefore, it does not appear to be a very promising candidate.

Solar–thermal reforming of natural gas is an interesting future prospect, but has yet to be fully demonstrated. Consequently, the practicalities and economics of the process have not been established. Thermochemical cycles for the splitting of water have been studied for many years, with new cycles emerging periodically. Very few, however, have been taken beyond the laboratory stage. Some scientists link this technology to high-temperature nuclear reactors for provision of the necessary heat input, others to solar–thermal towers. High-temperature nuclear reactors are not yet available, but would be of such a size that considerable quantities of heat would be liberated. This, in turn, would entail the construction of correspondingly large engineering plant to undertake the thermochemical processes. Such a situation may be seen as 'piling high technology on top of high technology'. Solar–thermal towers, by contrast, would have a much lower heat output, but still sufficient for hydrogen production. Nevertheless, the commercial viability of undertaking thermochemical cycles on a relatively small scale must be questionable. It is doubtful whether either approach will find major application in the near-term.

Finally, there is the possibility of splitting water directly by means of solar energy and photo-electrochemical or photo-biochemical reactions. These processes are still at the research stage and await scale-up and economic evaluation.

8.2.2 Distribution and Storage

Distribution and storage of hydrogen represent two of the barriers that impede a transition to the widespread adoption of hydrogen energy. As discussed above, centralized production leads to a requirement for a national grid of pipelines to supply the gas. It is doubtful whether such facilities will be built, both because of the high capital cost and because of the amount of energy consumed in pumping long distances in relation to the calorific value of the delivered gas. Rather, regional production and local distribution is all that can be projected. Bulk storage underground is feasible in principle and has been demonstrated in practice, but unless the gas is also consumed locally (as in a nearby electricity station) it still has to be dispensed to the end users. This is the fundamental weakness of centralized production. Conveyance by road as compressed gas in cylinders is suitable for limited quantities, but is highly inefficient given the petroleum consumed per unit of energy delivered. Extending greatly this mode of distribution would create severe road congestion and would be unduly expensive. The delivery of larger amounts of hydrogen as a cryogenic liquid (LH_2) is perfectly possible and also offers short-term storage. The problems with LH_2 are the capital cost of the liquefaction and storage plant, the substantial input of electrical energy required to liquefy the gas, and the losses due to boil-off during transfer and on standing. Distribution via LH_2 has been proposed and demonstrated for use in hydrogen-fuelled vehicles, but a question mark remains over the practicality and safety of permitting members of the public to refuel their own cars.

Localized storage of hydrogen gas, whether on-board vehicles or to supply stationary fuel cells or CHP plant, also presents challenges. There are several possible modes of storage. To date, however, compressed gas in cylinders has proved to be the most practical and affordable method. This approach has been adopted by the principal developers of fuel cell vehicles, regardless of the fact that the low volumetric energy density of the gas is a major disadvantage for road transportation applications where space is generally at a premium. For private cars running on hydrogen, the range between refuelling stops has therefore been restricted. Notable recent improvements in cylinder technology have permitted the storage of gas at higher pressures so that the driving range is becoming more acceptable.

Metal hydride storage has the drawbacks of excessive mass and high cost, as well as very slow refuelling due to the time required to dissipate the heat produced during the hydriding reaction. Storage in the form of a chemical compound that releases hydrogen by catalytic decomposition or by reaction with water sounds to be more promising, although this route has not been subjected to thorough vehicle trials. In particular, the use of sodium borohydride ($NaBH_4$) as a water-splitting reagent is worth further investigation, even though this chemical is currently very expensive. The evolution of hydrogen leaves behind sodium borate ($NaBO_2$), which would have to be collected during refuelling and then returned to the factory for re-conversion to $NaBH_4$. There may also be difficulties in controlling the reaction so as to produce hydrogen at the variable rate set by the vehicle's driving schedule.

8.2.3 Fuel Cells

Hydrogen energy and the fuel cell are closely linked. Most types of fuel cell use hydrogen as a fuel, at a level of purity that differs according to the specific system under consideration. For localized electricity generation, hydrogen is converted into electricity (together with some heat) in a fuel cell, whereas for large-scale operations a gas turbine would be employed. Stationary fuel cells are seen as clean and quiet sources of electricity for dispersed generation. It is often claimed that they are efficient at converting hydrogen into electricity, although in fact only about 45% of the original electrical energy used to generate the hydrogen by electrolysis, for example, may be recovered in stationary and mobile applications. This low electrical efficiency arises from losses in the electrolyzer and the fuel cell. Indeed, the figure of 45% is optimistic since it takes no account of the losses incurred in compressing and distributing the hydrogen. When these are considered, along with the losses incurred in generating the electricity in the first place, the efficiency for the overall process

$$\text{Fossil fuel} \xrightarrow{\text{generation}} \text{Electricity} \xrightarrow{\text{electrolysis}} \text{Hydrogen} \xrightarrow{\text{fuel cell}} \text{Electricity}$$

is likely to be under 20%. Energy efficiency is not, however, the sole criterion on which to judge fuel cells. The concept of distributed power generation without release of carbon dioxide is attractive, particularly when part of a CHP scheme, but (as mentioned previously) this is then dependent on the delivery of hydrogen from a central production facility that practises sequestration of carbon dioxide.

In the very long term, when renewable electricity may be the main source of energy, considerations of efficiency will dictate that the bulk of it should be transmitted directly by electrons and not by hydrogen or other chemical carriers. Electrolysis and fuel cell combinations may be employed for temporary storage of energy between generation and use, but only in situations or applications where convenience outweighs inefficiency. The use of fuel cells to propel electric vehicles is one such application, motivated primarily by the petroleum supply and security question and also by concerns over atmospheric pollution in urban environments and global carbon emissions. Electric road vehicles have long been considered an attractive mode of transportation, particularly in cities where their quiet, fume-free operation is greatly appreciated. The twin inconveniences of battery-powered vehicles are their short autonomous range and the unacceptably long times required for recharging the batteries. The mass, volume, cost and operating life of the batteries are further limiting factors. By comparison, fuel cells are claimed to provide a superior power source. This is a somewhat optimistic view, as they do introduce problems of their own, the most notable of which are the logistics of supplying hydrogen of suitable purity, its storage on-board the vehicle, and the overall cost and durability of the fuel-cell modules. The hard truth is that there is no power source as versatile and inexpensive as the internal combustion engine and, while petroleum is available at an acceptable price, it will be very difficult to dislodge the engine from its dominant position in road transportation.

Substantial improvements have been made in the technology of several of the different types of fuel cell. For stationary applications, phosphoric acid fuel cells have become commercially available in modules of 250 kW output that can be linked together to form 1 MW power plants. The development of high-temperature fuel cells, which are ideally suited to CHP applications, has progressed well. Internal-reforming molten carbonate fuel cells with outputs up to 2 MW are starting to enter the market. Work on solid oxide fuel cells is also proceeding steadily; 250 kW systems have been built and subjected to performance trials. Despite these advances, the target life set for stationary fuel cells (often quoted as 40 000 hours or 5 years) has yet to be fully demonstrated.

For road transportation applications, especially buses, proton-exchange membrane fuel cells have been shown to operate satisfactorily under realistic service conditions. Generally, these fuel cells are used in combination with a storage device such as a battery or an electrochemical capacitor. Fuel cell buses have now been operational in many cities around the world and it is reported that the staff involved (drivers, fleet managers), and also passengers, have found the vehicles to be acceptable. Prototypes of fuel cell cars have also been produced by several automotive manufacturers and have undergone road trials. The technology is advancing rapidly, but still has some way to go before a commercially viable product will be available. Many factors will influence the public's choice of fuel-cell vehicles and these include the rate of fuel excise duty that may be placed on hydrogen.

What does the future hold? At present, fuel cells are far too expensive for widespread deployment and more work is required to lower their costs to affordable levels – a reduction of one to two orders of magnitude is involved. While mass production always leads to falling prices, this does represent a very tough goal to meet for such a complex entity. As mentioned above, the overall efficiency ('well-to-wheels') for transportation applications is poor, possibly no better than modern diesel cars, and there are the logistical problems of hydrogen supply and on-board storage. In short, at the present time, fuel cells to power vehicles are not simple, not cheap and not particularly energy efficient. Nevertheless, should electricity from renewable sources become the primary source of indigenous energy, as in Iceland today or on the Japanese island of Yakushima, hydrogen would be a potential means of storing energy and fuel cells would be a prospective power source for road vehicles. The incentive for the Icelandic government to pursue this route is that of national energy independence rather than one of energy efficiency or cost minimization.

8.3   The Role of Government

Globalization, brought about by advances in mass transportation, communications and education, is leading to profound changes in world society. In particular, it is now possible to manufacture high-quality goods in countries with low labour rates rather than in expensive, developed nations. With rising levels of education, an increasing number of service activities (*e.g.*, insurance, banking, software writing, call centres) are also being transferred to low-cost countries. As this process continues, developed nations may begin to stagnate economically, while developing countries in Asia will flourish. This will lead to a slow equalization of status between these two groupings. There is, however, a danger that some other parts of the world, for instance much of Africa and regions of South America, will be left by the wayside. Undoubtedly, labour costs will rise progressively in Asia and, with increasing prosperity, the desire of their citizens for a better standard of living will become unstoppable. Already this is leading to an ever-greater demand for more energy in the industrial, commercial and domestic sectors.

The primary responsibilities of government in energy matters are:

- to safeguard the nation's future energy supplies;
- to abate carbon dioxide emissions to the greatest degree possible, in collaboration with other countries;
- to protect vulnerable members of society from unaffordable energy costs.

These targets are best achieved by (i) taking active steps to encourage energy savings and (ii) diversifying the nation's dependence on primary energy sources, with particular emphasis on low-carbon nuclear and renewable energy sources.

8.3.1   Energy Conservation Policies

The introduction of entirely new energy industries will encounter many difficulties, *e.g.*, the inevitable reluctance to change, the capital cost, the political problems and the time-scale necessary for implementation. When these are compounded by the limited global impact of a feasible programme for carbon dioxide sequestration, it is clear that far more urgent action is required to abate carbon emissions. The only short-term approach must lie in energy conservation and in the decoupling of energy use from living standards. This is where governments (national, regional and local) can play an immediate and major role. The use of propaganda and media advertising is a powerful means of influencing people's behaviour and has been well demonstrated in the environmental field by the success of recycling and composting schemes.

There are literally dozens of energy-conservation measures that can be encouraged, supported or even mandated without invoking untried technology. With respect to curtailing use of electricity or gas, savings can be made through:

- removal of excessive public lighting;
- change to low-wattage, long-life lamps;
- better insulation of buildings;
- replacement of old gas boilers by modern condensing boilers.

In the field of transportation, possible initiatives include:

- a requirement for vehicle manufacturers to promote energy-efficient, rather than high-powered, vehicles;
- a lowering of extra-urban speed limits to $100 \, \text{km h}^{-1}$;
- improvements to public transportation systems;
- promotion of car-sharing schemes;
- provision of more extensive networks of cycle tracks between residential centres and factories/office blocks;
- reduction in business travel by assisting people to work at home, particularly where reliable digital technology and video-conferencing facilities are available.

In the longer term, commuting by car can be reduced by encouraging people to take a job nearer home or live closer to their place of employment, as in former days. In addition to conserving petroleum, such actions will reduce traffic congestion. Many of these two sets of measures will require higher energy prices in addition to public exhortation before they make a major impact.

At the sociological level, there is scope for replacing the concept of 'standard of living', usually defined by material possessions and the possibility of unrestricted travel, with 'quality of life', as represented by a state of happiness. While it is true that increased wealth does contribute to happiness, especially for the underprivileged, it is by no means the only factor that makes for contentment. Local leisure pursuits that consume little energy – most types of sport, hobbies, gardening, handicrafts, recreational and charitable activities – can give great pleasure and satisfaction. It would appear that we shall have to place greater emphasis on spending our lives nearer to home, with increasing reliance on digital communications to interact with distant colleagues, friends and family, and on television and the internet to explore the world virtually.

In summary, governments have the capability to bring about major economies in energy use through persuasion, through legislation and regulations, and through taxes and subsidies. All the possible actions have to be examined and introduced where appropriate to bring about short-term reductions in carbon dioxide emissions. This is by far the best hope for restricting the build-up of this greenhouse gas in the atmosphere over the next decade or two.

8.3.2   Energy Diversification

8.3.2.1   *Electricity*

Electricity, with its attributes of cleanliness and versatility, is widely predicted to assume a greater share of the overall energy market. A major debating point is whether electricity will continue to be produced centrally in traditional power stations and transmitted via a national grid, or whether there will be a gradual transition to distributed generation. Some see the latter route as having significant advantages. Renewable sources of energy (wind, waves, solar, micro-hydro) are widespread in nature and are therefore appropriate to small-scale generation. While this electricity can be fed into a national grid up to a certain point (said to be around 15% of the total supply), there are technical issues and it may prove more practical and cost-effective to utilize the power close to where it is produced.

Fossil fuels will continue to provide much of the base-load generation for quite some time to come. Where natural gas is available at acceptable cost, it will be the preferred fuel on grounds of energy efficiency, convenience, cleanliness and the lower emissions of carbon dioxide relative to those from coal[†]. In countries where there is indigenous, low-cost coal, the resource will be used in order to minimize the import of other fuels. It is probable that political pressures will be such that future coal-fired power stations will be mandated to employ gasification procedures, once provision has been made for carbon capture and storage (*i.e.*, so-called 'clean coal' technology). Nevertheless, it could take two or three decades for all the component processes (coal gasification, hydrogen separation and purification, carbon dioxide sequestration) to be perfected and still longer (post-2050) for them to be widely introduced. At that point in time, hydrogen energy – if found to be practical – will have come into its own for electricity generation.

Meanwhile, in the short- to mid-term, the only major low-carbon energy source available (apart from renewables) is nuclear power. Many energy strategists now regard nuclear power as a partial answer to diversifying a nation's electricity supply and to mitigating carbon dioxide emissions, despite

---

[†] Compared with natural gas, coal liberates about twice as much carbon dioxide per mole of hydrogen produced.

the high costs of construction, decommissioning and radioactive waste disposal. If utilities are to be encouraged to invest in advanced types of reactor, then the playing field must be levelled – for example, by governments providing subsidies to defray the expense incurred with the secure containment of spent fuel and wastes and/or imposing heavy taxes on emissions from fossil-fuelled plants. It is unrealistic for authorities to expect the nuclear industry to safeguard its wastes for all time, but allow conventional generators freely to discharge carbon dioxide to the atmosphere, given that both by-products are deemed to be environmental threats.

In 2004, 357 GW of nuclear capacity was installed world-wide and provided 15.7% of the total global production of electricity (17 450 TWh). By comparison, the contribution from fossil fuels (coal, natural gas, oil) was 66.1%. Thus, to realize a significant impact on global carbon emissions, an appreciable expansion of nuclear operations would be required. For example, a world target of 1 TW of installed capacity by 2050 (*i.e.*, three times the present level) would necessitate the commissioning of a 1 GW nuclear station, somewhere, every 3–4 weeks. Obviously, such a task has profound implications in terms of reactor technology, capital finance, skilled construction teams, fuel processing (and possible reprocessing), radioactive waste disposal and public acceptability. The threat posed by terrorists is also a major cause of anxiety. Overall, this rate of expansion constitutes an implausible scenario. Enthusiasts for solar cells argue that there is no need for more nuclear plant. By extrapolating the rates at which the price of photovoltaic modules is falling and the production is rising, they conclude that within 10–20 years solar energy could supply much of the extra demand for electricity. This, too, seems implausible.

Fuel cells constitute a source of local electricity, but (as we have emphasized earlier) there is no benefit at present in converting renewable energy to hydrogen and then employing the latter to regenerate electricity in a stationary fuel cell. Compared with using 'green' electricity directly, this would be grossly inefficient and would increase the cost substantially.

Traditionally, most countries have produced their electricity in centralized power stations coupled to a common grid. In some nations, generation and transmission are undertaken by a single state-owned industry. In others, a number of independent companies either operate as regional monopolies or compete in an open market to sell their product. Many generators and governments now have to face up to a number of difficult decisions concerning the sustainability of future electricity supplies. Some of the important questions to be addressed are as follows.

- Is it better to remain with centralized plants or should a move towards distributed units be implemented?
- If the latter, then what technology should be used and what percentage of local generation should be targeted and over what period?
- If more large power utilities are to be built, should these be gas-fired, coal-fired, nuclear, or some combination of all three?

- To what extent should carbon capture and storage be mandated for new fossil-fuel stations, bearing in mind that such facilities generally have an expected life of ~ 40 years?
- What provisions should be made for the identification and licensing of underground storage sites for carbon dioxide and the installation of pipelines for conveying the gas to the chosen locations?

From these broad topics, many subordinate matters will arise with respect to the passing of legislation and regulations to ensure implementation of government policy. Fiscal measures, including taxes and subsidies, will be required to ensure the commercial viability of new technologies, together with the establishment of a business environment in which all forms of electricity generation can compete fairly.

Whereas countries with nationalized industries will find it difficult to resolve the above complex set of issues, the challenge will be much greater for those with private utilities. In the latter instance, there might be a natural desire for government to bury its head in the sand and leave the choice of power systems to the free market. This simply will not do. Private generators are looking to governments for a national strategy within which they can operate and compete. In Europe, the situation is complicated still further by the import/export of electricity between countries and by regulations that may be set by the European Union. Many power stations throughout the world are approaching their end of life and decisions will have to be reached relatively soon. Once made, however, the consequences will be borne for several decades.

8.3.2.2 Transportation

The difficulty that hydrogen faces as a transportation fuel is that of competing with established liquid fuels. The latter (petrol and diesel) are readily available and, as yet, comparatively cheap, especially in the USA. Distribution of fuel to service stations is straightforward and inexpensive. Storage there poses no problems. Refuelling a vehicle is both rapid and easy and can be undertaken by the public. The on-board fuel tank is low cost to construct and, given the high energy content of the fuel, is of an acceptable size while still providing the vehicle with a long driving range. The internal combustion engine benefits from over 100 years of development, is powerful and generally affordable when mass-produced. Compared with these attributes, present-generation hydrogen-powered transportation fails on all counts and it will be an uphill struggle to persuade the consumer to accept an inferior product at a considerably higher price. Only an acute shortage of liquid fuels or stringent restrictions on carbon dioxide emissions will bring about a change in this situation.

The increasing prosperity of China and India, which together account for a high proportion of the world population, will inevitably lead to an escalation in the demand for liquid hydrocarbon fuels – just as supplies may be faltering. When native petroleum becomes in short supply and the price rises accordingly, synthetic liquid hydrocarbons (manufactured from bitumen, natural gas or coal) and biofuels (alcohols and bio-diesel derived from agricultural products and waste materials) will probably serve as alternatives. Both of these classes of fuel are produced by established technologies, which can readily be implemented. With crude oil trading at around $US60 a barrel, the processes are now thought to be economically viable. For instance, oil sands are being mined extensively in Canada. South Africa has existing plant to convert coal to petrol, but it would take substantial capital and a considerable period of time to build many more such plants. Of course, these approaches (with the exception of biofuels) do nothing to address the problem of carbon emissions.

Compared with electricity generation (discussed above), a government's role in directing future strategies for transportation may be less complicated. The objectives must be to reduce petroleum consumption (and therefore emissions) and to minimize road traffic congestion. Some short-term measures that might be adopted have been addressed earlier under the topic of 'Energy Conservation Policies'. In contrast to electricity generation, long-term actions can probably be left to the market once the government has established a policy framework. Automotive manufacturing and petroleum refining are highly developed industries that can operate without government direction. Vehicles usually have a lifespan of 12–15 years, unlike the 40+ years of power stations, and so decisions that prove to have been mistaken can be remedied sooner. The main responsibility of governments is to set fiscal policy and regulations that will lead to the desired goals. There may be a requirement to impose still higher taxes on over-powered vehicles, to encourage the purchase of energy-efficient vehicles (such as diesel–electric hybrids) and to improve public transportation. Many of the controls designed to solve the problems may fall to local authorities to implement rather than to central government. For example, traffic lanes reserved for buses and taxis, or priority for commuter cars taking a passenger, all encourage energy saving and reduce congestion. In central London, electric vehicles are exempt from the congestion charge. A further incentive for the purchase of these vehicles might be provided by offering free, reserved parking spaces and equipping them with battery-recharging points.

The prognosis, then, for road transportation is for a varied mix of propulsion systems and fuels as conventional petroleum becomes less plentiful and prices rise. Internal combustion engines operating with synthetic liquid fuels (hydrocarbons, alcohols and bio-diesel) are expected to appear in increasing numbers, often supported by hybrid electric drives to reduce fuel consumption. 'Plug-in' hybrid electric vehicles may become popular when more durable and affordable batteries become available. This would allow the power source to be part liquid fuel and part mains electricity. Some of the major obstacles to be overcome with hydrogen-fuelled vehicles are:

- the production and distribution of pure hydrogen at an acceptable cost;
- the on-board storage of sufficient hydrogen to give a reasonable vehicle driving range and a fast re-fuelling time (of the order of a few minutes);
- the mass production of fuel cells at a substantially reduced cost (by at least a factor of 20);

- the development of fuel-cell systems with greater reliability and durability;
- the formulation of international standards for fuel cell vehicles, together with regulations governing their operation;
- public acceptance of hydrogen as a fuel.

This is an intimidating list of challenges to be faced. Heavy vehicles (buses and some trucks) may be better suited to hydrogen than cars, given that there is greater capability to accommodate the hydrogen containers, regular scheduled services are undertaken between fixed points and the vehicles are operated and refuelled by professional, trained crews. Nevertheless, it seems likely that the widespread uptake of such vehicles is destined for the second half of the 21st century or later. Even optimists are speaking in terms of at least 20 years for fuel cell vehicles to achieve appreciable penetration of the car market.

The combined fuel-cell stack, balance-of-plant and hydrogen store must not impose a heavy mass or volume penalty. Fuel cells should operate satisfactorily in ambient temperatures from about -20 to $+45\,°C$. Consumers are looking for vehicles that are competitive with present-day models as regards performance, efficiency, cost, range, comfort, convenience, reliability, life, refuelling time and safety, while using fuel that is no more expensive than petroleum. Frankly, the prospects of meeting these requirements in full are slim and serious compromises will almost certainly have to be made. These will only be acceptable to the consumer as a last resort when liquid fuels are no longer readily available.

The future of aircraft fuelled by liquid hydrogen is still more speculative. Such a profound departure from present practice would take decades to introduce on a global basis.

8.3.3 Carbon Emissions

Carbon emissions from large stationary sources are more readily controlled than those from smaller, dispersed units. In Europe, the European Union Emission Trading Scheme (ETS) that aims to limit carbon emissions from these 'point sources' was implemented on 1 January 2005[‡]. Individual plants are issued with permits which specify the level of carbon that each may emit annually. Companies with surplus permits may sell these to others who wish to exceed their respective allowances. At first, the scheme did not work well because the initial allocations were generous and the market price of permits was too low to be effective in curtailing emissions. In the next phase of the programme, permits could be restricted and their market value would then rise. As these regulations begin to bite and companies have to purchase additional permits on the open market at realistic prices, it may become financially attractive – if not actually required by law – to fit carbon-capture equipment to their exhaust stacks.

[‡] Not to be confused with the Large Combustion Plant Directive that specifies the quantities of sulfur dioxide, nitrogen oxides and dust that may be emitted.

New coal-fired power stations may well opt for gasification technology to produce hydrogen (with or without capture of carbon dioxide), followed by combusting the fuel in combined-cycle gas turbines to generate electricity. Although coal gasification has been practised, it is still novel for the electricity industry and therefore carries risks. The addition of carbon dioxide capture and storage will not only heighten these risks, but will also result in reduced generation efficiency and increased cost that, in turn, will lead to more expensive electricity. Hence there will be a natural reluctance to set out on this path. Moreover, it should be noted that the adoption of carbon sequestration is dependent on the technology being perfected, on the ready availability and proximity of storage sites, and the resolution of all the attendant logistical and legal problems. Given the time normally needed for the development of major new energy technologies and for the design and building of power stations based upon them, one has to be looking to the second half of the 21st century, and possibly beyond, to see hydrogen derived from coal as a key player in the world-wide generation of electricity.

The outlook for transportation fuels is similarly difficult from the standpoint of greenhouse gas emissions and climate change. Whereas it is generally held that conventional petroleum will become increasingly scarce and expensive, there is less immediate concern over the availability of natural gas, oil sands and, particularly, coal for conversion to synthetic liquid fuels. Already, as previously discussed, this option is becoming economically viable. It is possible that a vast new petroleum synthesis and refining industry, based on bitumen and coal, will be built up progressively over the course of this century. True, it will take time and huge capital investment, but there is no global shortage of primary fossil fuels for conversion to petroleum and at prices not too far removed from present levels. It is pointless appealing to developing countries to economize in the use of petroleum when it is consumed so profligately in the developed world. There are only a few ways to curb emissions. The first would be through securing international agreement on carbon emission taxes. These will not, however, be easy to implement in an era when energy prices are rising and would unfairly penalize the poor to the extent that personal transportation becomes beyond their reach. An alternative, and perhaps fairer, approach would be a system of petroleum rationing similar to that introduced during World War II, but probably in an updated form as tradable, personal carbon-emission permits. Low-income people would not be disadvantaged and they could either use or sell their allocations. These could apply solely to petroleum products or to all fossil fuels. The issuing of personal emission permits to all the world's drivers (or citizens if applied to all fuels) would, of course, be an immensely difficult undertaking.

8.3.4 Renewable Energy

It is anticipated that the harnessing of renewable energy will expand rapidly, in the light of widespread concern over global warming and the remedial actions that governments are taking. Despite such good intentions, there is every indication that, overall, sustainable energy will still make only a modest contribution in 2020. It is vital, therefore, that society continues to develop the various technologies and gains experience in their operation as a step towards growth later in the 21st century. An important feature of renewable energy is that up to 80% of it will be available in the form of electricity, with the remainder as heat from the burning of biomass. By contrast, most energy is currently consumed as heat and less than 20% as electricity. This fact alone will necessitate a massive change in the world's energy distribution and consumption patterns and that will be a protracted affair.

Most renewable forms of energy now rely on government subsidies to be competitive. With rising prices of fossil fuels and the introduction of carbon emission permits, such subsidies may in time become unnecessary. People generally, especially the young, are positive and enthusiastic about renewables, but cannot be expected to invest in enterprises where the financial return might take decades. This is especially true of solar photovoltaic installations in high latitudes, although the cost of modules is declining fairly rapidly. In the long term, many enthusiasts see solar energy as the central plank of a sustainable energy future, with wind energy also playing an important role in temperate regions of the world. If society is serious about this vision, then it is imperative that these sustainable forms of energy are cost competitive. Otherwise, their deployment will be well below that required to make a significant impact on the present consumption of fossil fuels. Even then, as an example, it would be necessary to install many hundreds of wind turbines to produce as much electricity annually as one conventional power station and there is doubt about whether public opinion will consent to the erection of so many wind farms to despoil the countryside. Off-shore installations are more acceptable, but are also more costly to build and maintain. Solar arrays in sub-tropical deserts are less controversial, but often are not viable because they are sited too far from centres of population. Unfortunately, much of the advocacy for renewable energy is based upon sentiment rather than a thorough analysis of the prospects and associated costs of producing the quantities of electricity that a country needs. In any event, as pointed out earlier, too great a proportion of intermittent power generation from these sources would destabilize the national grid. Furthermore, a back-up facility would almost certainly have to be installed to regulate the system and guarantee supply, especially during adverse weather conditions (*e.g.*, cloudy/windless days) and during periods of heavy demand.

8.4   Hydrogen Energy: The Prospects

In a true Hydrogen Economy, the gas will be produced from diverse and distributed sources and then used to power vehicles, homes and industry. A transition to this future is uncertain and by no means inevitable. The extent to which hydrogen might be adopted will depend, variously, on technological advances and on developments in primary energy markets, electricity generation and transportation. The uptake will vary from country to country according to political, sociological and geographical considerations. These interwoven factors are highly complex and it is therefore difficult to forecast the overall outcome with any degree of confidence, still less the time-scale for implementation. The latter will be strongly influenced by the difficulty, cost and time needed to set up an appropriate infrastructure for supplying and distributing the hydrogen. Widespread use of hydrogen is projected only for the long term, *i.e.*, after 2050. Nevertheless, it is prudent to continue with the research and development of hydrogen energy and to keep a watching brief as events unfold, with a view to a possible major investment in hydrogen as a new energy vector, if and when this is justified by the prevailing circumstances.

The benefits of a Hydrogen Economy, and also the barriers to its development, cut across national policies on energy, transportation, air pollution, regional planning, agriculture and waste. Before hydrogen energy can be implemented in a major way, it may be necessary for governments to draw together these diverse areas of public responsibility and formulate a strategic plan that is acceptable to the various stakeholders. For entities like the European Union, it will be vital to agree internationally upon codes and standards of operation. All these matters will take a considerable time to resolve. Meanwhile, considerably more research and development will be required to overcome the formidable technical hurdles that currently stand in the way of hydrogen. For example, finding a practical solution to the problem of storing hydrogen on board vehicles is a critical challenge. In addition, large reductions in unit costs, notably in bulk transport and storage and in fuel cells, are essential for hydrogen to become competitive with existing energy systems.

The best prospect for bulk use of hydrogen rests on the successful evolution of clean coal technology with carbon dioxide sequestration. This would produce large quantities of hydrogen for the generation of electricity in combined-cycle gas turbines. Alternatively, the hydrogen might be piped around the country to fuel local heat and power schemes based on gas turbines, gas engines, or high-temperature fuel cells. Yet another option is to generate the hydrogen locally in dedicated steam reformers, thereby avoiding the necessity of a distribution system. All these possibilities must be subjected to careful evaluation.

In the interests of energy sustainability, it is preferable, where possible, to replace conventional fuels by electricity. The extent to which this is possible in the field of transportation is limited. In due course, it is probable that all railways will be electrified, but aircraft will still rely exclusively on liquid fuels, possibly even liquid hydrogen. The future of electrically-propelled road vehicles depends upon developing better traction batteries for urban use and on much cheaper fuel cells for long-distance travel. The combination of the high cost of fuel cells, hydrogen supply problems and on-board storage limitations makes it difficult for fuel cell vehicles to compete with internal combustion engines burning synthetic liquid fuels. Urban buses may prove to be the exception to this generalization; their feasibility has already been demonstrated and there are significant advantages. Again, a major cost barrier has to be overcome.

The mass use of hydrogen to power private cars is seen, realistically, to be not until 2050 or later, although there may be localized 'hydrogen highways' before then. The penalties compared with internal combustion engines are just too great, as are the investment costs and the time that it would take to construct the infrastructure. In the medium term, the best hope for transportation in confronting oil shortages lies in the use of synthetic liquid fuels and biofuels. For city driving, battery electric vehicles may finally become acceptable, particularly if the performance of traction batteries is improved. Hybrid electric vehicles, notably hybrid diesels, should grow in numbers. To the extent that mains electricity is used to recharge plug-in hybrids, the fuel base of transportation will be partially diversified. The large-scale adoption of hydrogen energy for this application should be left to the private sector to pursue as it sees fit, with some public funding of research and demonstration projects, coupled with fiscal incentives for operators of vehicles with low carbon emissions. Governments are expected to play a role in promoting awareness of the benefits of hydrogen. Public appreciation of these benefits and the provision of information on safety are likely to be the keys to acceptance of the technology. Projects involving fuel cell buses are contributing to this spread of understanding, but it is important not to disillusion the public by promising too much before it is certain that fuel cell vehicles will be fully practical, commercially available and cost competitive.

What of the more distant future? Here, inevitably, the crystal ball is even cloudier. So many unforeseeable events and developments can arise during the course of 50 years. Who in 1950, for instance, could have foreseen all the changes in society and all the technological advances of the second half of the 20th century? Many analysts expect that petroleum, and perhaps natural gas, will be past their peak production rates by 2050. Prices for refined or processed energy will surely rise more than inflation if clean coal conversion and carbon sequestration become established procedures. By that time, many forms of renewable energy should have become economically competitive. It is doubtful, however, whether collectively they would be able to satisfy the bulk of the world's energy demand. Nuclear power may well make a comeback, with the realization of advanced designs of reactor, possibly even sustainable fast-fission reactors. Late in the 21st century, nuclear fusion reactors may also be available.

There are many daunting challenges to be solved before the world can achieve true energy sustainability, that is, the zero use of fossil fuels. For the present, it is important for governments and industry to focus their attention and funding on:

- coal gasification processes;
- carbon capture and storage schemes;
- synthetic liquid fuels and biofuels;
- the development of renewable energy in all its forms;
- research and assessment projects on every aspect of hydrogen energy.

These activities will occupy the next few decades at least.

The world's current energy practices are so entrenched that it will prove difficult, prolonged and expensive to achieve true sustainability. Much capital plant (power stations, oil refineries, pipelines) are built to last for decades. To retire these units prematurely would be very costly. The widespread deployment of carbon sequestration from large emission sources will be a massive undertaking. Furthermore, raising the investment required to overhaul the entire existing energy system within a decade or two would be an impossible task for the financial community, even supposing society were prepared to pay the enormous cost of phasing out present energy facilities early. If competitive hydrogen technologies emerge within the next 20 years, it would probably take most of the remainder of this century to complete the transition to a Hydrogen Economy – a stage at which only hydrogen and electricity are used to deliver energy services.

Meanwhile, governments will need to start preparing for hydrogen energy by working with regulatory agencies to establish appropriate standards and codes for designing, building, testing and ultimately marketing hydrogen-related equipment. In this respect, industrial stakeholders who manufacture such equipment will have to participate in public–private sector partnerships. International harmonization of standards will also be an important area for development. Such activities will be crucial in ensuring safety. The advancement of hydrogen technologies – to improve practicability, raise efficiency and lower costs – will inevitably require world-wide collaboration to lever resources through cooperation. The establishment of the International Partnership for the Hydrogen Economy (IPHE) is a first step in this direction.

In this book, an attempt has been made to analyze in outline the *challenges* faced in implementing the various facets of a Hydrogen Economy. From this analysis, we have advanced some opinions on the *prospects* for hydrogen in the world energy scene over the next 40–50 years. A critical factor will be the outcome of the current debate on climate change and how urgently and universally it is addressed through reducing the anthropogenic emission of greenhouse gases. Our broad conclusion is that in the latter half of this century or beyond, hydrogen as an energy carrier may play an important role in a sustainable future based on renewable energy. At that time, energy demand will be determined by strict conservation measures, by less consumption of oil in the transportation sector, and by the use of 'green' electricity wherever and whenever possible. Others will have different views and see hydrogen occupying a more dynamic and immediate role.

Subject Index

Printed in the United Kingdom
by Lightning Source UK Ltd.
135985UK00001B/118-138/P